Mathematics and Visualization

Series Editors

Gerald Farin
Hans-Christian Hege
David Hoffman
Christopher R. Johnson
Konrad Polthier
Martin Rumpf

Mohamed Elkadi
Bernard Mourrain
Ragni Piene

Editors

Algebraic Geometry and Geometric Modeling

With 52 Figures, 27 in Color and 7 Tables, 1 in Color

Mohamed Elkadi
University of Nice Sophia Antipolis
Laboratory J.A. Dieudonné
Parc Valrose
FR-06108 Nice, France
E-mail: elkadi@math.unice.fr

Bernard Mourrain
GALAAD, INRIA, BP 93
FR-06920 Sophia Antipolis
France
E-mail: mourrain@sophia.inria.fr

Ragni Piene
Centre of Mathematics for Applications and Department of Mathematics
University of Oslo
P.O. Box 1053 Blindern
NO-0316 Oslo, Norway
E-mail: ragnip@math.uio.no

Library of Congress Control Number: 2006929216

Mathematics Subject Classification: 12Y05, 13C15, 13D02, 13P05, 13P10, 14J10, 14J17, 14P25, 65D17

ISBN-10 3-540-33274-X Springer Berlin Heidelberg New York
ISBN-13 978-3-540-332749 Springer Berlin Heidelberg New York

This work is subject to copyright. All rights are reserved, whether the whole or part of the material is concerned, specifically the rights of translation, reprinting, reuse of illustrations, recitation, broadcasting, reproduction on microfilm or in any other way, and storage in data banks. Duplication of this publication or parts thereof is permitted only under the provisions of the German Copyright Law of September 9, 1965, in its current version, and permission for use must always be obtained from Springer. Violations are liable for prosecution under the German Copyright Law.

Springer is a part of Springer Science+Business Media
springer.com
© Springer-Verlag Berlin Heidelberg 2006

The use of general descriptive names, registered names, trademarks, etc. in this publication does not imply, even in the absence of a specific statement, that such names are exempt from the relevant protective laws and regulations and therefore free for general use.

Typesetting by the authors and SPi using a Springer LaTeX macro package
Cover picture by Oliver Labs
Cover design: *design & production* GmbH, Heidelberg

Printed on acid-free paper   SPIN: 11693703   46/SPi/3100   5 4 3 2 1 0

Preface

Algebraic Geometry and Geometric Modeling are two distinct domains of research, with few interactions up to now, though closely linked. On the one hand, Algebraic Geometry has developed an impressive theory targeting the understanding of geometric objects defined algebraically. On the other hand, Geometric Modeling is using every day, in practical and difficult problems, virtual shapes based on algebraic models. Could these two domains benefit from each other? Recent and interesting developments in this direction are about to convince us to answer yes. In this book, we have collected articles which reinforce, in some way, the natural bridge which exists between these two areas. The confrontation of the different points of view should result in a better analysis of the key problems and related methods to solve them. This was the aim of the workshop entitled *Algebraic Geometry and Geometric Modeling*, held from September 27 to September 29, 2004, at the University of Nice-Sophia Antipolis. This workshop was organized, in the context of the European project GAIA II (IST-2002-35512).

The articles of this book are grouped into three main families:

- Implicitization problems
- Classification problems
- Intersection problems

The first group of articles is about Implicitization problems, namely the conversion of a parametric representation of an algebraic variety into an implicit one. This problem is fundamental in Computer Aided Geometric Design (CAGD), where geometric objects are often given parametrically and some operations like testing if a point is in a variety, intersecting two varieties, determining the singular locus of a variety, is better understandood via implicit representations.

In his contribution, Goldman claims that the main contribution of Algebraic Geometry to Geometric Modeling is insight, not computation, and that we should not confuse concrete theoretical tools given by Algebraic Geometry with efficient computational methods needed by Geometric Modeling.

He defends this thesis using some examples, as solving polynomial systems, intersecting curves and surfaces, and implicitization problem.

Chardin presents in his paper a survey of the recent method of approximation complexes and its application to the implicitization problem of a parameterization given by a rational map from $\mathbb{P}^{n-1}$ to $\mathbb{P}^n$.

Shalaby, Thomassen, Wurm, Dokken and Jüttler compare four methods for approximate implicitization by piecewise polynomials. This comparison is performed by applying algorithms to academic surfaces and industrial ones provided by the CAGD vendor, Think3, a partner in the European project GAIA II.

Lazard in his paper shows that, using the Gb-RS software, Gröbner basis computations allow efficient computation in order to solve some problems like testing if a point is on a variety, computing the intersection of parameterized varieties, or studying the singularities of parametric varieties without implicitization.

Aigner, Szilágyi, Jüttler and Schicho study from a numerical point of view the computation of the implicit representation of a curve that approximates a given parametric curve in a domain of interest. They show that for any approximate parameterization of the given curve, the curve obtained by an approximate implicitization with a given precision is contained within a perturbation region.

The theme of the second string of papers in this volume is, loosely speaking, the theory of classification of algebraic surfaces. Traditionally in algebraic geometry, this theory has been developed mainly in the complex, projective setting. The important issue for CAGD applications, however, is an understanding of *real* algebraic curves and surfaces in affine real 3-space, or in bounded regions of this space. The understanding of the complex projective case helps understand the real case, for example one can obtain bounds, coming from the complex case, for the number of connected components and the number and type of singularities.

The paper by Dimca discusses how the presence of singularities affects the geometry and topology, or shape, of complex projective hypersurfaces. Relevant theory of deformations and singularities is surveyed, and examples are given in the curve and surface cases.

The topology of real algebraic surfaces is the topic of Kharlamov's contribution. Applying Smith theory and Hodge theory to real surfaces and higher dimensional varieties, he provides various prohibitions on the total Betti number, the Euler characteristic, and the signature. As an example, he discusses surfaces of low degree. The possible topological types of cubic and of (nonsingular) quartic surfaces are known, but already for quintics the knowledge is very limited.

Holzer and Labs illustrate the classification of real projective cubic surfaces. They show how to choose affine equations so that no singularity or real line lies in the plane at infinity. Using their new visualization tool SURFEX, the images of all types can be shown. A useful feature of SURFEX is that it allows visualizations of deformations of surfaces.

Most surfaces used in CAGD are rational, in fact given by a parameterization. In his paper, Krasauskas shows how to use toric surfaces to find Bezier patches of optimal degree on a given surface with boundary curves. This is done by studying all parameterizations of the given surface and applying the theory of Universal Rational Parameterization.

Elkadi, Galligo, and Lê also study parameterized surfaces, more precisely tensor surfaces of low bidegree. They work in the complex projective case, and consider equivalences of parameterizations and the corresponding moduli space. The number of moduli is determined and normal forms are provided; also the implicit equations and the nature of the singularities are studied.

The third group of papers deals with intersection problems in geometry. This is the corner-stone of many operations on shapes. Numerical difficulties have to be tackled, together with purely geometric questions such as computing the topology of the constructed objects. Behind these geometric problems, are hidden questions about the resolution of polynomial systems of equations and their simplification.

The paper by Fioravanti, Gonzalez-Vega and Necula deals with the intersection problems of rational surfaces of a special type, namely revolution and canal surfaces. These surfaces appear naturally in modeling problems, where the characterizations of shapes by distances are used.

The problem handled in the next paper by Galligo and Pavone is about computing the self-intersecting points of a parametric surface, that is the points where this parameterization is not injective. A subdivision approach is described and its practical efficiency illustrated on real CAGD models.

An important problem in geometric modeling concerns the assembling of objects by imposing some constraints on basic geometric primitives. The paper by Peter, Sitharam, Zhou, and Fan considers the 3D case and describes techniques to analyze it, from a combinatorial point of view.

The paper by Emiris and Tsigaridas is of another flavor. The problem that they consider is the decomposition of polynomials as products, approached from a convex geometry point of view. Namely, the decomposition of a polytope as the Minkowski sum of two other polytopes is analyzed in detail for polygons, in dimension two.

Finally the paper by Carlini handles the problem of reducing the number of variables in the presentation of a polynomial. Tools from effective algebraic geometry are used for this purpose, illustrating this interaction between algebra and geometry.

Nice/Oslo,
September 2005

Mohamed Elkadi
Bernard Mourrain
Ragni Piene

Contents

Algebraic geometry and geometric modeling: insight and computation
Ron Goldman 1

Implicitization using approximation complexes
Marc Chardin 23

Piecewise approximate implicitization: experiments using industrial data
Mohamed F. Shalaby, Jan B. Thomassen, Elmar M. Wurm, Tor Dokken, Bert Jüttler.. 37

Computing with parameterized varieties
Daniel Lazard 53

Implicitization and Distance Bounds
Martin Aigner, Ibolya Szilágyi, Bert Jüttler, Josef Schicho 71

Singularities and their deformations: how they change the shape and view of objects
Alexandru Dimca 87

Overview of topological properties of real algebraic surfaces
Viatcheslav Kharlamov 103

Illustrating the classification of real cubic surfaces
Stephan Holzer, Oliver Labs 119

Bézier patches on almost toric surfaces
Rimvydas Krasauskas...135

On parametric surfaces of low degree in $\mathbb{P}^3(\mathbb{C})$
Mohamed Elkadi, André Galligo, Thi Ha Lê 151

On the intersection with revolution and canal surfaces
Mario Fioravanti, Laureano Gonzalez–Vega, Ioana Necula 169

A sampling algorithm computing self-intersections of parametric surfaces
A. Galligo, J.P. Pavone 185

Elimination in generically rigid 3D geometric constraint systems
Jörg Peters, Meera Sitharam, Yong Zhou, JianHua Fan 205

Minkowski decomposition of convex lattice polygons
Ioannis Z. Emiris, Elias P. Tsigaridas 217

Reducing the number of variables of a polynomial
Enrico Carlini 237

Index 249

Algebraic geometry and geometric modeling: insight and computation

Ron Goldman

Department of Computer Science
Rice University
6100 Main Street
Houston TX 77005
rng@rice.edu

Summary. Recent trends in algebraic geometry emphasize effective computation over transcendent theory. The theme of this paper is that from the perspective of geometric modeling this trend is largely misguided – that for the purpose of geometric modeling the true role of algebraic geometry is insight not computation.

1 Introduction

The conventional wisdom expressed at several recent joint conferences and workshops on algebraic geometry and geometric modeling – *Algebraic Geometry and Geometric Modeling 2002*, Vilnius, Lithuania, *The MSRI Workshop on Real Algebraic Geometry in Applications*, Berkeley, 2004, and *Algebraic Geometry and Geometric Modeling 2004*, Nice, France – is that algebraic geometry has much to offer geometric modeling. Algebraic geometry and geometric modeling both deal with polynomials: algebraic geometry investigates the algebraic and geometric properties of polynomials; geometric modeling uses polynomials to build computer models for industrial design and manufacture. Since algebraic geometry is by far the older, more mature discipline, geometric modeling should have a lot to learn from algebraic geometry.

Contemporary trends in algebraic geometry emphasize effective computation. This current fashion in algebraic geometry is largely an outgrowth of two recent developments from outside the discipline: the ubiquity and utility of modern digital computers and the tastes and predilections of the funding agencies.

Computers demand effective computation, but the ideal of effective computation was present at the creation. Descartes first merged geometry with algebra; Descartes' dream was to invoke algebra to mechanize geometry. Descartes envisioned a day when a machine using algebra could take over all of the difficult and tedious analysis of geometry [18]. Today we are still far from realizing Descartes' dream, but modern computers certainly accentuate this ambition.

Funding agencies emphasize applications. There are many reasons for this emphasis. To justify their work within contemporary capitalist economies, funding agencies need to show practical returns for their disbursements. The public and private donors who put up the money for research also demand immediate returns. Bureaucrats and donors both prefer safe short term gains over risky long term investments. Thus the funding culture is naturally inclined to believe that with computers in hand, mathematicians should concentrate on developing immediately useful computational tools to solve pressing engineering problems rather than pursue difficult long term theoretical goals.

Here I shall argue against this trend toward effective computation in algebraic geometry. My experience is that *the main contribution of algebraic geometry to geometric modeling is insight not computation*. I shall give many examples to support this thesis, and I shall argue for theoretical and practical reasons that despite the ubiquity of computers and the aspirations of funding agencies, this relationship between algebraic geometry and geometric modeling is not likely to change.

The claims and examples in this paper are based mostly on my personal research. Perhaps my experience is too narrow, or perhaps the current imbalance between insight and computation is about to change due to recent improvements in computational techniques. I am aware of at least some of these developments and I will return to these hopeful possibilities in the final section of this paper, for despite what I am about to say in the body of the text the outlook for constructive algebraic geometry is not unremittingly negative. Nevertheless, I believe that currently the burden of proof still lies overwhelmingly with those who insist on the practical utility of effective computational techniques from algebraic geometry. Geometric modeling would certainly benefit a great deal if these constructive techniques ever turn out to be practical. At present I remain skeptical of this prospect, but lest this paper seem relentlessly negative, I will revisit this contingency in the final section of this paper.

Section 2 begins the discussion with the canonical computational problem in both algebraic geometry and geometric modeling: solving systems of polynomial equations. Although algebraic geometry offers several constructive tools such as Gröbner bases, resultants, and homotopy continuation, for solving systems of polynomial equations, these techniques are not currently competitive with numerical methods. What algebraic geometry offers is insight into the difficulty of finding the solutions rather than practical computational techniques for calculating solutions.

Intersection algorithms for algebraic curves and surfaces are the focus of Section 3. For low degree curves and surfaces algebraic geometry does indeed provide the most effective computational tools for calculating intersections. But such low degree curves and surfaces are not the major focus of geometric modeling. For the curves and surfaces most common in geometric modeling, numerical methods are, at present, far superior to algebraic techniques.

Section 4 investigates implicitization. Recently the implicitization problem has received a good deal of attention from the algebraic geometry community, ostensibly because of its practical applications. We shall see that the practical advantages of implicitization are less than generally supposed and that, in any event, the computational tools available from algebraic geometry are not particularly efficient. Again what algebraic geometry offers, at present, is insight into the nature of implicitization rather than practical computational techniques for calculating the implicit equation.

Resultants are the focus of Section 5. Though algebraic geometry provides lots of insight into different resultant formulations, almost nothing is currently known about the relative speed or numerical stability of various resultant techniques. Thus the utility and viability of resultants in practical applications is still an open question.

The method of moving surfaces is discussed in Section 6. This implicitization technique is an area of conscious collaboration between algebraic geometry and geometric modeling: the method – based on linear algebra – was introduced in geometric modeling, the proofs – based on sheaf cohomology – were provided by algebraic geometry. Thus algebraic geometry supplies the insight underlying the computations.

Efficient resolvents are the subject of Section 7. In principle, resolvents can be used to intersect non-planar curves and to find base points on rational surfaces. In practice, current resolvent methods are way too inefficient to be applied in practical geometric modeling applications. Thus, like resultants, resolvents are an engaging and effective theoretical tool, but much work remains to be done on efficiency and numerical stability before resolvents can be used in practical applications.

Section 8 is devoted to surface classification and feature extraction. Much insight is gained in geometric modeling from the classification of different surface types in algebraic geometry. But when we want to extract geometric features – axes, centers of symmetry, foci – from rational parameterizations, algebraic geometry seems to lack effective computational tools. Nor does algebraic geometry supply any intuition into curvature formulas for algebraic or rational surfaces. Geometric modeling is often interested in metric properties of surfaces; algebraic geometry is mainly concerned with affine or projective invariants. Thus, whereas algebraic geometry provides insight into the different kinds of surfaces that are available as tools for geometric modeling, the computational interests of the two disciplines concerning surfaces do not at all coincide.

All of my concerns and conclusions are summarized in Section 9, where I provide as well some hope and some challenges for future research that may help to resolve the role played by effective algebraic geometry in geometric modeling.

2 The canonical problem: solving systems of polynomial equations

Since we want to investigate the efficacy of computational algebraic geometry in geometric modeling, it is natural to ask: what important computational problems in geometric modeling can algebraic geometry solve more efficiently than standard numerical methods?

Many computational problems in geometric modeling can be reduced to solving systems of polynomial equations. For example, in geometric modeling curves and surfaces are typically represented by polynomial equations; therefore intersection algorithms for curves and surfaces reduce to finding common roots of polynomials. Thus finding common roots of polynomial equations is one of the canonical problem in geometric modeling. Polynomial root finding is also basic to many other branches of computational science and engineering, including robotics, computational biology, and coding theory.

What then is the best way to find the roots for a system of polynomial equations? There are many variants of this problem, depending on which roots are of interest in the application. For example, we might be interested in finding all the complex roots, or all the real roots, or all the real roots inside a bounding box, or even in just one real root inside a bounding box.

In geometric modeling, we are usually interested in all the real roots inside some bounded domain. What does algebraic geometry have to say about the solution to this problem? The main insight from algebraic geometry concerning roots of polynomial equations is provided by Bernstein's theorem [15, 47].

Bernstein's Theorem
Consider a collection of polynomials $F_1(x_1, ..., x_N), ..., F_{N+1}(x_1, ..., x_N)$. If the number of roots is finite, then the number of complex roots is

$$N! \times (Mixed\ Volume\ of\ the\ associated\ Newton\ Polygons) \qquad (2.1)$$

If there are infinitely many roots, then the right hand side of Equation (2.1) provides a bound on the number of connected components of the roots.

As a consequence of Bernstein's theorem, even if both the degree and the number of variables is of moderate size, say of order 10, the number of isolated complex roots can be very large, roughly of order 10^8. Moreover, connected components may also appear. Thus algebraic geometry does indeed provide a good deal of insight into the complexity of the root finding problem.

What about computation? Effective algebraic geometry offers several techniques for solving systems of polynomial equations, including Gröbner bases, resultants, and homotopy continuation. To judge the effectiveness of these tools, we need to compare and contrast these techniques with standard numerical methods such as binary search, Bezier subdivision, and Newton iteration.

The current state of the art in solving for the bounded real roots of a large system of polynomial equations is due to Zhang, who can solve for the real roots inside a unit hypercube of 9 equations in 9 variables, each variable of degree at most 2 [61]. Zhang can solve these equations in less than 10 minutes on a standard PC using Bezier subdivision. Problems of this size arise in computer aided drug design, where the short term goal is to solve 15 equations in 15 variables, where each variable is of degree at most 2. Much larger polynomial systems with hundreds of equations in hundreds of unknowns appear in computational biology in protein folding problems.

At present no technique from computational algebraic geometry comes even close to solving such systems of equations. The problem with current methods from effective algebraic geometry such as Gröbner bases, resultants, and homotopy continuation is that these methods are built to find all the complex roots; at present these techniques, unlike Bezier subdivision, cannot be fine tuned to look only for bounded real roots. Since we know from Bernstein's theorem that the number of complex roots is often huge, these methods either consume way too much memory or they must run for a very long time.

Thus while algebraic geometry provides insight into the complexity of solving systems of polynomial equations, the computational tools developed by algebraic geometry are nowhere near competitive with numerical methods such as Bezier subdivision already available in geometric modeling. In effect, the computational approaches of algebraic geometry are in reality theoretical tools for proving theorems by concrete methods rather than abstract techniques. We should not confuse concrete theoretical tools with effective computational methods. To reconfirm this conclusion, let us look at more examples.

3 Intersecting rational curves and surfaces

A large part of geometric modeling depends on intersecting rational curves and surfaces. Yet, although a good deal of work has been done on this intersection problem both in algebraic geometry and in geometric modeling, very few comparative studies are available.

About twenty years ago, Sederberg and Parry compared resultant methods to Bezier subdivision for intersecting planar rational curves [52]. What they found was that for degree ≤ 3 resultant techniques are faster than Bezier subdivision, but for degree ≥ 4 Bezier subdivision is faster than resultants. These results are consistent with what we have already discussed about solving systems of polynomials equations, but here the conclusions are even more dramatic: algebraic techniques are more efficient than numerical methods for intersecting rational curves only for curves of very low degree. Perhaps in twenty years improvements in constructive algebraic methods have shifted these bounds and algebraic methods are now superior for higher degrees, but I know of no recent studies that either confirm or contradict these earlier conclusions.

The simplest curved surfaces that appear in geometric modeling are quadric surfaces, especially spheres, cylinders and cones. Many research groups have implemented algebraic methods to intersect quadric surfaces. From algebraic geometry (Bernstein's theorem), we know that three quadric surfaces can have up to eight real intersection points. In the early 1980's General Motors implemented homotopy continuation to compute these intersection points [44]. In the early 1990's [11] applied Macaulay's resultant to calculate the intersections of three quadric surfaces. Though both groups were successful, neither group reports any comparisons with numerical methods. Nor despite recent improvements in algebraic techniques such as homotopy continuation or Gröbner bases have I heard or seen any contemporary comparisons of algebraic techniques vs. numerical methods for intersecting three quadric surfaces.

Many individual researchers and research groups [40, 48, 43, 45, 25, 22, 55] have applied tools from algebraic geometry to compute the intersection curve between two arbitrary quadric surfaces. Here a complete classification of the geometry and topology of the intersection curve along with degenerate cases is available from algebraic geometry, and explicit parameterizations of the intersection curve are also provided. Thus the analysis of the intersection curve between two quadric surfaces is one of the great successes of algebraic geometry in geometric modeling.

Nevertheless, even this success has drawbacks. Quadric surfaces form a closed system. Techniques that work well for quadric surfaces do not generally extend to higher order surfaces. But quadric surfaces by themselves are far from sufficient for the purposes of geometric modeling. The geometric coverage of quadric surfaces is quite limited; higher order surfaces are typically required to model industrial parts.

For this reason many companies – Boeing, Volkswagen, SDRC, Alpha-1 – have developed geometric modeling systems based on rational Bezier or rational B-spline surfaces. Bicubic patches are the most popular surfaces in these systems. Generic bicubic patches have implicit degree 18, so the intersection curve between two generic bicubic patches has degree $18^2 = 324$. To my knowledge, no industrial or academic geometric modeling system has attempted to implement algebraic techniques to compute intersection curves for bicubic patches; the degrees are just too large. Typically to compute intersections, these systems use numerical methods such as subdivision for Bezier surfaces or knot insertion for B-splines. Thus once again algebraic geometry provides insight into the complexity of the intersection problem rather than practical computational tools for calculating these intersections.

4 Implicitization

One problem that has received a good deal of attention lately is implicitization [2, 3, 6, 4, 13, 14, 16, 20, 33, 50]: given a rational curve or surface, find the corresponding implicit equation. Applications of implicitization are purported to include algorithms for intersecting rational curves and surfaces, ray casting procedures for rendering rational surfaces, and inside-outside tests for solid modeling. We have already seen that intersection algorithms for rational curves and surfaces based on algebraic techniques are generally much less efficient than numerical methods, but the other two applications may have some merit.

Algebraic geometry provides several implicitization methods, including resultants – classical, sparse, and residual – Gröbner bases, syzygies (known in geometric modeling as the method of moving surfaces [49, 62]), and the method of undetermined coefficients. Algebraic geometry also provides insight by telling us that if there are no base points, then the expected degree of the implicit equation is the mixed volume of the Newton polygons of the parameterizing polynomials. Base points are parameter values where all the rational coordinate functions evaluate to 0/0. Base points reduce the expected implicit degree, and may also cause an implicitization method to fail.

In terms of actual computation, Gröbner bases are generally way too slow (at least for bicubic patches) and undetermined coefficients are way too inefficient. Thus we are left with resultants for which there are many possible formulations, and syzygies for which the theoretical foundations are not yet complete. Approximate implicitization methods have also been considered [20, 33], but these methods are numerical rather than algebraic.

Though several computational tools for implicitization are provided by algebraic geometry, precise quantitative measurements concerning the relative speed and numerical stability of these techniques are not yet available. Apparently much work remains to be done to determine the reliability and utility of these implicitization methods. Nevertheless, in the next two sections we shall look more closely both at resultants and at syzygies in order to determine what algebraic geometry has contributed to geometric modeling concerning these two potential computational techniques.

5 Resultants

A resultant is a polynomial in the coefficients of a polynomial system that vanishes if and only if the polynomials in the system have a common root. Resultants have many formulations, but typically resultants are represented either by determinants [57] or by ratios of determinants [17] or by GCD's of determinants [15, 27].

The two most common representations for the resultant of two polynomials in one variable are the Sylvester resultant and the Bezout resultant [9]. Each entry of the Sylvester resultant is either zero or a coefficient of one of the original polynomial equations; the entries of the Bezout resultant are more complicated, but the Bezout resultant employs a much smaller matrix. Resultants for two polynomials in one variable can also be represented by hybrids of the Sylvester and Bezout resultants [51] as well as by companion matrices [58].

Resultants for multivariate polynomials have at least a half dozen different representations many of which are valid only in special cases. The Sylvester formulation represents the resultant by a single determinant each of whose entries is either zero or one of the coefficients of the original polynomial equations [38]. The Dixon formulation uses smaller matrices than the Sylvester representation, but employs more complicated Bezoutian entries [19]. Macaulay represents the resultant as the ratio of two determinants whose entries are similar to the Sylvester resultant [41]. There are also hybrids representations that consist of a mixture of Sylvester and Bezout type entries [60, 37]. For low degree polynomial systems there are special formulations such as the Jacobian representation [10]. And there exist other residual formulations as well to handle other general cases including base points [5].

The polynomial representing the resultant is often quite large. Therefore to compute the resultant, Manocha and Canny provide a term expansion technique involving multivariate polynomial interpolation together with modular arithmetic [42]. Matrices representing the resultant are also often quite large.

Explicit formulas as well as efficient recursive computations for the entries of the Bezout and Dixon resultants are provided by [12].

Despite all this theory, there are often more questions than answers regarding actual computations with resultants. Natural questions include: what are the relative speed and numerical stability of different resultant formulations? Speed and stability are clearly key issues for practical implementations. Specific resultant formulations also raise theoretical questions. For example, it is known that when the Dixon determinant vanishes identically, the maximal minor is a multiple of the resultant [23]. But it is not known how to find the maximal minor; nor is it known how to predict the extraneous factors in the maximal minor.

The only work so far on numerical stability for resultants is [59]. Winkler and Goldman show that in the univariate setting Sylvester's resultant is numerically unstable; Bezout's resultant is inherently more reliable than Sylvester's resultant. These facts about numerical stability are crucial in practice for geometric modeling applications.

Conditions under which the Dixon determinant generates the exact resultant are provided by [8]. In cases where the Dixon matrix does not generate the exact resultant, the maximal minor of the Dixon resultant for some simple Newton polygons is found by [26]. Foo and Chionh can also predict the Dixon extraneous factors for these simple Newton polygons.

These three papers – one on numerical stability in the univariate setting, one on specific conditions for when the Dixon determinant generates the exact resultant, and one on finding the maximal minor and the extraneous factor for the Dixon resultant in the multivariate setting – together with the paper of Manocha and Canny on term expansion seem to be all that is known about numerical computations with resultants. Surely a lot more study on computational issues needs to be done before resultants can be considered as a practical tool for numerical computation in geometric modeling applications. Yet these investigations are not part of the research agenda of algebraic geometry.

Rather than an analysis of the numerical properties of resultants, what algebraic geometry provides is insight into when certain specific resultant formulations actually exist. For example, [37] uses sheaf cohomology to establish that a determinant formulation for the resultant always exists for unmixed systems of bivariate polynomials with arbitrary Newton polygons. These resultants are hybrids with complicated structures: some rows and columns are in the style of Sylvester, whereas others are in the style of Bezout.

Pure Sylvester style resultants – resultants where each entry of the resultant matrix is either zero or a coefficient of one of the polynomial equations

– do not always exist, even in the case of polynomials of a fixed total degree where the Newton polygon is a triangle. Nevertheless, because of their simple structure, it would be desirable to know when such pure Sylvester style resultants do exist. Moreover, whenever pure Sylvester style resultants exist, pure Bezout style resultants – resultants generated by Bezoutian matrices (see, for example, [5]) – must also exist. Bezout resultants are more compact than the hybrid resultants of Khetan, and their structure may be simpler as well. Thus we seek Newton polygons with pure Sylvester style resultants.

For which Newton polygons do Sylvester style resultants exist? The general answer is not known even in the bivariate setting. What is known is that for triangles (total degree) the answer is never for degree>1; for rectangles (bidegree), the answer is always [19]; and for hexagons the answer is sometimes [38].

To construct a Sylvester style resultant for three bivariate polynomials f, g, h with the same Newton polygon A, we typically seek a collection of monomials M such that the determinant of the matrix formed by the coefficients of the polynomials Mf, Mg, Mh is the resultant. The monomials M are called a m*ultiplying set* for the Newton polygon A. Thus we ask: for which Newton polygons A do multiplying sets M exist?

The multiplying set must satisfy the following three constraints [38]:

$$\#M = 2\text{Area}(\text{ConvexHull}(A)),$$

$$\#(M \oplus A) = 3(\#M) = 6\text{Area}(\text{ConvexHull}(A)), \oplus \text{ denotes Minkowski sum},$$

$$\text{Det}(\text{Mf}, \text{Mg}, \text{Mh}) \neq 0.$$

The first constraint says that the degree of the resultant in the coefficients of any of the original polynomials must be $2Area(ConvexHull(A))$. The second constraint is simply that the matrix formed by the coefficients of the polynomials Mf, Mg, Mh must be a square matrix (#rows=#columns). The final constraint is that the determinant does not vanish identically. When the Newton polygon of M has the same shape as the Newton polygon of f, g, h – that is, when the sides of the Newton polygon of M are parallel to the sides of the Newton polygon A – the first two constraints reduce to two quadratic Diophantine equations [Khetan et al 2004]. These equations can often be solved by a computer using a simple search. The difficult part is finding explicit examples to show that $Det(Mf, Mg, Mh) \neq 0$. Using sheaf cohomology, [38], show that $Det(Mf, Mg, Mh) \neq 0$ can be verified without constructing canonical examples. Thus once again, algebraic geometry provides insight into a theoretical question that may have practical applications in geometric modeling.

6 Implicitization revisited: the method of moving surfaces

Resultants can be applied to find the implicit equation of a rational surface. But resultants vanish in the presence of base points. The method of moving surfaces was introduced by [49] to find the implicit equation for a rational surface in the presence of base points. We shall give a brief description of this method here; for further details and insights, see [13, 14, 62].

A moving plane of bidegree (m, n) is an expression of the form

$$(A_{m,n}s^m t^n + \cdots + A_{0,0})x + (B_{m,n}s^m t^n + \cdots + B_{0,0})y$$
$$+ (C_{m,n}s^m t^n + \cdots + C_{0,0})z + (D_{m,n}s^m t^n + \cdots + D_{0,0})w = 0 \quad (6.2)$$

or equivalently

$$(A_{m,n}x + B_{m,n}y + C_{m,n}z + D_{m,n}w)s^m t^n + \cdots$$
$$+ (A_{0,0}x + B_{0,0}y + C_{0,0}z + D_{0,0}w) = 0.$$

Similarly a moving quadric of bidegree (m, n) is an expression of the form

$$(A_{m,n}s^m t^n + \cdots + A_{0,0})x^2 + \cdots + (J_{m,n}s^m t^n + \cdots + J_{0,0})w^2 = 0. \quad (6.3)$$

or equivalently

$$(A_{m,n}x^2 + \cdots + J_{m,n}w^2)s^m t^n + \cdots + (A_{0,0}x^2 + \cdots + J_{0,0}w^2) = 0.$$

For each value of (s, t), a moving plane is the implicit equation of a plane; different values of (s, t) represent different planes. Similarly for each value of (s, t), a moving quadric is the implicit equation of a quadric surface; again different values of (s, t) represent different quadric surfaces.

Moving surfaces are related to syzygies. A moving plane (6.2) is said to follow a rational surface

$$x = \frac{x(s,t)}{w(s,t)}, y = \frac{y(s,t)}{w(s,t)}, z = \frac{z(s,t)}{w(s,t)}, \quad (6.4)$$

if

$$(A_{m,n}s^m t^n + \cdots + A_{0,0})x(s,t) + (B_{m,n}s^m t^n + \cdots + B_{0,0})y(s,t)$$
$$+ (C_{m,n}s^m t^n + \cdots + C_{0,0})z(s,t) + (D_{m,n}s^m t^n + \cdots + D_{0,0})w(s,t) \equiv 0.$$

Thus a moving plane that follows a rational surface is equivalent to a syzygy of the polynomials $x(s,t), y(s,t), z(s,t), w(s,t)$ representing the rational parameterization. Similarly a moving quadric (6.3) is said to follow the rational surface (6.4) if

$$(A_{m,n}s^m t^n + \cdots + A_{0,0})x^2(s,t) + \cdots + (J_{m,n}s^m t^n + \cdots + J_{0,0})w^2(s,t) \equiv 0.$$

Thus a moving quadric that follows a rational surface is a syzygy of the ten polynomials $x^2(s,t), \ldots, w^2(s,t)$.

To implicitize a rational surface of bidegree (m, n), consider moving planes of bidegree $(2m - 1, n - 1)$ that follow the rational surface:

$$(A_{2m-1,n-1}s^{2m-1}t^{n-1} + \cdots + A_{0,0})x(s,t) +$$
$$(B_{2m-1,n-1}s^{2m-1}t^{n-1} + \cdots + B_{0,0})y(s,t) +$$
$$(C_{2m-1,n-1}s^{2m-1}t^{n-1} + \cdots + C_{0,0})z(s,t) +$$
$$(D_{2m-1,n-1}s^{2m-1}t^{n-1} + \cdots + D_{0,0})w(s,t) \equiv 0.$$

Treating the parameters $A_{i,j}, B_{i,j}, C_{i,j}, D_{i,j}$ as undetermined coefficients leads to $6mn$ homogeneous linear equations in $8mn$ unknowns. Hence from linear algebra there exist at least $2mn$ linearly independent solutions. Placing these solutions in a matrix and taking the determinant generates

$$\begin{vmatrix} A^1_{2m-1,n-1}x + \cdots + D^1_{2m-1,n-1}w & \cdots & A^1_{0,0}x + \cdots + D^1_{0,0}w \\ \vdots & \vdots & \vdots \\ A^{2mn}_{2m-1,n-1}x + \cdots + D^{2mn}_{2m-1,n-1}w & \cdots & A^{2mn}_{0,0}x + \cdots + D^{2mn}_{0,0}w \end{vmatrix} = 0$$

a polynomial equation of degree $2mn$ in x, y, z, w. When the rational surface has no base points, this polynomial is the implicit equation of the rational surface (6.4).

When base points are present, we can apply the method of moving quadrics to find the implicit equation. To implicitize a rational surface of bidegree (m, n), consider moving quadrics of bidegree $(m - 1, n - 1)$ that follow the rational surface:

$$(A_{m-1,n-1}s^{m-1}t^{n-1} + \cdots + A_{0,0})x^2(s,t) + \cdots$$
$$+(J_{m-1,n-1}s^{m-1}t^{n-1} + \cdots + J_{0,0})w^2(s,t) \equiv 0.$$

Now treating the parameters $A_{i,j}, \cdots, J_{i,j}$ as undetermined coefficients leads to $9mn$ homogeneous linear equations in $10mn$ unknowns. Hence there exist at least mn linearly independent solutions. Placing these solutions in a matrix and taking the determinant generates

$$\begin{vmatrix} A^1_{m-1,n-1}x^2 + \cdots + J^1_{m-1,n-1}w^2 & \cdots & A^1_{0,0}x^2 + \cdots + J^1_{0,0}w^2 \\ \vdots & \vdots & \vdots \\ A^{mn}_{m-1,n-1}x^2 + \cdots + J^{mn}_{m-1,n-1}w^2 & \cdots & A^{mn}_{0,0}x^2 + \cdots + J^{mn}_{0,0}w^2 \end{vmatrix} = 0,$$

a polynomial equation of degree $2mn$ in x, y, z, w. When the rational surface has no base points, this polynomial is the implicit equation of the rational surface. When base points are present, some of the quadratic rows reduce to linear expressions – that is, some of the moving quadrics become moving planes – and in many cases the determinant still represents the implicit equation of the rational surface.

The method of moving quadrics has certain advantages over resultant techniques. First, the method employs smaller matrices than resultants – $mn \times mn$ vs. $2mn \times 2mn$ for surfaces of bidegree (m, n). Second, the method often works even in the presence of base points. Here for each base point counted with proper multiplicity, one row of the moving quadric matrix reduces to a row of moving planes. Moreover, the method of moving quadrics is valid for total degree as well as for bidegree rational surfaces, though for surfaces of a fixed total degree some of the rows must always be filled in with moving planes. Finally, the method of moving quadrics uses only straightforward tools from linear algebra. Therefore the method of moving quadrics is easier for members of the geometric modeling community to understand than competing techniques such as residual resultants [5] or approximation complexes [6].

The method of moving quadrics is a method of undetermined coefficients. But moving quadrics is faster than naive undetermined coefficients, since to use the methods of moving quadrics we need to solve $O(mn)$ equations in $O(mn)$ unknowns, whereas to use naive undetermined coefficients we would have to solve $O(m^3n^3)$ equations in $O(m^3n^3)$ unknowns.

Why do these moving surface implicitization techniques work? Again algebraic geometry provides the key insights. When there are no base points, both methods – moving planes and moving quadrics – generate a polynomial in x, y, z of the same degree as the implicit equation, and by construction this polynomial vanishes on the rational surface. The key point is that this expression is not identically zero.

The method of moving planes can be shown to be equivalent to implicitization by resultants [51]. The method of moving quadrics is more subtle. When there are no base points, [16] use sheaf cohomology to show that the method will always generate the implicit equation provided there are no moving planes of bidegree $(m-1, n-1)$ that follow the surface. Moreover, [4, 6, 14] extend these arguments to show that the method of moving quadrics works for surfaces of fixed total degree even in the presence of base points provided that the base points are local complete intersections. Thus algebraic geometry provides the insight underlying the computations developed by geometric modeling.

7 Resolvents

A *resultant* is a single polynomial condition that vanishes when $N+1$ polynomials in N variables have a common root. *Resolvents* are a collection of polynomial conditions that vanish when M polynomials in N variables ($M > N+1$) have a common root.

Resolvents have several potential applications in geometric modeling. For example, we could use resolvents to represent non-planar rational curves

$$x = \frac{x(t)}{w(t)}, \quad y = \frac{y(t)}{w(t)}, \quad z = \frac{z(t)}{w(t)}$$

as the intersection of implicit algebraic surfaces by computing the resolvents of the three polynomials $x(t) - xw(t)$, $y(t) - yw(t)$, $z(t) - zw(t)$. Thus resolvents could be applied to implicitize and intersect non-planar rational curves [29, 31]. In the bivariate setting, resolvents could be employed to detect base points on a rational surface

$$x = \frac{x(s,t)}{w(s,t)}, \quad y = \frac{y(s,t)}{w(s,t)}, \quad z = \frac{z(s,t)}{w(s,t)}$$

by detecting common roots of the polynomials $x(s,t), y(s,t), z(s,t), w(s,t)$ representing the parameterization. For example, if we want to know if a rational quadratic parameterization represents a quadric surface, then we need to determine if the parameterization has two base points.

Kronecker's method is a technique that uses resultants for computing resolvents. To compute the resolvents for three univariate polynomials $F(t), G(t), H(t)$ of degree D, Kronecker computes the resultant $Res(F(t), G(t) + uH(t))$, a polynomial in u of degree D. This polynomial in u vanishes identically if and only if $F(t), G(t), H(t)$ have a common root. Hence the coefficients of $u^k, k = 0, \cdots, D$, are resolvents for the polynomials $F(t), G(t), H(t)$. Thus Kronecker's method generates $D+1$ resolvent conditions for three univariate polynomials of degree D. Similarly Kronecker's method can be applied to find resolvents in the multivariate setting. Kronecker's methods is effective, but not efficient. Alternative resolvent methods due to [34, 21] and others often lead to fewer conditions.

Consider, for example, the simple case of three quadratic polynomials

$$F(t) = a_2 t^2 + a_1 t + a_0,$$
$$G(t) = b_2 t^2 + b_1 t + b_0,$$
$$H(t) = c_2 t^2 + c_1 t + c_0.$$

Kronecker's method generates 3 conditions from $Res(F(t), G(t) + uH(t)) \equiv 0$:

1. $\operatorname{Res}(F(t), G(t)) = 0 \Leftrightarrow F$ and G have a common root
2. $\operatorname{Res}(F(t), H(t)) = 0 \Leftrightarrow F$ and H have a common root
3. $2a_2^2 b_0 c_0 - a_1 a_2 b_1 c_0 + a_1^2 b_2 c_0 - 2a_0 a_2 b_2 c_0 - a_1 a_2 b_0 c_1$
   $+ 2a_0 a_2 b_1 c_1 - a_0 a_1 b_2 c_1 + a_1^2 b_0 c_2 - 2a_0 a_2 b_0 c_2 - a_0 a_1 b_1 c_2 + 2a_0^2 b_2 c_2 = 0.$

On the other hand, two conditions suffice to determine if three quadratic polynomials F, G, H have a common root.

1. $\operatorname{Res}(F(t), G(t)) \equiv \operatorname{Det} \begin{pmatrix} a_2 & 0 & b_2 & 0 \\ a_1 & a_2 & b_1 & b_2 \\ a_0 & a_1 & b_0 & b_1 \\ 0 & a_0 & 0 & b_0 \end{pmatrix} = 0 \Leftrightarrow F$ and G have a common root

2. $\operatorname{Det}(F(t), G(t), H(t)) \equiv \operatorname{Det} \begin{pmatrix} a_2 & b_2 & c_2 \\ a_1 & b_1 & c_1 \\ a_0 & b_0 & c_0 \end{pmatrix} = 0 \Leftrightarrow F, G, H$ are linearly dependent.

In this simple example, it may be possible to recover the two simpler conditions from the three Kronecker conditions, but, in general, algebraic geometry does not seem to provide efficient computational methods for expressing resolvents. Surely we do not need $D + 1$ algebraic surfaces to implicitize a non-planar rational curve of degree D. What then is the minimum number of surfaces required? What are their minimal degrees and how do we find them? The existence of resolvents is guaranteed by the theory, but before resolvents can be used effectively in geometric modeling applications, additional research needs to be done to determine the fastest, most efficient, most compact, most numerically stable approach to resolvents.

8 Surface classifications and surface features

Algebraic geometry distinguishes many different types of surfaces: rational, total degree, bidegree, toric, del Pezzo, and many others. Each of these surface types incorporates special geometric properties that have potential applications in geometric modeling. Tensor product bidegree surface patches are the mainstay of geometric modeling because these surface patches are easily integrated with rectangular meshes [24, 28]. Triangular total degree surface patches are the next most common surfaces in geometric modeling because these surface patches are well suited for handling scattered data [24, 28]. Recently toric surface patches with arbitrary Newton polygons have entered geometric modeling in order to build multisided patches and to fill n-sided holes [36, 39]. Geometric insights – Segre, Veronese, and toric embeddings – and algebraic tools – Macaulay, Dixon, and toric resultants – originate from within algebraic geometry, but the computational techniques – Bezier representations as well as subdivision, evaluation, and differentiation algorithms – arise from inside geometric modeling.

Surfaces have many geometric features such as curvatures, axes and centers of symmetry, and foci. For quadric surfaces, schemes are known for extracting standard features such as the type, center, axes and foci from their implicit representation [46], but in geometric modeling the standard representation is most often rational parametric rather than implicit polynomial. To extract these quadric surface features from rational representations, we must implicitize and then compute. But as we have seen, implicitization for surfaces with base points can be complicated. Recently, straightforward algorithms have been developed for extracting the geometric features of conic sections – type, center, axes, foci – from their rational parameterizations [55], but direct computational tools for extracting geometric features for quadric surfaces from rational representations are at present unavailable.

Gaussian and mean curvature formulas for surfaces both for implicit representations and for parametric representations are known from differential geometry [53], [54]. Algebraic geometry provides neither geometric insight nor computational tools for extracting these basic features.

Thus, whereas algebraic geometry provides insight into the different surface types that are available as tools for geometric modeling, the computational interests of the two disciplines concerning surfaces are often distinct. What is the reason for this dichotomy? Geometric modeling is frequently interested in metric properties of surfaces, but algebraic geometry is concerned primarily with affine or projective invariants. Perhaps this distinction helps to explain why algebraic geometry does not provide effective tools for computing many of the surface features of interest in geometric modeling, even though algebraic geometry does indicate effectively which surface types are worth further study.

9 Conclusions and some open problems for future research

The thesis of this paper is that it is a mistake to confound constructive theoretical tools with effective computational techniques. I have endeavored to provide many examples to substantiate this premise. We have seen that for problems of even moderate size constructive tools from algebraic geometry such as Gröbner bases, resultants, and homotopy continuation for solving systems of polynomial equations are not competitive with numerical methods from geometric modeling such as Bezier subdivision.

Rather than efficient computation, the main contribution of algebraic geometry to geometric modeling has been insight into the complexity of the computational problems at hand. For example, Bernstein's theorem provides the number of complex roots of a system of polynomial equations. This result bounds the number of real intersections of two rational curves and the degree of the real intersection curve of two rational surfaces. Bernstein's theorem also furnishes the degree of the implicit equation of a rational curve or surface.

Algebraic geometry has also provided a service to geometric modeling by supplying proofs for numerical techniques such as implicitization by the method of moving surfaces developed from within geometric modeling. What algebraic geometry has not provided to geometric modeling is efficient numerical techniques. Fashions in the philosophy of mathematics vary from transcendent to constructive, but constructive philosophical trends in mathematics do not necessarily translate into effective computational tools for engineers: there is a big difference between constructive theoretical tools and effective computational techniques.

Nevertheless, these conclusions should not be taken to mean that constructive methods are not at all useful. To the contrary, computers are potentially powerful theoretical tools. Computation can be allied to intuition; even inefficient computations can lead to novel theoretical insights.

There may also be isolated areas where algebraic geometry furnishes computation as well as insight to geometric modeling. Finding singularities [7] and rendering algebraic curves and surfaces based on an analysis of singularities [1] are two such potential applications. Perhaps too the proof of the validity of the method of moving quadrics for the special case presented in [4] (total degree surfaces with base points that are local complete intersections) provides a more efficient implicitization algorithm than the method of moving quadrics based on standard linear algebra.

Moreover, new, more efficient computational approaches to Gröbner bases or homotopy continuation or resultants may yet prove to be effective in geometric modeling. Much of the original work and many of the original comparative studies in geometric modeling are now dated – resultants [52], homotopy continuation [44], Gröbner bases [32] – and it may well be time to reevaluate newer, more efficient computational methods embodied in more modern software. In addition, almost no comparative studies exist on the efficiency or numerical stability of the various resultant and resolvent formulations.

I will close then with some problems for future research that may help to fill in some holes in the current theory as well as bridge the gap between the constructive tools of algebraic geometry and the effective numerical methods required by geometric modeling.

1. *Generic Resultants.* Explore the relative speed, efficiency, and numerical stability of the different resultant formulations, especially in the bivariate setting.
2. *Sylvester Resultants.* Classify the Newton polygons for which exact Sylvester resultants (and exact Bezout resultants) exist.
3. *Dixon Resultants.* When the Dixon resultant vanishes identically, determine both the maximal minor and the extraneous factor.
4. *Resolvents.* Find an efficient, compact, numerically stable representation for resolvents. Show how to use resolvents to determine efficiently the number and multiplicity of the base points for a rational surface.
5. *Implicitization.* Investigate the relative speed, efficiency, and numerical stability of the different implicitization techniques for rational surfaces, including Gröbner bases, resultants and syzygies.
6. *Extraction of Surface Invariants.* Develop techniques for finding axes, centers of symmetry, foci and other geometric features of rational surfaces directly from their rational parameterizations.
7. *Floating Point Computations.* For speed, most computations in geometric modeling are performed using floating point rather than exact arithmetic. Analyze the effect of floating point computations on various techniques from constructive algebraic geometry, including resultants, resolvents, Gröbner bases, and homotopy continuation. Determine whether these floating point computations are sufficiently stable for geometric modeling applications.
8. *Comparative Studies.* Compare constructive techniques based on the latest implementations of constructive algebraic geometry – Gröbner bases, homotopy continuation, resultants – to numerical methods such as Bezier subdivision from geometric modeling. Determine which type and which size problems are most amenable to techniques from constructive algebraic geometry and which problems are best suited to numerical methods from geometric modeling. Consider, in particular, implicitizing and intersecting rational curves and surfaces, as well as finding the real roots of a system of polynomials inside a bounded domain.

Acknowledgements

This work is partially supported by NSF grants CCF-0203318 and INT-0421771. I would also like to thank David Cox and Falai Chen for reading a preliminary version of this manuscript and making some helpful comments and suggestions. All the opinions expressed in this paper as well as any errors that still remain are, of course, entirely my own.

References

1. Bodnar, G. and Schicho, J. (2000), A computer program for the resolution of singularities, in H. Hauser, editor, *Resolution of Singularities*, Vol. 181 of Progr. Math., pp. 231-238. Birkhauser.
2. Busé, L. (2001), Residual resultant over the projective plane and the implicitization problem, Proceedings of ISSAC 2001, pp. 48-55.
3. Busé, L. and Chardin, M. (2004), Implicitizing rational hypersurfaces using approximation complexes, submitted to *Journal of Symbolic Computation*.
4. Busé, L., Cox, D., and D'Andrea, C. (2003), Implicitization of surfaces in $\mathbb{P}^3$ in the presence of base points, *Journal of Algebra and its Applications*, Vol. 2, pp. 189-214.
5. Busé, L., Elkadi, M., and Mourrain, B. (2001), Resultant over the residual of a complete intersection, *Jour. of Pure and Applied Algebra*, Vol. 164, pp. 35-57.
6. Busé, L. and Jouanolou, J.P. (2003), On the closed image of a rational map and the implicitization problem, *Journal of Algebra*, Vol. 265, pp. 312-357
7. Chen, F. and Wang, W. (2004), Computing the singular points of a planar rational curve using the μ-basis, preprint.
8. Chtcherba, A. and Kapur, D. (2000), Conditions for exact resultants using the Dixon formulation, *International Symposium on Symbolic and Algebraic Computation (ISSAC 2000)*, St. Andrews, Scotland.
9. Chionh, E. and Goldman, R. (1995a), Elimination and resultants part 1: Elimination and bivariate resultants, *IEEE Computer Graphics and Applications*, Vol. 15, pp. 69-77.
10. Chionh, E. and Goldman, R. (1995b), Elimination and resultants part 2: Multivariate resultants, *IEEE Computer Graphics and Applications*, Vol. 15, pp. 60-69.
11. Chionh, E., Miller, J. and Goldman, R. (1991), Using multivariate resultants to find the intersection of three quadric surfaces, *Transactions on Graphics*, Vol. 10, pp. 378-400.
12. Chionh, E., Zhang, M. and Goldman, R. (2002), Fast computation of the Bezout and Dixon resultant matrices, *Journal of Symbolic Computation*, Vol. 33, pp. 13-29.
13. Cox, D. (2001), Equations of parametric curves and surfaces via syzygies, in *Symbolic Computation: Solving Equations in Algebra, Geometry and Engineering*, AMS Contemporary Mathematics, Vol. 286, pp. 1-20.
14. Cox, D. (2003), Curves, surfaces and syzygies, in *Topics in Algebraic Geometry and Geometric Modeling*, edited by R. Goldman and R. Krasauskas, AMS Contemporary Mathematics, Vol. 334, pp. 131-150.
15. Cox, D., Little, J. and O'Shea, D. (1998), *Using Algebraic Geometry*, Springer, New York.
16. Cox, D., Zhang, M. and Goldman, R. (2000), On the validity of implicitization by moving quadrics for rational surfaces with no base points, *Journal of Symbolic Computation*, Vol. 29, pp. 419-440.
17. D'Andrea, C. and Dickenstein, A. (2001), Explicit formulas for the multivariate resultant, *Jour. Pure Appl. Algebra*, Vol. 164, pp. 59-86.
18. Davis, P. and Hersh, R. (1986). *Descartes' Dream: The World According to Mathematics*, Harcourt Brace Jovanovich, San Diego.
19. Dixon, A. (1908), The eliminant of three quantics in two independent variables, *Proc. London Mathematical Society*, Vol. 6, pp. 49-69.

20. Dokken, T. (2001), Approximate implicitization, in *Mathematical Methods in CAGD*, T. Lyche and L. Schumaker (eds.), Vanderbilt University Press, Nashville.
21. Du, H. (1991), *New Resolvent Methods with Applications to Curves and Surfaces in Geometric Modeling*, Ph.D. Thesis, Department of Computer Science, University of Waterloo.
22. Dupont, L., Lazard, D., Lazard, S., and Petitjean, S. (2003), Near-optimal parameterization of the intersection of quadrics, *Symposium on Computational Geometry 2003*, pp. 246-255.
23. Emiris, I. and Mourrain, B. (1999), Matrices in elimination theory, *Journal of Symbolic Computation*, Vol. 28, pp. 3-44.
24. Farin, G. (2002), *Curves and Surfaces for Computer Aided Geometric Design: A Practical Guide*, Fifth Edition, Academic Press, Inc., San Diego.
25. Farouki, R., Neff, C., and O'Connor, M. (1989), Automatic parsing of degenerate quadric surface intersections, *ACM Transactions on Graphics*, Vol. 8, pp. 174-203.
26. Foo, M. and Chionh, E. (2004), Corner edge cutting and Dixon A-resultant quotients, *Journal of Symbolic Computation*, Vol. 37, pp. 101-119.
27. Gelfand, I. M., Kapranov, M. M. and Zelevinsky, A. V. (1994), Discriminants, Resultants and Multidimensional Determinants, Boston, Birkhauser.
28. Goldman, R. (2002), *Pyramid Algorithms: A Dynamic Programming Approach to Curves and Surfaces for Geometric Modeling*, Morgan Kaufmann.
29. Goldman, R. (1985), The method of resolvents: A technique for the implicitization, inversion, and intersection of non-planar, parametric, rational cubic curves, *Computer Aided Geometric Design*, Vol. 2, pp. 237-255.
30. Goldman, R. and Sederberg, T. (1986), Algebraic geometry for computer-aided geometric design, *IEEE Computer Graphics and Applications*, Vol. 6, pp. 52-59.
31. Goldman, R. and Sederberg, T. (1987), An analytic approach to the intersection of all piecewise parametric rational cubic curves, *Computer-Aided Design*, Vol. 19, pp. 282-292.
32. Hoffmann, C. (1989), *Geometric and Solid Modeling: An Introduction*, Morgan Kaufmann.
33. Juettler, B. and Wurm, E. (2003), Approximate implicitization via curve fitting, in L. Kobbelt, P. Schroder, H. Hoppe (eds.), *Symposium on Geometry Processing, Eurographics/ACM Siggraph*, New York, pp. 240-247.
34. Kakie, K. (1976), The resultant of several homogeneous polynomials in two indeterminates, *Proceedings of AMS*, Vol. 54, pp. 1-7.
35. Karciauskas, K. (2003), Rational M-patches and tensor border patches, in *Topics in Algebraic Geometry and Geometric Modeling*, edited by R. Goldman and R. Krasauskas, AMS Contemporary Mathematics, Vol. 334, pp. 101-128.
36. Karciauskas, K. and Krasauskas, R. (2000), Comparison of different multisided patches using algebraic geometry, *Curve and Surface Design: Saint-Malo 1999*, P-J. Laurent, P. Sablonniere, and L. Schumaker (eds.), pp. 163-172, Vanderbilt University Press, Nashville.
37. Khetan, A. (2003), The resultant of an unmixed bivariate system, *Journal of Symbolic Computation*, Vol. 36, pp. 425-442.
38. Khetan, A., Song, N., and Goldman, R. (2004), Sylvester A-resultants for bivariate polynomials with planar Newton polygons, *Proceedings of the 2004 International Symposium on Symbolic and Algebraic Computation (ISSAC)*, Santander, Spain.

39. Krasauskas, R. (2000), Toric surface patches, *Advances in Computational Mathematics*, Vol. 21, pp. 1-25.
40. Levin, J. (1979), Mathematical models for determining the intersections of quadric surfaces, *Computer Graphics and Image Processing*, Vol. 1, pp. 73-87.
41. Macaulay, F. (1902), Some formulae in elimination, *Proceedings London Math. Soc.*, Vol. 1, pp. 3-27.
42. Manocha, D. and Canny, J. (1993), Multipolynomial resultant algorithms, *Journal of Symbolic Computation*, Vol. 15, pp. 99-122.
43. Miller, J. (1987), Geometric approaches to nonplanar quadric surface intersection curves, *ACM Transactions on Graphics*, Vol. 6, pp. 274-307.
44. Morgan, A, (1983), A method for computing all solutions to systems of polynomial equations, *ACM Trans. on Math. Software*, Vol. 9, pp. 1-17.
45. Ocken, S., Schwartz, J. and Sharir, M. (1987), Precise implementation of CAD primitives using rational parametrizations of standard surfaces, Schwartz, Hopcroft, and Sharir, eds., *Planning, Geometry, and Complexity of Robot Motion*, Ablex Publishing Corporation, pp. 245-266.
46. Roe, J. (1993), *Elementary Geometry*, Oxford University Press.
47. Rojas, J. M. (2003), Why polyhedra matter in non-linear equation solving, in *Topics in Algebraic Geometry and Geometric Modeling*, edited by R. Goldman and R. Krasauskas, AMS Contemporary Mathematics, Vol. 334, pp. 293-320.
48. Sarraga, R. (1983), Algebraic methods for intersections of quadric surfaces in GMSOLID, *Computer Vision, Graphics and Image Processing*, Vol. 22, pp. 222-238.
49. Sederberg, T. and Chen, F. (1995), Implicitization using moving curves and surfaces, *Proceedings of Siggraph'95*.
50. Sederberg, T., Goldman, R., and Anderson, D. (1984), Implicit representation of parametric curves and surfaces, *Computer Vision, Graphics, and Image Processing*, Vol. 28, pp. 72-84.
51. Sederberg, T., Goldman, R., and Du, H. (1997), Implicitizing rational curves by the method of moving algebraic curves, *Journal of Symbolic Computation*, Vol. 23, pp. 153-175.
52. Sederberg, T. and Parry, S. (1986), A comparison of curve-curve intersection algorithms, *Computer-Aided Design*, Vol. 18, pp. 58–63.
53. Spivak, M. (1975), *A Comprehensive Introduction to Differential Geometry*, Vol. 3, Publish or Perish, Inc., Boston.
54. Stoker, J. (1969), *Differential Geometry*, John Wiley and Sons, New York.
55. Wang, W. and Goldman, R. (2004), Using invariants to extract geometric characteristics of conic sections from rational quadratic parametrizations, to appear in *The International Journal of Computational Geometry and Applications*.
56. Wang, W., Joe, B. and Goldman, R. (2003), Computing quadric surface intersections based on an analysis of planar cubic curves, *Graphical Models*, Vol. 64, pp. 335-367.
57. Weyman, J. and Zelevinsky, A. (1994), Determinantal formulas for multigraded resultants, *Jour. Algebraic Geometry*, Vol. 3, pp. 569-597.
58. Winkler, J. (2003), A companion matrix resultant for Bernstein polynomials, *Linear Algebra and Its Applications*, 362, pp. 153-175.
59. Winkler, J. and Goldman, R. (2003), The Sylvester resultant matrix for Bernstein polynomials, in *Curve and Surface Design: Saint-Malo 2002*, T. Lyche, M. Mazure and L. Schumaker (eds.), Nashboro Press, Brentwood, Tennessee, pp. 407-416.

60. Zhang, M., Chionh, E. and Goldman, R. (1998), Hybrid Dixon resultants, *Proceedings of the Eighth IMA Conference on the Mathematics of Surfaces*, August 1998, pp. 193-212.
61. Zhang, M., White, R., Wang, L., Kavraki, L., Goldman, R. and Hassett, B. (2005), Improving conformational searches by geometric screening, accepted to appear in the *Journal of Bioinformatics*.
62. Zheng, J., Sederberg, T., Chionh, E. and Cox, D. (2003), Implicitizing rational surfaces with base points using the method of moving surfaces, in *Topics in Algebraic Geometry and Geometric Modeling*, edited by R. Goldman and R. Krasauskas, AMS Contemporary Mathematics, Vol. 334, pp. 151-168.

Implicitization using approximation complexes

Marc Chardin

Institut de Mathématiques de Jussieu,
4 place Jussieu, Paris,
France
chardin@math.jussieu.fr

Summary. This paper describes a method, exploiting approximation complexes, for computing the implicit equation of a parameterized hypersurface. The fundamental ingredients and properties used in this approach are recalled and illustrated on simple examples.

1 Introduction

We present in this short account a method for computing the image of a rational map from $\mathbb{P}^{n-1}$ to $\mathbb{P}^n$, under suitable hypotheses on the base locus and on the image.

The formalism we use is due to Jean-Pierre Jouanolou, who gave a course on this approach at the University of Strasbourg during the academic year 2000–2001. In his joint article with Laurent Busé [5], this formalism is explained in details and applications to the implicitization problem are given.

The idea of using a matrix of syzygies for the implicitization problem goes back to the work of Sederberg and Chen [17] and was at the origin of several important contributions to this approach (see for instance [7], [8], [9], [10] and the articles on this subject in the volume of the 2002 conference on Algebraic Geometry and Geometric Modeling [1]).

Most of this note is dedicated to presenting the method, the geometric ideas behind it and the tools from commutative algebra that are needed. Some references to classical textbooks are given for the concepts and theorems we use for the presentation. In the last section, we give the most advanced results we know related to this approach. We illustrate this technique on an example that we carry out in details throughout the article.

References are given to the publication that fits best our statments. They may not be the first place where a similar result appeared. For instance, many results were first proved for $n = 2$ or for $n = 3$.

2 General setting

Given
$$\phi : \mathbb{P}^{n-1} \dashrightarrow \mathbb{P}^n$$
a rational map defined by $f := (f_0, \ldots, f_n)$, $f_i \in R := k[X_0, \ldots, X_n]$ homogeneous of degree $d \geq 1$, such that the closure of its image is a hypersurface $\mathcal{H}$, the goal is to compute the equation H of this hypersurface.

No hypothesis on the field k is needed. When we compute for instance the image of a map, in the sense of algebraic k-schemes, it should be remembered that it corresponds geometrically to the image over the algebraic closure of the field (the image is itself defined over k if the scheme is).

We let:
- $I := (f_0, \ldots, f_n) \subset R$ be the ideal generated by the f_i's,
- $X := \mathrm{Proj}(R/I) \subset \mathbb{P}^n$ be the subscheme defined by I.

A specific example: We will illustrate in this article the different steps and constructions on an example, taken from [4, Example 3.2]:
$$\phi : \mathbb{P}^2 \dashrightarrow \mathbb{P}^3,$$
given by $f := (ac^2, b^2(a+c), ab(a+c), bc(a+c))$ with $R_E := \mathbb{Q}[a,b,c]$. The ring of the target will be $R'_E := \mathbb{Q}[x,y,z,t]$.

We will refer to this example several times during the remainder of the article.

3 The algebro-geometric intuition

If $\Gamma_0 \subset \mathbb{P}^{n-1} \times \mathbb{P}^n$ is the graph of $\phi : (\mathbb{P}^{n-1} \setminus X) \longrightarrow \mathbb{P}^n$ and Γ the Zariski closure of Γ_0, one has:
$$\mathcal{H} = \overline{\pi(\Gamma_0)} = \pi(\Gamma),$$
where $\pi : \mathbb{P}^{n-1} \times \mathbb{P}^n \longrightarrow \mathbb{P}^n$ is the projection, and the bar denotes the Zariski closure (or equivalently the closure for the usual topology in the case $k = \mathbb{C}$).

The first equality directly follows from the definition of $\mathcal{H}$, and the second from the fact that π is a projective morphism (so that the image of a variety is a variety).

On the algebraic side [15, II §7], one has
$$\Gamma = \mathrm{Proj}(\mathcal{R}_I),$$
with $\mathcal{R}_I := R \oplus I \oplus I^2 \oplus \cdots$ and the embedding $\Gamma \subset \mathbb{P}^{n-1} \times \mathbb{P}^n$ corresponds to the natural graded map:

$$S := R[T_0, \ldots, T_n] \xrightarrow{s} \mathcal{R}_I$$
$$T_i \longmapsto f_i \in I = (\mathcal{R}_I)_1.$$

If $\mathfrak{P} := \ker(s)$, $\mathfrak{P}_1$ (the degree 1 part of $\mathfrak{P}$) is the module of syzygies of the f_i's:
$$a_0 T_0 + \cdots + a_n T_n \in \mathfrak{P}_1 \iff a_0 f_0 + \cdots + a_n f_n = 0.$$

The ideal $\mathfrak{P}$ is called by the geometric modeling community the moving hypersurface ideal and its elements of degree 1 (the syzygies) are called moving hyperplanes.

Setting $\mathcal{S}_I := \operatorname{Sym}_R(I)$ and $V := \operatorname{Proj}(\mathcal{S}_I)$, we have natural onto maps

$$S \longrightarrow S/(\mathfrak{P}_1) \quad \text{and} \quad \mathcal{S}_I \simeq S/(\mathfrak{P}_1) \longrightarrow S/\mathfrak{P} \simeq \mathcal{R}_I$$

which correspond to the embeddings

$$\Gamma \subseteq V \subset \mathbb{P}^{n-1} \times \mathbb{P}^n.$$

As $\mathcal{R}_I$ is the bigraded domain defining Γ, the projection $\pi(\Gamma)$ is defined by the graded domain $\mathcal{R}_I \cap k[T_0, \ldots, T_n]$. We have assumed that $\pi(\Gamma)$ is the hypersurface $H = 0$, so that this may be rewritten:

$$(H) = \mathfrak{P} \cap k[T_0, \ldots, T_n].$$

In our example, with $S_E := R_E[x,y,z,t] = \mathbb{Q}[a,b,c,x,y,z,t]$:

$$\mathfrak{P} = \bigl(ay - bz, at - cz, bt - cy, act - b(a+c)x\bigr) + \bigl(bx(z+t) - at^2\bigr) + \bigl(xy(z+t) - zt^2\bigr)$$

where we have separated the (minimal) generators of degrees 1, 2 and 3 for simplicity. Of course it follows that $H = xy(z+t) - zt^2$. Also, by definition,

$$\mathcal{S}_I \simeq S_E/(\mathfrak{P}_1) = S_E/\bigl(ay - bz, at - cz, bt - cy, act - b(a+c)x\bigr).$$

The fact that $\mathcal{R}_I$ and $\mathcal{S}_I$, as well as the canonical map $\mathcal{S}_I \longrightarrow \mathcal{R}_I$, do not depend on generators of I is useful in proving the following:

Theorem 1. *We have $\Gamma = V$ if X is locally a complete intersection.*

In our example, the saturation of I is the complete intersection ideal $\bigl(ac^2, b(a+c)\bigr)$ because $b(a+c) \cdot (a,b,c) \subset I$ so that $I \subseteq \bigl(ac^2, b(a+c)\bigr) \subseteq I^{sat}$, and $\bigl(ac^2, b(a+c)\bigr)$ is saturated. Therefore X is locally a complete intersection (it is even globally a complete intersection). See the next section for the definition and elementary properties of the saturation of an ideal.

More refined criteria exist to insure that $\Gamma = V$, but we will stick here to this one. This is partly justified by the following result:

Proposition 2. *If $\dim X = 0$, $\Gamma = V$ if and only if X is locally a complete intersection.*

The theorem above explains the key role of syzygies in computing H: they are equations of definition of Γ when X is locally a complete intersection.

A more algebraic way to state the theorem is the following:

Theorem 3. *The prime ideal $\mathfrak{P}$ is the saturation of the ideal generated by its elements of degree 1 in the T_i's (the syzygies) if X is locally a complete intersection.*

In our example, $\mathfrak{P}_2 \subset (\mathfrak{P}_1) : (a,b,c)$ and $\mathfrak{P} = (\mathfrak{P}_1) : (a,b,c)^2$.

Nevertheless, as it is clear from this formulation of the theorem, one should not forget that even if $\Gamma = V$, it need not be the case that $\mathcal{R}_I = \mathcal{S}_I$. In fact, the equality $\mathcal{R}_I = \mathcal{S}_I$ may only hold in trivial cases in our context, because H is a minimal generator of $\mathfrak{P}$. The difference between these algebras (which is the torsion part of $\mathcal{S}_I$, because $\mathcal{R}_I$ is a domain) is a key point when one uses the syzygies to compute H. This is very much similar to the fact that a homogeneous ideal defining a variety in the projective space need not be saturated.

The way the method proceeds is somehow parallel to determinantal methods for computing resultants: it uses graded pieces of a resolution of $\mathcal{S}_I$ to compute $\pi(V)$.

The connection between the elimination theory viewpoint, which looks at H as the generator of $\mathfrak{P} \cap k[T_0, \ldots, T_n]$, and the determinantal approach that computes H from graded pieces of a resolution of $\mathcal{S}_I$ is shown by the following:

Proposition 4. *[5, 5.1] Assume that $\Gamma = V$ and let η be such that $H^0_{\mathfrak{m}}(\mathcal{S}_I)_\mu = 0$ for all $\mu \geq \eta$. Then,*

$$\mathrm{ann}_{k[T_0,\ldots,T_n]}(\mathcal{S}_I^\eta) = \mathfrak{P} \cap k[T_0,\ldots,T_n].$$

Here $\mathfrak{m} := (X_1, \ldots, X_n)$, and if M is a R-module, $H^0_{\mathfrak{m}}(M) := \{m \in M \mid \exists \ell, \ X_i^\ell m = 0 \ \forall i\}$. The graded pieces of $\mathcal{S}_I$ will be described below, and we will provide estimates for η satifying the vanishing condition. Notice that $H^0_{\mathfrak{m}}(\mathcal{S}_I)$ is the torsion part of $\mathcal{S}_I$ when $\Gamma = V$.

Remark 5. The choices of gradings are one of the delicate points in this approach. For instance, the hypothesis $H^0_{\mathfrak{m}}(\mathcal{S}_I)_\mu = 0$ is equivalent to

$$H^0_{\mathfrak{m}}\big(\mathrm{Sym}^j_R(I)\big)_{\mu+dj} = 0, \quad \forall j,$$

if we adopt the natural grading of $\mathrm{Sym}^j_R(I)$ making the canonical map

$$\mathrm{Sym}^j_R(I) \longrightarrow I^j \subset R$$

a homogeneous map of degree zero.

A candidate for a resolution of $\mathcal{S}_I$ is the $\mathcal{Z}$-complex introduced and studied by Herzog, Simis and Vasconcelos. We will decribe this complex in the next section.

4 Tools from commutative algebra

The saturation of an ideal — An homogeneous ideal I in a polynomial ring $R := k[X_1, \ldots, X_n]$ (where k is any commutative ring) is *saturated* (or, more precisely $\mathfrak{m}$-saturated) if $I : \mathfrak{m} = I$, where $\mathfrak{m} := (X_1, \ldots, X_n)$. In other words, I is saturated if: $X_i f \in I, \forall i \Rightarrow f \in I$.

The ideal $I^{sat} := \bigcup_j (I : \mathfrak{m}^j)$ is saturated, it is the smallest saturated ideal containing I and is called the saturation of I.

Another way of seeing the saturation of an ideal, that directly extends to modules, is given by the remark that:

$$I^{sat} = I + H^0_{\mathfrak{m}}(R/I)$$

that one can also write $R/I^{sat} = (R/I)/H^0_{\mathfrak{m}}(R/I)$.

The saturation of a module M will be $M/H^0_{\mathfrak{m}}(M)$. As usual, one should be careful about the fact that the saturation of an ideal I corresponds to saturating the module R/I and not the ideal considered as a module over the ring.

Seeing the saturation operation in relation with the left exact functor $H^0_{\mathfrak{m}}(-)$ naturally leads to the consideration of the derived functors $H^i_{\mathfrak{m}}(-)$, and to the cohomological approach of algebraic geometry.

There is a one-to-one correspondance between the subschemes of a projective space $\mathbb{P}^{n-1}$ and the *saturated* homogeneous ideals of the polynomial ring $R := k[X_1, \ldots, X_n]$. To see this notice that, by definition, two subschemes of $\mathbb{P}^{n-1}$ are the same if they coincide on all the affine charts $X_i = 1$. If ϕ_i is the specialization homomorphism $X_i \mapsto 1$ then the homogenization of $\phi_i(I)$ is the ideal $I_{(i)} := \bigcup_j \left(I : (X_i^j)\right)$. It follows that I and J define the same scheme if and only if $I_{(i)} = J_{(i)}$ for all i, which is easily seen to be equivalent to the equality of their saturation as $I^{sat} = \bigcap_i I_{(i)}$.

When considering multigraded ideals, with respect to a set of variables that are generating ideals $\mathfrak{m}_1, \ldots, \mathfrak{m}_t$ (these ideals are never maximal unless $t = 1$ and k is a field), the operations of saturation with respect to the different ideals naturally appears. The subschemes of the corresponding product of projective spaces corresponds one-to-one to ideals that are saturated with respect to all the ideals $\mathfrak{m}_i$, or equivalently with respect to the product of these ideals.

The ring of sections [15, II §5],[11, §A4.1] — If $R := k[X_1, \ldots, X_n]$ and $B := R/I$ is the quotient of R by the homogeneous ideal I, an interesting object to consider is:

$$\Gamma B := \ker\Big(\bigoplus_i B_{X_i} \longrightarrow \bigoplus_{i<j} B_{X_i X_j}\Big),$$

where $B_{(f)} := \{\frac{x}{f^j} \mid x \in B, \ j \in \mathbb{N}\}$ and the maps are the evident ones up to a sign chosen so that $(1, \ldots, 1)$ maps to 0. One has a natural isomorphism $B_{X_i} \simeq B/(X_i - 1)$ and ΓB should be interpreted as the applications that are defined on each affine chart $X_i = 1$ and matches on the intersection of any two of these charts. Notice that it is clear from the definition that replacing I by its saturation do not affect ΓB.

In a sheaf theoretic language, one has

$$\Gamma B = \bigoplus_{\mu \in \mathbb{Z}} H^0(\mathbb{P}^{n-1}, \mathcal{O}_X(\mu)),$$

with $X := \mathrm{Proj}(B)$, and the natural grading of ΓB coincides with the grading of the section ring on the right hand side.

These considerations extend to modules along the same lines. Also, the map we used to define ΓB fits into a complex, called the Čech complex,

$$0 \longrightarrow B \xrightarrow{\phi} \bigoplus_i B_{X_i} \xrightarrow{\psi} \bigoplus_{i<j} B_{X_i X_j} \longrightarrow \cdots \longrightarrow B_{X_1 \cdots X_n} \longrightarrow 0.$$
$$\parallel \qquad \parallel \qquad \qquad \parallel \qquad \qquad \qquad \parallel$$
$$\mathcal{C}_\mathfrak{m}^0(B) \quad \mathcal{C}_\mathfrak{m}^1(B) \qquad \mathcal{C}_\mathfrak{m}^2(B) \qquad \qquad \mathcal{C}_\mathfrak{m}^n(B)$$

One has $H_\mathfrak{m}^0(B) = \ker(\phi)$ and $\Gamma B = \ker(\psi)$. It is a standard fact that $H_\mathfrak{m}^i(B)$ is isomorphic to the i-th cohomology module of this complex. This in particular gives an exact sequence:

$$0 \longrightarrow H_\mathfrak{m}^0(B) \longrightarrow B \longrightarrow \Gamma B \longrightarrow H_\mathfrak{m}^1(B) \longrightarrow 0,$$

which splits into two parts, the difference between the homogeneous quotient B and the more geometric notion of the section ring attached to $X := \mathrm{Proj}(R/I) \subseteq \mathbb{P}^{n-1}$.

If k is a field and X is of dimension zero, $\Gamma B_\mu = H^0(\mathbb{P}^{n-1}, \mathcal{O}_X(\mu))$ is a k-vector space of dimension the degree of X for any μ. In particular, when $\dim X = 0$, ΓB is not finitely generated. In any dimension, it can be shown that $\Gamma(R/I)$ is finitely generated if and only if I has no associated prime $\mathfrak{p}$ such that $\mathrm{Proj}(R/\mathfrak{p})$ is of dimension zero (*i.e.* $\dim(R/\mathfrak{p}) = 1$).

Castelnuovo-Mumford regularity [11, §20.5] — The Castelnuovo-Mumford regularity is an invariant that measures the algebraic complexity of a graded ideal or module over a polynomial ring $R := k[X_1, \ldots, X_n]$. The two most standard definitions are given either in terms of a minimal finite free R-resolution of the module (this resolution exists by Hilbert's theorem on syzygies) or in terms of the vanishing of the cohomology modules defined above (using a theorem of Serre [15, III 5.2] to show that this makes sense).

Definition 6. *Let $b_i(M)$ be the maximal degree of a minimal i^{th} syzygy of M and $a_i(M) := \inf\{\mu \in \mathbb{Z} \mid H_\mathfrak{m}^i(M)_\nu = 0, \ \forall \nu > \mu\}$, then*

$$\mathrm{reg}(M) = \max_i \{a_i(M) + i\} = \max_i \{b_i(M) - i\}.$$

Notice that if $M = R/I$, minimal 0th syzygies of M are minimal generators of M (namely, the element 1), minimal 1st syzygies of M are minimal generators of I, and 2nd syzygies of M are syzygies between the chosen (minimal) generators of I. If one looks at I as a module, these modules are the same up to a shift in the labeling, except 0th module for R/I, and one has $\operatorname{reg}(I) = \operatorname{reg}(R/I) + 1$.

The existence of different interpretations of the regularity is a key to many results on this invariant. It is for instance immediate from the cohomological definition that $\operatorname{reg}(I^{sat}) \leq \operatorname{reg}(I)$, but this is not easy to see using the definition in terms of syzygies. Also, when $\dim X = 0$ ($X := \operatorname{Proj}(R/I)$, as above), it easily follows from the cohomological definition and the fact that $H^i_{\mathfrak{m}}(M) = 0$ for $i > \dim M$ (Grothendieck's vanishing theorem) that $\operatorname{reg}(I)$ is the smallest integer μ such that:

(1) $I_\mu = (I^{sat})_\mu$ (recall that I^{sat} is the saturation of I),
(2) $\dim(R/I^{sat})_{\mu-1} = \deg(X)$.

In case X is a set of simple points, condition (2) says that passing through the $\deg(X)$ different points of X impose linearly independant conditions on polynomials of degree $\mu - 1$. An elementary account on regularity in this context is given in [1, §4, D. Cox, Curves, surfaces and syzygies].

The fact that $\operatorname{reg}(I)$ bounds the degrees of the syzygies of I shows the naturality of considering this invariant in the implicitization problem using the syzygy matrix.

On the computational side, the degrees of generators of a Gröbner basis of the ideal for the degree-reverse-lex order, under a quite weak conditions on the coordinates, is bounded by $\operatorname{reg}(I)$. This is another way of understanding the regularity as a measure of the complexity of the ideal.

The Koszul complex [11, §17] — Let $x = (x_1, \ldots, x_r)$ be a r-tuple of elements in a ring A. The (homological) Koszul complex $K_\bullet(x; A)$ is the complex with modules
$$K_p(x; A) := \bigwedge^p A^r \simeq A^{\binom{p}{r}}$$
and maps $d_p : K_p(x; A) \longrightarrow K_{p-1}(x; A)$ defined by:
$$g.e_{i_1} \wedge \cdots \wedge e_{i_p} \longmapsto g.\sum_{j=1}^p (-1)^{j+1} x_{i_j} e_{i_1} \wedge \cdots \widehat{\wedge e_{i_j}} \cdots \wedge e_{i_p}.$$

We set $Z_p(x; A) := \ker(d_p)$ and $H_p(x; A) := Z_p(x; A)/\operatorname{im}(d_{p+1})$.

The $\mathcal{Z}$-complex. [18, Ch. 3] — We consider $f_i \in R \subset S$ as elements of S and the two complexes $K_\bullet(f; S)$ and $K_\bullet(T; S)$ where $T := (T_0, \ldots, T_n)$. These complexes have the same modules $K_p = \bigwedge^p S^{n+1} \simeq S^{\binom{p}{n+1}}$ and differentials $d^f_\bullet$ and $d^T_\bullet$.

- It directly follows from the definitions that $d^f_{p-1} \circ d^T_p + d^T_{p-1} \circ d^f_p = 0$, so that $d^T_p(Z_p(f;S)) \subset Z_{p-1}(f;S)$. The complex $\mathcal{Z}_\bullet := (Z_\bullet(f;S), d^T_\bullet)$ is the called $\mathcal{Z}$-complex associated to the f_i's.
- Notice that $Z_p(f;S) = S \otimes_R Z_p(f;R)$ and
  — $Z_0(f;R) = R$,
  — $Z_1(f;R) = \mathrm{Syz}_R(f_0, \ldots, f_n)$,
  — the map $d^T_1 : S \otimes_R \mathrm{Syz}_R(f_0, \ldots, f_n) \longrightarrow S$ is defined by:

$$(a_0, \ldots, a_n) \longmapsto a_0 T_0 + \cdots + a_n T_n.$$

The following result shows the intrinsic nature of the homology of the $\mathcal{Z}$-complex, it is a key point in proving results on its acyclicity.

Theorem 7. *We have $H_0(\mathcal{Z}_\bullet) \simeq \mathcal{S}_I$ and the homology modules $H_i(\mathcal{Z}_\bullet)$ are $\mathcal{S}_I$-modules that only depend on $I \subset R$, up to isomorphism.*

- We let $R' := k[T_0, \ldots, T_n]$ and look at graded pieces:

$$\mathcal{Z}^\mu_\bullet : \cdots \longrightarrow R' \otimes_k Z_2(f;R)_\mu \xrightarrow{d^T_2} R' \otimes_k Z_1(f;R)_\mu \xrightarrow{d^T_1} R' \otimes_k Z_0(f;R)_\mu \longrightarrow 0$$

where $Z_p(f;R)_\mu$ is the part of $Z_p(f;R)$ consisting of elements of the form $\sum a_{i_1 \ldots i_p} e_{i_1} \wedge \cdots \wedge e_{i_p}$ with the $a_{i_1 \ldots i_p}$ are of the same degree μ.

Nota Bene. *This is not the usual convention for the grading of these modules, however we chose it here for simplicity. The usual grading (used for instance in [3] or [6]) makes the Koszul maps homogeneous of degree 0, so they require that $a_{i_1 \ldots i_p}$ be homogeneous of degree $\mu - pd$ in place of being of degree μ.*

We will denote the cokernel of the last map by $\mathcal{S}^\mu_I$.

Determinants of complexes [16, §3.6], [13, App. A] — Let A be a commutative domain, for simplicity.

If $A^n \xrightarrow{\alpha} A^n$ is A-linear we can define $\det(\alpha) \in A$.

If we have a complex $C_\bullet$ with three terms:

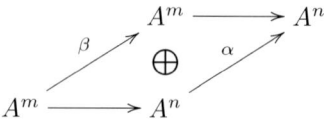

such that $\det(\beta) \neq 0$, we set $\det(C_\bullet) := \frac{\det(\alpha)}{\det(\beta)}$. In fact $\det(C_\bullet)$ is independent of the decomposition of C_1 as a direct sum $A^m \oplus A^n$.

More generally a bounded complex $C_\bullet$ of free A-modules such that $\mathrm{Frac}A \otimes_A C_\bullet$ is exact may always be decomposed in the following way

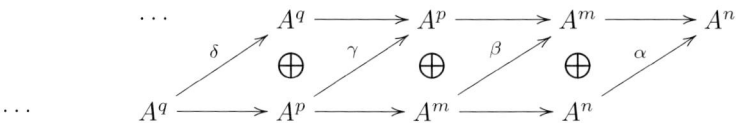

with $\alpha, \beta, \ldots$ having non-zero determinants. Then $\det(C_\bullet) := \frac{\det(\alpha).\det(\gamma)\cdots}{\det(\beta).\det(\delta)\cdots}$ is independent of the decomposition.

Performing the decomposition of a given complex is easy: decompose first C_1 into $A^n \bigoplus A^m$ so that $\det(\alpha) \neq 0$ (this amounts to choose a non-zero maximal minor of the map $C_1 \longrightarrow C_0 = A^n$), and apply the procedure recursively to the complex:

$$\cdots \longrightarrow C_3 \longrightarrow C_2 \longrightarrow A^m \longrightarrow 0.$$

Fitting ideals [16, §3.1], [11, §20.2] — If A is a ring and M is a module represented as the cokernel of a map $\psi : A^m \longrightarrow A^n$, the ideal generated by minors of size $n-i$ of ψ only depends on M and i. This ideal is called the i-th Fitting ideal of the A-module M. One of the most important of these ideals associated to the A-module M is the 0-th Fitting ideal (*i.e.* the one generated by the maximal minors of ψ), denoted by $\mathrm{Fitt}^0_A(M)$.

5 The method and main results

Recall that $R := k[X_1, \ldots, X_n]$, $R' := k[T_0, \ldots, T_n]$, $S = R \otimes_k R'$, $\Gamma = \mathrm{Proj}(\mathcal{R}_I) \subseteq \mathbb{P}^{n-1} \times \mathbb{P}^n$ (see §3 for the definition of $\mathcal{R}_I$) and $\pi : \mathbb{P}^{n-1} \times \mathbb{P}^n \longrightarrow \mathbb{P}^n$ is the natural projection.

We assume hereafter that $\pi(\Gamma)$ is of codimension 1 in $\mathbb{P}^n$ defined by the equation $H = 0$ and denote by δ the degree of the map π from Γ onto its image.

If J is a R'-ideal, we will denote by $[J]$ the gcd of the elements in J. It represents the component of codimension one of the scheme defined by J (its divisorial component) because R' is factorial.

With this notation, one has:

Proposition 8. *[5, 5.2] If $X = \emptyset$, $\mathcal{Z}_\bullet$ is acyclic and*

$$[\mathrm{Fitt}^0_{R'}(\mathcal{S}^\mu_I)] = \det(\mathcal{Z}^\mu_\bullet) = H^\delta,$$

for every $\mu \geq (n-1)(d-1)$.

The identities above are identities of principal ideals in R', therefore they correspond to equalities of elements of R' up to units. Recall that $[\mathrm{Fitt}^0_{R'}(\mathcal{S}^\mu_I)]$ is the gcd of maximal minors of the map $d_1^T : (a_0, \ldots, a_n) \longmapsto a_0 T_0 + \cdots + a_n T_n$ from the syzygies of degree μ (each a_i is of degree μ) seen as a vector space over k to $R' \otimes_k R_\mu$. The entries of this matrix are therefore linear forms in the T_i's with coefficients in k.

This proposition shows that the determinant of this graded part of $\mathcal{Z}_\bullet$ actually computes the divisor $\pi_*(\Gamma) = \delta \cdot \pi(\Gamma)$ obtained as direct image of the cycle Γ (see [12, §1.4] for the defintion of the direct image $\pi_*(\Gamma)$ of the cycle Γ).

In the case X is of dimension zero, the situation is slightly more complicated:

Proposition 9. *[5, 5.7, 5.10], [3, 4.1] If* $\dim X = 0$,

i) *The following are equivalent:*
   a) X *is locally defined by at most* n *equations,*
   b) $\mathcal{Z}_\bullet$ *is acyclic,*
   c) $\mathcal{Z}_\bullet^\mu$ *is acyclic for* $\mu \gg 0$.
ii) *If* $\mathcal{Z}_\bullet$ *is acyclic, then*

$$[\mathrm{Fitt}^0_{R'}(\mathcal{S}^\mu_I)] = \det(\mathcal{Z}^\mu_\bullet) = H^\delta G,$$

for every $\mu \geq (n-1)(d-1) - \varepsilon_X$, *where* $1 \leq \varepsilon_X \leq d$ *is the minimal degree of a hypersurface containing* X *and* $G \neq 0$ *is a homogeneous polynomial which is a unit if and only if* X *is locally a complete intersection.*

Remark 10. In fact $[\mathrm{Fitt}^0_{R'}(\mathcal{S}^\mu_I)] = \det(\mathcal{Z}^\mu_\bullet) = \pi_* V$ for $\mu \geq (n-1)(d-1) - \varepsilon_X$, and the degree of G is the sum of numbers measuring how far X is from a complete intersection at each point of X.

Remark 11. It is very fast to compute the ideal I^{sat} with a dedicated computer algebra system (like Macaulay 2, Singular or Cocoa), and *a fortiori* to compute ε_X which is the smallest degree of an element in I^{sat}. Moreover the following result actually implies a good bound on the complexity of this task.

Proposition 12. *[6, 3.3] If* $J \subset R$ *is a homogeneous ideal generated in degree at most* d *with* $\dim(R/J) = 1$ *and* J' *its saturation (in other words, the defining ideal of the zero-dimensional scheme* $X := \mathrm{Proj}(R/J)$), *then*

$$\mathrm{reg}(J) \leq n(d-1) + 1 \quad \text{and} \quad \mathrm{reg}(J') \leq (n-1)(d-1) + 1.$$

In our example, $\mathrm{reg}(I) = \mathrm{reg}(I^{sat}) = 4$, while the general bound above gives $\mathrm{reg}(I) \leq 7$ and $\mathrm{reg}(I^{sat}) \leq 5$. A minimal free R_E-resolution of I gives a resolution of Z_1:

$$R_E[-2] \xrightarrow{\begin{bmatrix} c \\ -b \\ -a \\ 0 \end{bmatrix}} R_E[-1]^3 \oplus R_E[-2] \xrightarrow{\begin{bmatrix} 0 & 0 & 0 & -b(a+c) \\ a & 0 & c & 0 \\ -b & -c & 0 & 0 \\ 0 & a & -b & ac \end{bmatrix}} Z_1 \subset R_E^4 = (S_E)_1$$

and we have seen that $I^{sat} = (ac^2, b(a+c))$, so that $\mathrm{indeg}(I^{sat}) = 2$ and therefore $(n-1)(d-1) - \varepsilon_X = 2 \times (3-1) - 2 = 2$. The syzygies of degree 2 are of the form:

$$\ell_1(ay - bz) + \ell_2(at - cz) + \ell_3(cy - bt) + \lambda_4\bigl(act - b(a+c)x\bigr)$$

with $\ell_i \in (R_E)_1$ and $\lambda_4 \in (R_E)_0 = k$. Notice that they are not linearly independent, and that the relation (unique in this degree) is given by the second syzygy:

$$c(ay - bz) - b(at - cz) - a(cy - bt) = 0.$$

We may for instance choose as generators of syzygies of degree 2 the 9 syzygies, s_1 to s_9: $a(ay - bz)$, $b(ay - bz)$, $a(at - cz)$, $b(at - cz)$, $c(at - cz)$, $a(cy - bt)$, $b(cy - bt)$, $c(cy - bt)$, $act - b(a+c)x$ which gives the 6×9 matrix of linear forms (elements of $(R'_E)_1$) for the matrix of d_1^T in degree 2 (recall that $T = (x, y, z, t)$ with the notation of the example):

$$\begin{array}{c} a^2 \\ ab \\ ac \\ b^2 \\ bc \\ c^2 \end{array} \begin{bmatrix} y & 0 & t & 0 & 0 & 0 & 0 & 0 & 0 \\ -z & y & 0 & t & 0 & -t & 0 & 0 & -x \\ 0 & 0 & -z & 0 & t & y & 0 & 0 & t \\ 0 & -z & 0 & 0 & 0 & 0 & -t & 0 & 0 \\ 0 & 0 & 0 & -z & 0 & 0 & y & -t & -x \\ 0 & 0 & 0 & 0 & -z & 0 & 0 & y & 0 \end{bmatrix}.$$

with columns labeled $s_1\ s_2\ s_3\ s_4\ s_5\ s_6\ s_7\ s_8\ s_9$.

Now, Z_2 has a free R_E-resolution of the form:

$$0 \longrightarrow R_E[-4] \longrightarrow R_E[-1] \oplus R_E[-3]^3 \longrightarrow Z_2 \subset \bigwedge^2(S_E)_1 = R_E^6.$$

In particular, Z_2 has one minimal generator of degree 1 and no minimal generator of degree 2. The element of degree 1

$$\Sigma := c\,y \wedge z - b\,z \wedge t - a\,y \wedge t \in \bigwedge^2(S_E)_1$$

satisfies $d_2^f(\Sigma) = b(a+c)[c(ay - bz) - b(at - cz) - a(cy - bt)] = 0$. Therefore $(Z_2)_1 = k\,\Sigma$ and $(Z_2)_2 = \Sigma(R_E)_1$. We have $d_2^T(\Sigma) = c(z \otimes y - y \otimes z) - b(z \otimes t - t \otimes z) - a(t \otimes y - y \otimes t) \in S_E \otimes_{R_E} \bigwedge^1(S_E)_1$, that we may rewrite

$$d_2^T(\Sigma) = -t \otimes (ay - bz) + y \otimes (at - cz) + z \otimes (cy - bt).$$

In degree 2, the matrix of $d_2^T : R'_E \otimes_k \Sigma(R_E)_1 \longrightarrow R'_E \otimes_k Z_1(f; R_E)_2$ on the bases $(a\Sigma, b\Sigma, c\Sigma)$ for the source and $(s_1, \ldots, s_9)$ for the target is therefore the transpose of

$$\begin{array}{c} a\Sigma \\ b\Sigma \\ c\Sigma \end{array} \begin{bmatrix} -t & 0 & y & 0 & 0 & z & 0 & 0 & 0 \\ 0 & -t & 0 & y & 0 & 0 & z & 0 & 0 \\ 0 & 0 & 0 & -t & y & t & 0 & z & 0 \end{bmatrix}.$$

with columns labeled $s_1\ s_2\ s_3\ s_4\ s_5\ s_6\ s_7\ s_8\ s_9$.

We now choose a maximal non-zero minor of this matrix, for instance the minor Δ_2 given by lines 3, 4 and 5 of the matrix of d_2^T, and the minor Δ_1 of the matrix of d_1^T obtained by erasing columns 3,4 and 5. We get the formula:

$$H = \frac{\Delta_1}{\Delta_2} = \frac{\begin{vmatrix} y & 0 & 0 & 0 & 0 & 0 \\ -z & y & t & 0 & 0 & -x \\ 0 & 0 & y & 0 & 0 & t \\ 0 & -z & 0 & -t & 0 & 0 \\ 0 & 0 & 0 & y & -t & -x \\ 0 & 0 & 0 & 0 & y & 0 \end{vmatrix}}{\begin{vmatrix} y & 0 & 0 \\ 0 & y & -t \\ 0 & 0 & y \end{vmatrix}} = \frac{-y^3(xyz + xyt - t^2z)}{y^3}.$$

Computations of the free R_E-resolutions of Z_1 and Z_2 were done using the dedicated software Macaulay 2 by Dan Grayson and Mike Stillman [14]. In the case $n = 3$, this computation goes very fast, even for pretty high degree d, and Macaulay 2 performs degree truncations to speed up the computation, if needed. The graded pieces that we need to know can also easily be computed using linear algebra routines, as explained in [3] and implemented in [2].

When the dimension of the base locus X of the map ϕ increases, the situation becomes harder to analyze. In dimension 1, the situation is pretty well understood:

Proposition 13. *[6, 8.2, 8.3] Assume that $\dim X = 1$ and let $\mathcal{C}$ be the union of components of dimension 1 of X (its "unmixed part"). Then,*

i) *The following are equivalent:*
   a) *X is locally defined by at most n equations and $\mathcal{C}$ is defined on a dense open subset by at most $n - 1$ equations,*
   b) *$\mathcal{Z}_\bullet^\mu$ is acyclic for $\mu \gg 0$.*

ii) *If $\mathcal{Z}_\bullet^\mu$ is acyclic for $\mu \gg 0$, the following are equivalent:*
   a) *$\mathcal{Z}_\bullet$ is acyclic,*
   b) *$\mathcal{C}$ is arithmetically Cohen-Macaulay,*
   b') *every section $f \in H^0(\mathcal{C}, \mathcal{O}_\mathcal{C}(\mu))$ is the restriction to $\mathcal{C}$ of a polynomial function of degree μ, for every $\mu \in \mathbb{Z}$.*

iii) *If $\mathcal{Z}_\bullet^\mu$ is acyclic for $\mu \gg 0$ and $H^0(\mathcal{C}, \mathcal{O}_\mathcal{C}(\mu)) = 0$ for all $\mu < -d$ —for instance if $\mathcal{C}$ is reduced— then $\mathcal{Z}_\bullet^\mu$ is acyclic for $\mu \geq (n-1)(d-1)$. If further X is defined by at most $n-1$ equations locally on the support of $\mathcal{C}$, then*

$$[\mathrm{Fitt}_{R'}^0(\mathcal{S}_I^\mu)] = \det(\mathcal{Z}_\bullet^\mu) = H^\delta G,$$

for every $\mu \geq (n-1)(d-1)$, where G is a homogeneous polynomial such that the support of $\pi(V)$ is the zero set of GH.

Here also, more precisely, $\det(\mathcal{Z}_\bullet^\mu)$ represents the divisor $\pi_* V$.

Remark 14. It is perhaps true that $\det(\mathcal{Z}_\bullet^\mu)$ represents the divisor $\pi_* V$ for $\mu \geq (n-1)(d-1)$ when $\mathcal{Z}_\bullet^\mu$ is acyclic for $\mu \gg 0$ and $H^0(\mathcal{C}, \mathcal{O}_\mathcal{C}(\mu)) = 0$ for all $\mu < -d$, but we needed the slightly stronger hypothesis above to prove it in [6].

References

1. R. Goldman, R. Krasaukas (Eds.). Topics in Algebraic Goemetry and Geometric Modeling. *Contemporary Mathematics* **334** (2003).
2. L. Busé. *Algorithms for the implicitization using approximation complexes.* (http://www-sop.inria.fr/galaad/personnel/Laurent.Buse/program.html).
3. L. Busé, M. Chardin. Implicitizing rational hypersurfaces using approximation complexes. *J. Symbolic Computation (to appear).*
4. L. Busé, D. Cox, C. D'Andréa. Implicitization of surfaces in $\mathbb{P}^3$ in the presence of base points. *J. Algebra Appl.* **2** (2003), 189–214.
5. L. Busé, J.-P. Jouanolou. On the closed image of a rational map and the implicitization problem. *J. Algebra* **265** (2003), 312–357.
6. M. Chardin. Regularity of ideals and their powers. Preprint **364** (Mars 2004), Institut de Mathématiques de Jussieu, Paris.
7. D. Cox. Equations of parametric curves and surfaces via syzygies. *Contemporary Mathematics* **286** (2001), 1–20.
8. D. Cox, T. Sederberg, F. Chen. The moving line ideal basis of planar rational curves. *Comp. Aid. Geom. Des.* **15** (1998), 803–827.
9. D. Cox, R. Goldman, M. Zhang. On the validity of implicitization by moving quadrics for rationnal surfaces with no base points. *J. Symbolic Computation* **29** (2000), 419–440.
10. C. D'Andréa. Resultants and moving surfaces. *J. of Symbolic Computation* **31** (2001), 585–602.
11. D. Eisenbud. *Commutative algebra. With a view toward algebraic geometry.* Graduate Texts in Mathematics **150**. Springer-Verlag, New York, 1995.
12. W. Fulton. *Intersection theory.* Second edition. Ergebnisse der Mathematik und ihrer Grenzgebiete **3**. Springer-Verlag, Berlin, 1998.
13. I. M. Gel'fand, M. Kapranov, A. Zelevinsky. *Discriminants, resultants, and multidimensional determinants.* Mathematics: Theory & Applications. Birkhäuser Boston, Inc., Boston, MA, 1994.
14. D. Grayson, M. Stillman. *Macaulay 2.* (http://www.math.uiuc.edu/Macaulay2/).
15. R. Hartshorne. *Algebraic geometry.* Graduate Texts in Mathematics **52**. Springer-Verlag, New York-Heidelberg, 1977.
16. D. G. Northcott. *Finite free resolutions.* Cambridge Tracts in Mathematics **71**. Cambridge University Press, Cambridge-New York-Melbourne, 1976.
17. T. Sederberg, F. Chen. Implicitization using moving curves and surfaces. *Proceedings of SIGGRAPH 95*, Addison Wesley, 1995, 301–308.
18. W. Vasconcelos. *The Arithmetic of Blowup Algebras.* London Math. Soc. Lecture Note Ser. **195**. Cambridge University Press, 1994.

Piecewise approximate implicitization: experiments using industrial data

Mohamed F. Shalaby[1], Jan B. Thomassen[2,3], Elmar M. Wurm[1], Tor Dokken[2] and Bert Jüttler[1]

[1] Institute of Applied Geometry, Johannes Kepler University, Altenberger Str. 69, 4040 Linz, Austria; firstname.lastname@jku.at
[2] SINTEF ICT, P.O. Box 124 Blindern, N–O314 Oslo, Norway; [Jan.B.Thomassen|Tor.Dokken]@sintef.no
[3] Center of Mathematics for Applications, P.O. Box 1053 Blindern, N–0316 Oslo, Norway; jan.b.thomassen@cma.uio.no

Summary. We compare several methods for approximate implicitization by piecewise polynomials which have been developed by the authors, and a linear-algebra-based numerical method for implicitization which is provided as a part of MAPLE. We investigate both quantitative criteria (such as computing time, memory use, and the error of the approximation) and qualitative criteria. As demonstrated by the results, piecewise approximate implicitization is able to handle surfaces arising in industrial applications. However, special care has to be taken to avoid additional branches and unwanted singularities.

1 Introduction

Parametric representations, such as NURBS surfaces, are used in most CAD systems, see [10]. The parametric representation offers a number of advantages, e.g., simple techniques for display and for analyzing the geometric properties. On the other hand, implicitly defined surfaces are better suited in many applications, such as surface fitting (possibly subject to shape constraints, such as convexity) and the possibility of defining solids. In many applications, such as the computation of intersection and the detection of possible self–intersections, it is advantageous to combine both types of representations [15, 16].

In order to exploit the potential of implicit representation, methods for conversion to and from implicit form are needed. The process of implicitization has been discussed by several authors. Various exact methods, such as resultants, Gröbner bases, moving curves and surfaces exist [3, 4, 5, 9, 12, 17, 19]. More recently, a number of approximate methods [1, 6, 7, 14] have emerged. They form a valuable alternative to the exact techniques.

Several computational techniques for approximate implicitization of surfaces by a single algebraic surface have been compared in [20]. However, for more complicated input surfaces, the approximation by a single algebraic surface is no longer feasible. In these cases, surfaces of relatively high degree would be needed, which may not be of much use for applications. In addition, the use of such surfaces would lead to numerical problems. Consequently, the use of piecewise polynomial representations (i.e., splines) is more appropriate.

This paper presents a feasibility study for two methods for approximate implicitization by piecewise polynomials. The two methods are the scattered data fitting method developed at Linz ('PPL') , and the sampling-based method developed at SINTEF ('PPS'). We restrict ourselves to the approximate implicitization of surfaces, since this case is more important for applications. Our results are based on work we have done in connection with the EU project GAIA II.

The feasibility study is performed by applying the algorithms to two sets of "academic" and industrial test surfaces. The "academic" surfaces were constructed in the course of testing and developing the algorithms, while the industrial surfaces were supplied by the CAD vendor think3, which is a partner in the GAIA II project.

We follow a procedure similar to [20], where benchmarking for implicitization by polynomials has been presented. The input was a single parametric patch while the output was a single polynomial implicit function. In this paper, we have generalized this to the case of piecewise polynomials. Consequently, we can now deal with much more complicated surfaces. The approach enables piecewise surfaces – i.e., NURBS surfaces – as input, and piecewise polynomial implicit functions as output. In addition to studying the feasibility, the two methods are qualitatively compared to each other.

Some of the test surfaces are given in terms of a single polynomial patch, and for this data it is possible to apply the implicitize routine implemented in MAPLE ('ML') and the previously studied single polynomial implicitization algorithm from SINTEF ('PS') [20]. For these test cases we have been able to compare all four methods (PPL, PPS, ML, PS), and thereby study the effects of using implicit functions that are piecewise polynomial.

This paper is organized as follows. Section 2 gives a short outline of the four methods of implicitization. Section 3 describes the test surfaces we have used in our experiments. In section 4, we comment on the parameters we are measuring in our experiments – time, memory usage, and approximation error – and specify how we made the measurements. The results are presented in section 5. In section 6 we give a qualitative analysis, and finally, in Section 7, we give a conclusion.

2 The methods

In this Section, we describe the four methods used for approximate implicitization. Only the first two of them (PPL and PPS) are relevant for piecewise implicitization. The two additional methods (PS and ML) are included in order to study some effects of using piecewise polynomials instead of just a single polynomial.

All methods produce an approximate implicit representation of the form

$$f(x,y,z) = \sum_{j \in \mathcal{J}} \phi_j(x,y,z) c_j \qquad (2.1)$$

with certain coefficients $c_j \in \mathbb{R}$ and finite index set $\mathcal{J}$. In the case of PPL and PPS, the basis functions $\phi_j(x, y, z)$ are trivariate tensor-product B-splines. Consequently, the output is a three–dimensional array of B-spline coefficients, along with three knot vectors. In the case of PPL, the array is sparse, and dynamical data structures may therefore be used in order to exploit the sparsity. PS and ML use a suitable basis of trivariate polynomials, such as Bernstein polynomials on a tetrahedron or tensor-product Bernstein polynomials.

Note that none of these methods takes care of approximating the boundary curves (e.g., by clipping surfaces). This may be a subject of further research. Also, the use of hierarchical spline spaces (such as T-splines) should lead to a further improvement of the method.

2.1 PPL

This method for approximate implicitization by piecewise polynomials has been developed at Linz [13]. It is characterized by the simultaneous approximation of sampled point data $\mathbf{p}_i = (x_i, y_i, z_i)$, $i \in \mathcal{I}$, and estimated unit normals $\mathbf{n}_i$ at these points.

If no other information is available (e.g., from a given parametric or procedural description of the surface), then each normal vector $\mathbf{n}_i$ is estimated from the nearest neighbors of the point $\mathbf{p}_i = (x_i, y_i, z_i)$. In addition, a region-growing-type algorithm is used for adjusting the orientation of the normal vectors.

In order to control the shape of the resulting surface, an additional tension term is optimized. It pulls the approximating surface towards a simpler shape. A possible quadratic tension term is

$$T(\mathbf{c}) = \iiint_{\Omega} \left(f_{xx}^2 + 2 f_{xy}^2 + f_{yy}^2 + 2 f_{xz}^2 + 2 f_{yz}^2 + f_{zz}^2 \right) \, \mathrm{d}x \, \mathrm{d}y \, \mathrm{d}z. \qquad (2.2)$$

By increasing the influence of this tension term, the resulting surface becomes more similar to a plane.

The implicit function is obtained as the minimum of a convex quadratic objective function

$$\sum_{i \in \mathcal{I}} f(x_i, y_i, z_i)^2 + w||\nabla f(x_i, y_i, z_i) - \mathbf{n}_i||^2 + \text{``tension''}, \tag{2.3}$$

where w is a positive weight. The weight controls the influence of the estimated normal vectors $\mathbf{n}_i$ to the resulting surface. As observed in our experiments, increasing the weight can be used to 'push away' unwanted branches of the surface.

This method is fully general, i.e., it can be applied to any space of functions, not only to piecewise polynomials. For practical applications, however, fast evaluation of basis functions is important. For this reason we implemented the algorithm for trivariate tensor-product B-splines. In this case, the basis functions ensure global smoothness, and the resulting system of linear equations is sparse.

The domain of interest is divided into cubes of the same size. This is done by specifying a cell-size. In order to guarantee an integer number of cells, the bounding box of the input surface is enlarged a bit. We consider only the cubes which contain data, and its neighbors. We named it as "active cells". The domain of the spline functions consists only of the active cells.

Another input parameter is the degree d of the spline function. We choose the knot vector with simple knots in the interior, so the continuity is C^{d-1}. The choice of the degree d depends on the singularities of the given curve/surface. Singularities can be reproduced by the algebraic approximation, provided that a sufficiently high degree is employed in the algebraic approximation. For example, in 2D, in order to represent a double point, one has to use degree three or higher. See also the discussion in Section 6.2, which addresses the conflict between pushing away unwanted branches and avoiding singularities.

Due to the compact support of the B-splines, the implementation is relatively fast because of the sparsity of the resulting linear system of equations. Consequently, even complicated singular surfaces can be implicitized. In this case, it is an important issue to create a consistent orientation of the (estimated) normals $\mathbf{n}_i$, and this can be achieved by a propagation technique [14]: First, an initial approximation is computed using only information from one part of the data, where a consistent orientation could be created without ambiguities. The result is then used to propagate the orientation of the normals to the neighboring segments of the data. The details of this method are described in [13, 20, 18]. A fast implementation of the method is important, since is then feasible to iterate this process, for orientation propagation and for adjusting the estimated normals.

2.2 PPS

The piecewise polynomial approximate implicitization algorithm made at SINTEF is a generalization of the polynomial SINTEF algorithm from [7, 20]. Essentially, it is a generalization in two ways: First, it produces a piecewise polynomial implicit function, more specifically a trivariate tensor product B-spline function. The three knot vectors of the B-spline function are obtained by dividing the bounding box of the input surface into $n \times n \times n$ cells, where n is a user-defined integer, and adding d additional knots at the boundaries (where d is the degree of the spline function). This gives interpolating boundary conditions. The knot vector is chosen with simple knots in the interior, hence the continuity equals C^{d-1}.

The second generalization is the use of point sampling on the input surface. Thus a linear system of equations is set up such that each point defines a row in the matrix of the equation,

$$f(x_i, y_i, z_i) = \sum_{j \in \mathcal{J}} \phi_j(x_i, y_i, z_i) c_j = 0, \quad i \in \mathcal{I}. \tag{2.4}$$

Points are sampled according to a uniform grid in the parameter domain. The density of the points in the parameter domain is chosen large enough such that the matrix equation is over-determined. From experimenting with the algorithm, a good choice for the number of points in the algorithm was found to be roughly two times the number of coefficients in the implicit function. Thus, if $M \times N$ denotes the dimensions of the matrix $\Phi_{ij} = \phi_j(x_i, y_i, z_i)$, we have $M \approx 2N$.

The rest of the algorithm proceeds essentially in the same way as the polynomial version [7, 20]: We get a matrix, and we want to find a suitable vector in an approximate null space of this matrix. This is done by SVD, and the resulting approximate nullvector will be the vector of coefficients for the implicit spline function. Special care must be taken to ignore vectors belonging to singular values that are exactly zero. Such singular values appear in general because some cells in the rectangular space grid are outside the support of the relevant basis functions, and this produces columns of zeros in the matrix. We remove these columns before we run the SVD. We also take the square of the matrix Φ, i.e. we use $\Phi^T \Phi$, which further reduces the size of the matrix from $M \times N$ to $N \times N$ before SVD is applied. The coefficients corresponding to the original zero columns are subsequently arbitrarily set to zero.

The input parameters to this algorithm, in addition to the surface, are the degree of the spline approximation and the number of grid cells in each direction.

2.3 PS

This method is the single polynomial approximate implicitization method developed at SINTEF, which is also based on singular value decomposition (SVD). A description can be found in [7].

In brief, we insert the parametric surface into an implicit polynomial function of chosen degree and with unknown coefficients. This results in the factorization

$$f(\mathbf{x}(u,v)) = \Phi(u,v)^T D \mathbf{c}, \tag{2.5}$$

where Φ is a vector of basis functions (bivariate tensor-product Bernstein polynomials) and $\mathbf{c}$ contains the unknown coefficients of f. In addition, it contains a certain matrix D, which depends on the given surface. The coefficients of the implicit representation are then found as the eigenvector corresponding to the smallest singular value of that matrix. The only input parameter is the degree of the implicit function. For this we have either used the exact degree, or, if the exact degree is too high, a relatively low degree to produce an approximation. We have found $d = 5$ to be a good choice.

2.4 ML

The 'implicitize' routine in MAPLE is based on the algorithm described in [2], which relies on numerical linear algebra. Again, the parametric surface is inserted into an implicit function of a chosen degree and unknown coefficients. The implicit approximation then is found by minimizing the integral of the resulting parametric function over the given parametric domain, with respect to the unknown coefficients. The eigenvector associated with the smallest eigenvalue of a certain matrix has to be computed. In order to generate this matrix, several integrations have to be performed.

3 Test cases

Piecewise approximate implicitization methods work for both curves and surfaces. In the sequel we consider only the surface case, since it is more interesting for applications, and – unlike curves – many surfaces are not accessible for exact implicitization techniques.

The test cases used in benchmarking the algorithms can be divided in two groups. The first group contains "academic" examples, which were constructed at SINTEF and Linz in order to help developing the algorithms and to display essential features like singularities and self-intersections in a relatively simple setting. The second group consists of industrial examples, provided by CAD vendor think3 (a partner in the GAIA II project).

The surfaces are visualized and described in Fig. 1 and Fig. 2. The tables in these figures give a short description of the various test cases, along with short motivations for choosing these examples.

The industrial surfaces have been generated using geometrical operations such as sweeping, which may create severe problems with self-intersections. Some of these examples have already been used in [20]. In this paper, we focus on the more complicated industrial examples, which could not be dealt with previously.

4 Quantitative criteria

The three main criteria for the comparison are the computing time, the memory needed by the algorithms being tested, and errors measured for the resulting implicit surface.

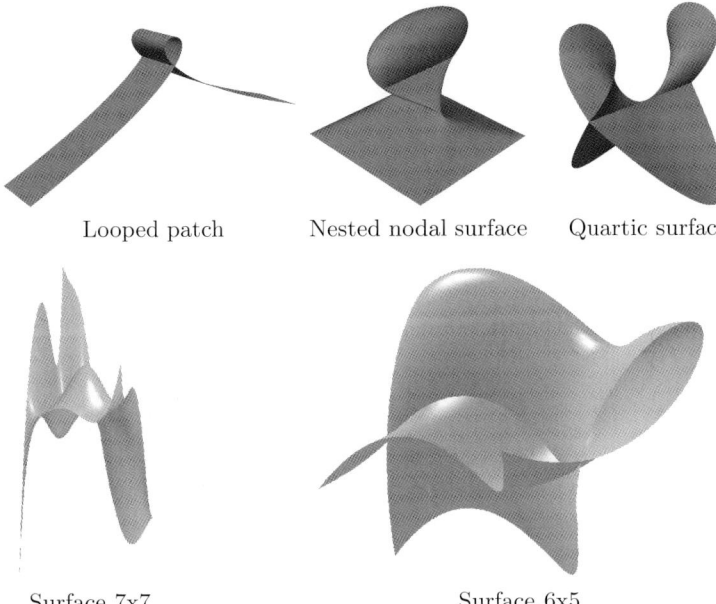

Surface	Degree	Description
Looped patch	$(3,1)$	Self-intersecting. It is the simplest type of surface with this property.
Nested nodal surface	$(3,3)$	Doubly self-intersecting. The bi-degree of the Bézier representation suggests an exact implicit representation of degree $2 \times 3 \times 3 = 18$, but due to reflection symmetries the correct degree is 6.
Quartic surface	$(3,2)$	Surface with self-intersection curves and a cusp-like singular point. As with the 'nested nodal surface' the correct implicit degree is reduced, in this case to 4.
Surface_7x7	$(7,7)$	High degree oscillating patch without singularities. The theoretical algebraic degree of this surface is $2 \times 7 \times 7 = 98$.
Surface_6x5	$(6,5)$	Self-intersecting patch. A self-intersecting curve is moved in space and at the same time bent. The theoretical algebraic degree of this surface is $2 \times 6 \times 5 = 60$.

Fig. 1. Academic test surfaces

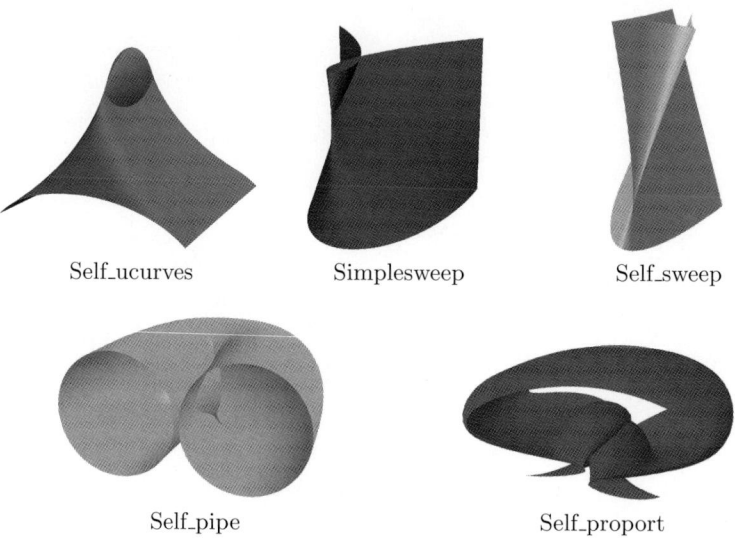

Surface	Degree	Description
Self_ucurves	(4, 2)	Surface with a closed self-intersection curve. Two points on this curve are cusp-like singularities. Effectively, the degree of the exact implicit representation for this surface is 8. The surface is obtained by blending a curve segment with a node between two non-self-intersecting curves.
Simplesweep	(4, 3)	Self-intersecting. The exact implicit representation has degree 6. The surface is obtained by "sweeping" between a piece of a node and a piece of a parabola.
Self_sweep	(12, 1)	The surface is generated by sweeping a line with constant draft angle along a planar curve. It consists of 8 patches each of degree (12, 1) connected with C^2 continuity.
Self_pipe	(3, 3)	The surface is generated by sweeping a circular section along a curve. It consists of 2310 bicubic patches connected with C^2 continuity.
Self_proport	(3, 3)	The surface is a general sweeping surface. It is obtained by sweeping a curve (drive) along another curve (boundary). It consists of 1824 bicubic patches connected with C^2 continuity.

Fig. 2. Industrial test surfaces (data courtesy of think3)

The quoted time is an approximation of the processor time used by the program, and similarly for the memory. These quantities, in the case of PPL, PPS and PS, were measured with standard Linux tools (time and top). In the case of ML the 'profile' procedure of MAPLE was used.

For the error measurements, we consider a set of points $\mathbf{p}_i$ on the parametric surface and compute their Euclidean distances with respect to the implicit surface, using a Newton-like iteration procedure. In order to get a scale independent measurement, we divide this by the length L of the shortest side of the bounding box of the parameterized surface. In a nutshell, the value listed for the error is a scaled version of the average Euclidean distances of the sampled points $\mathbf{p}_i$ from the implicit surface. The computation via footpoint computation is conservative, hence it will overestimate the real error.

Clearly, the average footpoint distance does not reflect the reproduction of singularities. A suitable error measure for singularities, which seems to be currently unknown, would therefore be interesting.

All tests were performed on standard hardware. Both PPL and ML were run on a computer with an Intel(R) Xeon(TM) CPU 2.40GHz, while the PPS and PS were tested on an Intel Pentium IV 2.80 GHz processor. Both systems had (approximately) the same operating system (Linux) and working memory. According to the CPU benchmark tests [11], both systems have almost the same performance.

5 Results

For implicitizing multi-patch parametric surfaces, the two piecewise approximate implicitization methods that are relevant are PPL and PPS. Even in the case of single patch parametric surfaces, which can be handled with single polynomial methods (ML, PL), these piecewise methods are able to produce an algebraic representation of lower degree.

Both the PPL and PPS take as input the degree of the B-spline and information about the number of cells in the input data bounding box. In the case of PPL, the cell size is specified and the input data bounding box is enlarged a bit in order to fit an integer number of cubic cells. In the case of PPS, the input data bounding box is subdivided into n^3 cells, where n is a user-defined integer.

The results given in Table 1 were computed using cubic B-splines. For both algorithms, almost the same "total" number of cells were used. However, as described in Section 2.3, PPL considers only the "active cells". Their number is listed separately in the table. For PPL we list also the number of sampled points for each test case.

The results of PPL and PPS, see Table 1, show that piecewise approximate implicitization is feasible. For all test cases, a tensor B–spline of degree ($3 \times 3 \times$

Example (#Points for PPL)	Degree for PPL&PPS	PPL # Active Cells	PPL Time Memory	PPL Error	# Cells	PPS Time Memory	PPS Error	Degree for ML&PS	ML Time Memory	PS Time Memory	PS Error
Looped patch (2250)	3	84	3.72 sec. 18.41 MB	0.000168	216	9.7 sec. 11 MB	0.02	3	4.64 sec. 11.75 MB	~ 0 0.9 MB	~ 0
Nested nodal surface (2500)	3	101	6.96 sec. 17.90 MB	0.005170	216	6.9 sec. 5.2 MB	0.029	6	48.76 sec. 45.79 MB	0.16 sec. 1.3 MB	~ 0
Quartic surface (2500)	3	168	5.52 sec. 31.32 MB	0.000455	216	10 sec. 9.3 MB	0.003	4	9.64 sec. 15.31 MB	0.01 sec. 0.9 MB	~ 0
Surface_7x7 (2500)	3	93	4.70 sec. 24.40 MB	0.000235	216	9.9 sec. 12 MB	0.008	5	failed	0.6 sec. 1.5 MB	0.0249
Surface_6x5 (2500)	3	111	4.35 sec. 12.44 MB	0.000572	216	8.7 sec. 8.5 MB	0.013	5	126.99 sec. 30.35 MB	0.2 sec. 1.3 MB	0.0161
Self_ucurves (2500)	3	154	4.77 sec. 30.70 MB	0.000210	216	7.8 sec. 8.6 MB	0.01	8	206.99 sec. 129.95 MB	1.38 sec. 1.8 MB	~ 0
Simplesweep (2500)	3	178	5.91 sec. 38.02 MB	0.000026	216	12 sec. 13 MB	0.008	6	30.14 sec. 47.98 MB	0.25 sec. 1.2 MB	~ 0
Self_sweep (2500)	3	168	6.34 sec. 36.57 MB	0.002690	216	13.3 sec. 14 MB	0.025		failed	failed	
Self_pipe (2500)	3	174	5.28 sec. 30.64 MB	0.002740	216	16.7 sec. 17 MB	0.055		failed	failed	
Self_proport (2500)	3	172	6.29 sec. 37.68 MB	0.004991	216	16.9 sec. 16 MB	0.035		failed	failed	

Table 1. Result for the academic (upper part) and industrial (lower part) surfaces.

3) are used. For both algorithms, the implicitization is performed in reasonable time, with reasonable memory usage, and with relatively small error. There are some small differences between the algorithms.

The results of ML and PS are also shown in the same table. Both methods take a parametric surface in terms of a single polynomial patch and a chosen degree of the implicit representation as input. Hence some analysis of the problem is required prior to performing an implicitization. The methods may fail if the chosen degree was too low/high compared with the exact one.

The first three academic test cases and the first two industrial cases in Table 1 were computed using the exact algebraic degree. In this case, PS computes the exact implicit equation. ML, using the 'numerical' option, was able to compute an approximation within tolerances.

For the test cases Surface_7x7 and Surface_6x5 the exact algebraic degrees are too high. We computed these two examples using degree 5, for PS & ML, as the degree of the implicit representation. ML, in the case of Surface_7x7, failed to give an implicit representation due to the low degree specified. Both methods, PS & ML cannot handle multi-patch parametric surfaces, and they therefore fail to compute the last three test cases in Table 1.

In order to demonstrate the feasibility of the methods, we plot the implicit surface for two selected test cases: Quartic surface and Self_pipe, see Fig. 3. PPL is able to generate surfaces without additional branches (cf. Fig. 3).

6 Qualitative comparisons

Finally we discuss some properties of the four techniques. Since pushing away unwanted branches is the most important issue, we dedicate a separate subsection to this subject.

6.1 General criteria

- Both PPL and PPS are able to handle general surfaces. Consequently, they can also be used to implicitize procedurally defined surfaces, since they only need samples of points. In contrast, PS and ML can be applied only to parametrically defined rational surfaces (patches of NURBS).
- Both PPL and PPS are able to compute an approximate implicitization for all test cases with reasonable error.
- In the case of one-patch parametric surfaces, PS (and similarly PPS) is able to reproduce the exact implicitization (within tolerances) if the exact degree is chosen. ML reproduces the exact implicitization if the symbolic integration option is used. Using this option the computation is extremely slow. For our test cases, we used the numerical option.

  PPL does not reproduce the exact implicit representation, since it approximates not only the points, but also the estimated unit normals.

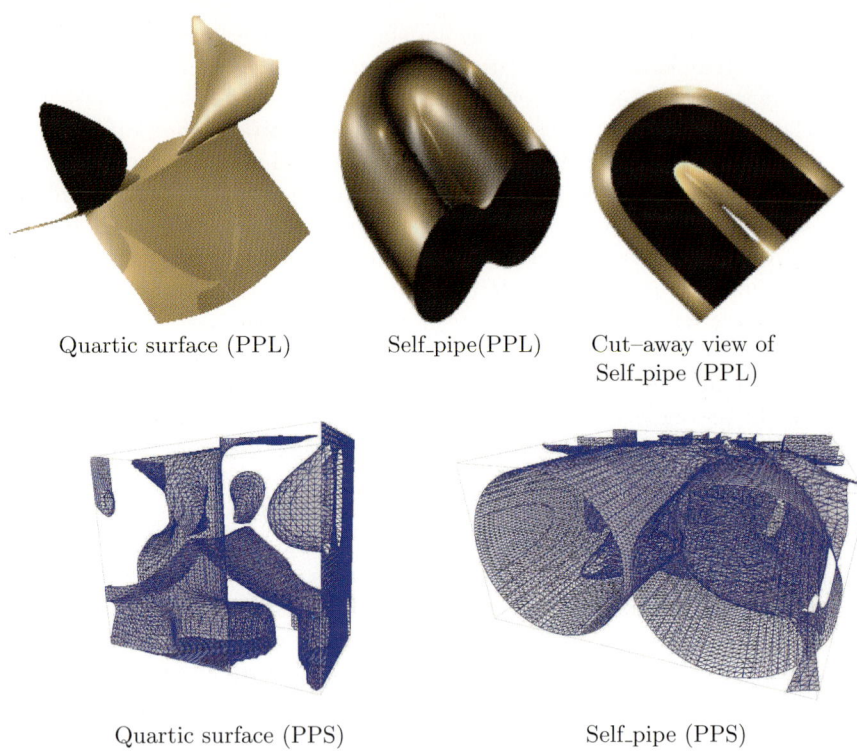

Fig. 3. Results (PPL, PPS)

- In the case of spline surfaces, the notion of an exact implicitization does not make much sense. In the case of one-patch parametric surface, PS is the fastest and the most accurate method.
- For both PPL and PPS one may increase the number of segments and use a low degree, while maintaining the same level of accuracy. A trivariate tensor-product spline function with k inner knots in each direction has $(k+1)^3$ cells/segments and $(k+d+1)^3$ scalar coefficients. In practice, however, it is more important how many of the cells are aligned with the surface ('active cells'), and the number of degrees of freedom will be more in the order of $(k+d+1)^2$ (but depends on the geometry of the surface).
- All methods need the degree as an input parameter. In addition, both PPL and PPS need information about the number (or size) of the cells.
- Currently, PPL is faster and more accurate than PPS for all test cases.

- In order to implicitize with piecewise polynomials, it is necessary to regularize the problem, since cells without or with only a small number of data may cause numerical problems. In the case of PPL, this was achieved by introducing an additional tension term. It is possible to develop (semi–) automatic techniques for adjusting the influence of this term.

6.2 Avoiding unwanted branches and the reproduction of singularities

All methods may produce unwanted branches of the surfaces, and possible additional singularities. The latter ones can be detected by analyzing the singular points of the implicitly defined surface, along the parametric surface. If one of these singular points corresponds to only one point on the parametric surface, which is not singularly parameterized, then it is an additional singularity, which was produced by the implicitization process.

PPL provides two possibilities for pushing away these branches away from the desired part of the surface. First, they can be avoided by exploiting the simultaneous approximation of points sampled from the surface and the associated unit normals. By increasing the number of sampled points and/or the weight w in the objective function, the result of the approximation can be modified so as to become more similar to the signed distance function of the surface. Second, the tension term (2.3) can be used to avoid these unwanted branches. However, this term tends to flatten the implicitly defined surface, and it may therefore lead to poorer approximations.

It is hoped that an improved version of PPS can achieve similar effects by taking the results for more than one singular value into account. This is subject of on–going research.

The two goals of reproducing singularities and avoiding unwanted branches are often in conflict with each other. This is illustrated by Fig. 4, which shows two approximate implicitizations of the 'nested nodal surface'. The piecewise polynomial approximation (left) has no unwanted branches, but the reproduction of the singularity is rather poor. On the other hand, the implicitization by a single high degree polynomial produces many unwanted branches, but it reproduces the double lines very well.

7 Conclusion

We have compared several algorithms for approximate implicitization by polynomials and by piecewise polynomials. As demonstrated by the results, industrial data needs piecewise polynomials, in order to generate well–defined implicit representations of low degree. It was shown that approximate implicitization of real–world surfaces is possible in reasonable time on standard hardware. Further research will focus on the important issue of avoiding unwanted branches and additional singularities.

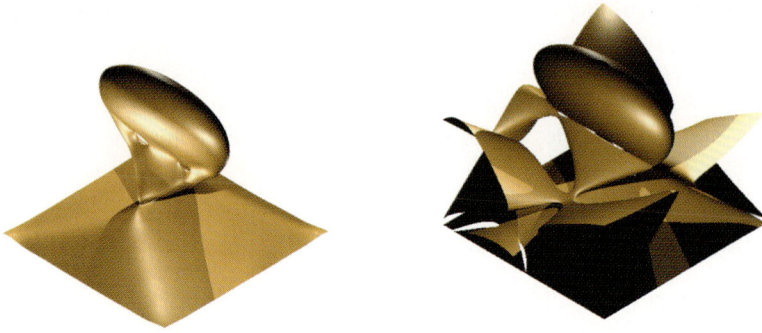

Fig. 4. Reproducing singularities vs. avoiding unwanted branches.

Acknowledgment. This research has been supported by the European Commission through project IST-2001-35512 'Intersection algorithms for geometry based IT-applications using approximate algebraic methods' (GAIA II).

References

1. Chuang, J., Hoffmann, C.: On local implicit approximation and its applications. ACM Trans. Graphics **8**, 4:298–324, (1989)
2. Corless, R., Giesbrecht, M., Kotsireas, I., Watt., S.: Numerical implicitization of parametric hypersurfaces with linear algebra. In: AISC'2000 Proceedings, Springer, LNAI 1930.
3. Cox, D., Little, J., O'Shea, D.: Ideals, Varieties and Algorithms, Springer, New York 1997.
4. Cox, D., Little, J., O'Shea, D,: Using algebraic geometry, Springer Verlag, New York 1998.
5. Cox, D., Goldman, R., Zhang, M.: On the validity of implicitization by moving quadrics for rational surfaces with no base points, J. Symbolic Computation, **11**, (1999)
6. Dokken, T.: Approximate Implicitization, in: Lyche, T., Schumaker, L. (eds.), Mathematical methods in CAGD, Nashboro Press, 2001, 1-25.
7. Dokken, T., Thomassen, J.: Overview of Approximate Implicitization, in: Topics in Algebraic Geometry and Geometric Modeling, AMS Cont. Math. 334 (2003), 169–184.
8. Dokken, T., Thomassen, J.: Weak approximate implicitization, preprint, 2005.
9. Elkadi, M., Mourrain B.: Residue and Implicitization Problem for Rational Surfaces, Applicable Algebra in Engineering, Communication and Computing, (2004), 361-379.
10. Farin, G.: Curves and Surfaces for Computer Aided Geometric Design, Academic Press, 2002.
11. PoVRAY Benchmarks, http://new.haveland.com/povbench/graph.php
12. Hoffmann, C.: Implicit Curves and Surfaces in CAGD, Comp. Graphics and Applics. **13**:79–88, (1993)

13. Jüttler, B., Felis, A.: Least-squares fitting of algebraic spline surfaces, Advances in Computational Mathematics **17**:135–152, (2002)
14. Jüttler, B., Wurm, E.: Approximate implicitization via curve fitting, in L. Kobbelt, P. Schröder, H. Hoppe (eds.), Symposium on Geometry Processing, Eurographics / ACM Siggraph, New York 2003, 240-247.
15. Lee, K.W.: Principles of CAD/CAM/CAE Systems, Prentice Hall, 1999.
16. Thomassen, J., Self-intersection problems and approximate implicitization, in: Dokken, T., Jüttler, B. (eds.), Computational Methods for Algebraic Spline Surfaces, Springer, Berlin 2005, 155-170.
17. Sederberg, T., Chen F.: Implicitization using moving curves and surfaces. Siggraph 1995, **29**:301–308, (1995)
18. Šír, Z.: Fitting of Piecewise Polynomial Implicit Surfaces, Proc. of the 24rd Conference on Geometry and Computer Graphics, University of Ostrava 2004, 202-206.
19. Gonzalez-Vega, L.: Implicitization of parametric curves and surfaces by using multidimensional Newton formulae. J. Symb. Comput. 23(2-3), 137-151 (1997)
20. Wurm, E., Thomassen, J., Jüttler, B., Dokken, T.: Comparative Benchmarking of Methods for Approximate Implicitization, in: Neamtu, M., and Lucian, M. (eds.), Geometric Modeling and Computing: Seattle 2003, Nashboro Press, Brentwood 2004, 537-548.

Computing with parameterized varieties

Daniel Lazard

LIP6, Université Paris VI and SALSA project, INRIA
8 rue de Capitaine Scott, 75015 Paris
Daniel.Lazard@lip6.fr

Summary. In this paper, it is shown that Gröbner bases allow efficient computation with real parametric varieties without implicitizing them. The differences between a real parametric variety and its complex and implicit counterparts are made explicit. The fundamental operations described are the intersection of two parametric varieties, the computation of the singular locus and the distance between a point and the variety. The latter operation provides a robust algorithm for identifying the parameters values of an approximately given point on the variety.

1 Introduction

Parameterized surfaces and, more generally, parameterized varieties are widely used in geometrical design, and it is generally recognized that their implicitization is necessary for answering several questions, such as testing if a point is on the surface, computing the intersection of parameterized curves and surfaces or studying the singularities of the surface.

The first aim of this paper is to show how all these questions may be solved efficiently without implicitization, by using Gröbner basis computations.

Moreover, it appears that, in most applications, one considers real surfaces, defined in some subdomain of the space of parameters, and that these real constraints are lost by the implicitization. On the other hand, the parameters are kept by our methods, which allows us to take into account at any stage of the computation the real aspect of the problem and the constraints induced by the inequalities.

While this paper was being written, it appeared to us that real parameterized varieties are specific mathematical objects whose properties are rather different from those of their associated complex parameterized varieties and implicit varieties. We do not know any place where these particular properties appear, and we have thus devoted a section of this paper to a precise definition of these notions and to the statement of these specific properties.

Another rather surprising fact appeared when writing this paper. We had already used most of the methods described here in some applications, but they appeared as *ad hoc* tools. To abstract them into general algorithms, we had to provide proofs which were not needed before (when working on a specific example, it is not necessary to prove that the algorithm will succeed: it suffices to run it and to look at the result). These proofs make explicit the geometric meaning of the operations which are done. The outcome of this is that the resulting algorithms are much more efficient than the previous *ad hoc* computations.

The structure of the paper is as follows. In Section 2 we describe the algorithmic tools and the corresponding software that we use: Gröbner bases, RUR, real roots isolation, ... and also the mathematical operations which they allow to implement, such as projection, localization and multivariate solving.

Section 3 is devoted to the precise definition of the real parametric varieties and to the description of their main properties. Most of this section is not really new, but we do not know of any reference for it.

Section 4 deals with the computations which may be done independently of the implicitization. It is divided into three subsections, devoted to the computation of the graph of the map associated to the parameterization, the membership test of a point to the variety and the computation of intersections.

The second subsection is probably the main result of this paper, as we give a stable algorithm for identifying the parameters values of an approximately given point on the variety.

In the last section, we show that the Gröbner basis computation which gives the implicit equation(s) also gives the singular locus. This provides information (directly or after more computation) on the topological structure of the real variety; especially this allows to detect the singular points which become regular when restricting the variety to a domain in the real space of the parameter. The way to obtain this information is only sketched, as it needs some tools which are yet under development and are beyond of the scope of this paper.

2 Computational tools

In this section, we describe the few software operations which will be widely used through this paper.

For each of them, there are several existing implementations, but the only available software which implement all of them efficiently and robustly (without errors due to numerical instability) is Gb-RS, written by Jean-Charles Faugère and Fabrice Rouillier, which is available from http://fgbrs.lip6.fr. This software may be used as a stand-alone program or through a Maple interface. In this paper we only consider this latter way to use Gb-RS. It should be noted that, with the current version of the interface, the Gröbner basis computations are usually much faster if the command advance("fast") is issued after loading Gb-RS. However a new version and a new interface are in preparation, which will be more efficient and include new built-in functions, such as a function similar to `mult_solve` (defined below).

Multivariate solve

The first operation we need is `mult_solve` which takes as input a list of polynomials with rational coefficients and outputs either the list of their common real zeroes or the message *"infinitely many real or complex solutions"*. With the current MAPLE interface of Gb-RS, this may be implemented as

```
mult_solve:=proc(list_pols) local vars;
            vars:=[op(indets(list_pols))];
            gbasis(list_pol,DRL(vars));
            if dimension(%)>0 then "infinitely many solutions"
            else isolecoords(rur(%)) fi
            end;
```

This function provides the values of the solutions as floating point numbers or as small intervals, depending on the version of the interface. The floating point output is usually more convenient for most applications and may easily be deduced from the interval output by taking the middle of the intervals. On the other hand, the interval output has the advantage of giving a bound on the rounding errors and signaling when very close values are different. As far as we know, there is no other implementation of a function with the same specifications which is both efficient and produces certified output.

Elimination and projection

The second operation that we need is the computation of Gröbner bases for a block ordering (called `lexdeg` in the `Groebner` package of Maple). Its syntax in Gb-RS is

```
basis_elim:=(list_pols, vars_to_elim, smaller_vars) ->
            op(1,gbasis(list_pols,DRL(vars_to_elim,smaller_vars)));
```

Frequently, one only needs the polynomials in the result which depend only on the variables in `smaller_vars`. Its may be obtained simply by adding the third argument `elim` to the function `gbasis`; we will call the resulting function `elim_grobner` in what follows. This operation is very important in geometrical applications, because its result is a set of implicit equations for the projection (on the space of the variables `smaller_vars`) of the algebraic variety defined by `list_pols`.

Localization

In computational algebraic geometry, very often we need to remove the components included in some hypersurface from the variety defined by a set of polynomials. This is especially useful for dealing with the difficulties due to the base points of a projective parameterization or the vanishing of the denominators of an affine one. This is obtained easily by the function

```
localize:=proc(list_pols,pol) local new_var,liste,vars;
          liste:=[op(list_pols), new_var*pol-1];
          vars:=[op(indets([op(list_pols), pol]))];
          elim_grobner(liste,[new_var],vars);
          end;
```

The proof that this function outputs the desired result is classical and may summarized as follows: The common zeroes of the result of `elim_grobner` are the closure of the projection of the common zeroes of its first argument. The equation `new_var*pol-1=0` implies that none of these zeroes is a zero of `pol`, and thus that the common zeroes of the result do not have a component contained in the zeroes of `pol`.

3 Parametric varieties

In this section, we examine the main differences between real and complex algebraic geometry in the context of parametric varieties. As most books and papers on real algebraic geometry do not consider parametric varieties explicitly and as most papers on parametric varieties do not describe the differences between the real and the complex cases, most of the the results in this section are not well known even when they are not new; it is therefore worthwhile to collect them together.

Terminology

In this paper we will only consider polynomials with real coefficients, and, for the computation, we will restrict ourselves to polynomials with rational coefficients.

Given a family $x_1, \ldots, x_n$ of unknowns, a set of complex (resp. real) values for them is called a *point* (resp. a *real point*) in the *affine space* $\mathbb{A}^n$. The set of common complex zeroes of a family of polynomials is the *(algebraic) variety* of the family. The subset of the common real zeroes is the real part of the variety or the *real variety* of the family. The *ideal* of a set of points is the set of the polynomials which vanish on these points; if the points are real, this ideal is generated by polynomials with real coefficients (because of its invariance by complex conjugation). The *Zariski closure* of a set of points is the variety of its ideal or equivalently the smallest variety which contains the points.

An *(affine) parameterization* of a variety of dimension k in the affine space $\mathbb{A}^n$ is a mapping

$$x_i = p_i/p_0, \ i = 1, \ldots, n,$$

where the $p_i \in \mathbb{R}[u_1, \ldots, u_k]$ are polynomials in k variables. The *domain* of the parameterization, where the mapping is defined, is the complement in $\mathbb{A}^k$ of the zeroes of p_0 (the hypersurface $p_0 = 0$). The *(real) image* of the parameterization is the image by the mapping of the real points of the domain. The *variety of the parameterization* is the Zariski closure of the image. Thus, what is usually called a *parametric variety* is a pair consisting of a parameterization and its variety.

In this paper we consider only parameterizations whose image has dimension k. This means that the $k \times n$ Jacobian matrix of the p_i/p_0 with respect to the u_i has rank k. We will say that a parameterization satisfying this property is *valid*.

The *index* of a parameterization is the number of complex points in the pre-image of a generic point of the image (in fact of any point outside some variety). A parameterization is said *proper* if it has index 1.

The *equations* of the parameterization are the polynomials $p_0 x_i - p_i$. Its *graph* is the set of the pairs $(u, x) \in \mathbb{A}^k \times \mathbb{A}^n$ such that u is in the domain of the parameterization and x is the image of u; it is also the set of the common zeroes of the equations which satisfy $p_0 \neq 0$. The variety defined by the equations is the union of the graph and of the variety defined by $p_0, \ldots, p_n$; therefore, the ideal of the graph is usually larger than the ideal generated by the equations. *The ideal of the graph is always prime* (there are many ways to prove this; the shortest one consists of remarking that it is the saturated ideal of the linear triangular set formed by the equations).

We will sometimes need the notion of *projective parameterization* of a variety of dimension k in the projective space $\mathbb{P}^n$; it is defined by a mapping

$$X_i = P_i, \ i = 0, \ldots, n,$$

where the $P_i \in \mathbb{R}[u_0, u_1, \ldots, u_k]$ are homogeneous polynomials of the same degree in $k + 1$ variables. Its *domain* is the complement of the *base points*, i.e. the common zeroes of the P_i. The other notions as *image, index, proper*, etc. extend in an obvious way from the affine case to the projective one. Without loss of generality we will suppose that the P_i do not have any non-constant common factor. One passes easily from an affine parameterization to a projective one by taking for the P_i the homogenization of the p_i in the same degree, and conversely by substituting 1 for u_0 in the P_i. Note that a projective parameterization may be viewed as the affine parameterization of a cone of dimension $k+1$ in the affine space $\mathbb{A}^{n+1}$. Thus any algorithm which works on affine parameterizations works also on projective parameterizations. However the homogenization introduces new variables and also the singularity at the vertex of the cone. Moreover, the fact that the homogenization should be done in the same degree often introduces highly singular base points: if, for example, one homogenizes Example 7 below, all but one of the resulting P_i are multiple of powers of u and of the homogenizing variable. Thus, when one starts from an affine parameterization, it is frequently much more efficient to make all the computation in an affine setting than to homogenize and work on the projective parameterization.

Basic properties of real parameterization

We state in the next theorem the main properties of the real parameterizations, which make them rather different from the complex ones. As we have said before, all these properties are probably not new, but we are unable to give a reference where they are explicitly stated.

The different items of this theorem are independent; from a usual mathematical point of view, it would be better to split them into several propositions. We state them together in order to give a synthetic view of the subject to a reader who is not necessary interested in the proofs.

Theorem 1. *Let us consider a real parameterization $\mathcal{P}$ of dimension k in $\mathbb{A}^n$, and its variety $\mathcal{V}$.*

  i. *If the parameterization is proper, the complement of the image of $\mathcal{P}$ in $\mathcal{V}$ is included in a variety of dimension lower than k (for the complex point of view this is true even without the hypothesis of properness).*
  ii. *The closure (for the usual topology) of the image of $\mathcal{P}$ is contained in $\mathcal{V}$, but may differ from it, even if the parameterization is proper. In this latter case, the difference lies in the real singular points of $\mathcal{V}$ such that the real part of $\mathcal{V}$ is of dimension lower than k in their neighborhood.*
  iii. *The image of a real projective parameterization is a connected set.*
  iv. *If the parameterization is not proper, the difference between the closure of its image and the real part of $\mathcal{V}$ may contain an open subset of $\mathcal{V}$ of dimension k.*
  v. *If $\mathcal{P}$ is not proper, it may be the case that there is no proper real parameterization which has $\mathcal{V}$ as a variety. In other words, there are parameterizable varieties without real proper parameterization.*

Proof. In [3], a notion of *discriminant variety* is defined which, when applied to the projection of the graph into $\mathbb{A}^n$, is a strict subvariety of $\mathcal{V}$ such that, outside it, the elements of $\mathcal{V}$ have a constant number of distinct complex preimages in $\mathbb{A}^k$ and the projection of the graph on the variety is a covering. Moreover, the discriminant variety has a dimension lower than k, because $\mathcal{V}$ is irreducible (it is the Zariski closure of the projection of the variety of the graph, which is itself irreducible, being linear in the x_i).

i. If the parameterization is proper, the preimage of a real point outside the discriminant variety is reduced to one complex value, which is necessarily real since it is invariant by complex conjugation. Therefore the complement of the image is contained in the discriminant variety.

ii. The inclusion is clear, the usual topology being finer than the Zariski one. Suppose w.l.o.g. that the parameterization is proper. Let P be a real regular point of $\mathcal{V}$; the set of real points of $\mathcal{V}$ in a neighborhood of P has dimension k, and there is a sequence of real points of $\mathcal{V}$, lying outside the discriminant variety and converging to P; thus, by a previous argument, these points are in the image and P is in the closure of the image. The same argument applies for a singular point in a neighborhood of which $\mathcal{V}$ has dimension k.

iii. The domain of a projective parameterization is connected, because its complement, the common zeroes of the P_i, has dimension at most $k-2$ (by hypothesis, the P_i have no common factor). Therefore, the same is true for the image.

iv and *v*. It is well known (cf. [4]) that any cubic surface is parameterizable and that there exist smooth projective cubic surfaces with two real connected components. If we consider an affine real parameterization of such a cubic surface, assertion *iii* shows that the affine part of one of the components is outside of the image, and assertion *i* shows that the parameterization is not proper.

The important meaning of this theorem is that the real variety of a parameterization is very different from its image and even from the closure of its image. This is enforced by the fact that, in most practical applications, one is not interested in the whole variety nor in the image of the parameterization, but in a subset of this image, defined by some inequalities (for example, bi-cubic patches are usually defined for parameters lying in the interval $[0, 1]$). It follows that using implicitization for solving problems on such surfaces like testing membership or computing the intersection with a curve or another surface usually gives wrong answers, like producing a non empty intersection when it is empty. This could be a last resort if there were no better solution to these problems. We will show now that computing directly with real parameterizations is nowadays possible, and even rather efficient.

4 Computing with parameterizations without implicitizing

In this section, we consider a real parameterization $x_i = p_i/p_0$ for $i = 1, \ldots, n$ with $p_i \in \mathbb{Q}[u_1, \ldots, u_k]$ (the typical case is the parameterization of a surface in $\mathbb{A}^3$, i.e. $n = 3$, $k = 2$, but most of the algorithms we describe work in any dimension).

In most applications, the u_i satisfy some polynomial inequalities. For example, for bi-cubic patches the constraints are $0 \leqslant u_i \leqslant 1$. In physical applications, the u_i may represent some lengths or some masses and they have to be positive; for example, in Example 7 below, the constraint, coming from the studied problem is $u > 0$. It follows that a set $\mathcal{C}$ of polynomial inequalities has to be added to the input of most algorithms.

4.1 Equations of the graph

Let us recall that our parameterization is defined by the *equations* $x_i p_0 - p_i$ for $i = 1, \ldots, n$, where p_0 and the p_i are in $\mathbb{Q}[u_1, \ldots, u_k]$. The set of the common zeroes of these polynomials is clearly the union of the *graph* of the parameterization and of the common zeroes of $p_0, p_1, \ldots, p_k$. If this extraneous component, similar to the base points in the projective setting, is kept, most algorithms will fail, by producing infinitely many solutions when a finite number is expected.

As the points of the graph are those of the variety defined by the equations, such that $p_0 \neq 0$, the Zariski closure of the graph is obtained by removing from the variety of the equations the components included in the hypersurface $p_0 = 0$. This is done by the following simple program which needs a fraction of a second on Example 7.

```
graph_eqns:=localize(equations,p_0):
```

As the result is a Gröbner basis of the ideal of the graph, the result may be rather large and it is worthwhile to not display it.

Remark 2. The variety of the graph is the union of the graph and of the limits of points of the graph when the parameters tend to a zero of p_0. It follows that the real part of the variety of the graph is included in the topological closure of the graph. Thus the replacement of the equations by the equations graph_eqns of the graph amounts to extend the domain and the image of the parameterization by continuity, exactly as simplifying a rational fraction amounts to extend the function it defines.

This extension of the image is not necessary closed: it differs from its closure by the images of infinite values of the parameters.

It may also be important to note that the points which are added to the image by this extension of the graph are exactly those which are not in the image of the corresponding projective parameterization, because of the base points. Therefore, the computation of the equations of the graph amounts to removing the base points.

4.2 Testing if a point lies on the variety – closest point on the variety – identification of the parameters

An important problem on parametric varieties consists of testing if a point lies on it, and, if so, finding the corresponding value(s) of the parameters.

If the point is exactly known, the problem is easy: *substitute the coordinates of the points in* equations *or in* graph_eqns, *solve by* mult_solve *and throw out the solutions which do not satisfy the inequalities.*

This algorithm does not work in practice because it is very unusual that one knows exactly a point on the surface. Thus the relevant problem, is *test if the point is close to the variety, and find the parameter values of the closest point of the variety.* The easiest way to do this consists of computing the critical values of the function which maps a point of the graph onto the squared distance from its image to the given point. Fortunately, in most cases this leads easily to a zero-dimensional system (i.e. a system with a finite number of complex zeroes) depending only on the parameters; in the rare cases where the system is not zero-dimensional, it may easily be transformed into a zero-dimensional one by a single localization.

The following theorem describes explicitly the conditions under which such zero-dimensional systems may be obtained. In fact, this theorem describes an algorithm and simultaneously establishes its correctness.

Theorem 3. *Let $p_0 x_i - p_i$, $i = 1, \ldots, n$ be a valid parameterization with $p_i \in \mathbb{Q}[u_1, \ldots, u_k]$ and let $A = (a_1, \ldots, a_n)$ be a point in $\mathbb{A}^n$. Then*

i. The local extrema of the squared distance $d = \sum_{i=1}^{n}(x_i - a_i)^2$ between A and a point of the image are obtained for the values of the u_i which are common zeroes of the k polynomials

$$\sum_{j=1}^{n}(p_j - a_j p_0)(p_j \partial p_0/\partial u_i - p_0 \partial p_j/\partial u_i),$$

or are reached for some values of the x_i which belong to the difference between the image of the parameterization and its closure. Let us denote S the system of polynomials obtained by localizing above polynomials with respect to p_0, in which the common zeroes of the p_i are removed (this system may be obtained by localize(T, p_0)*, where T is the list of the k above polynomials).*

ii. Let $\mathcal{V}_h$ be the sub-variety (possibly empty) of $\mathbb{A}^k$ where the $k \times n$ matrix of the $p_j \partial p_0/\partial u_i - p_0 \partial p_j/\partial u_i$ has rank at most h outside the hypersurface $p_0 = 0$. If the dimension of $\mathcal{V}_h$ is at most h for any h, then the system S has finitely many common zeroes for almost all A. More precisely, the set of points A such that this system of equations has infinitely many complex solutions is contained in some strict subvariety of $\mathbb{A}^n$.

iii. If some $\mathcal{V}_h$ has a dimension $> h$ then the system S has infinitely many solutions for almost all points $A = (a_1, \ldots, a_n)$. If it is the case, a zero-dimensional system may be obtained by localizing with respect to any non-zero minor of rank $k-1$ or k of the matrix of the $p_j \partial p_0/\partial u_i - p_0 \partial p_j/\partial u_i$ (command localize(T, D) *if D is this minor).*

iv. If $k = 1$, the local extrema are obtained without localization from the roots of the unique univariate polynomial

$$\sum_{j=1}^{n}(p_j - a_j p_0)(p_j \partial p_0/\partial u_1 - p_0 \partial p_j/\partial u_1).$$

If $k = 2$, the local extrema are obtained (again without localization) from the common zeroes of the quotients by their gcd of the 2 polynomials

$$\sum_{j=1}^{n}(p_j - a_j p_0)(p_j \partial p_0/\partial u_i - p_0 \partial p_j/\partial u_i), i = 1, 2.$$

Proof. *i.* If the minimal distance (or a local extremum) is reached for a point of the variety which is in the image of the map, then, when the parameters vary, the variation of the point should be orthogonal to the direction of A. Thus we have to write that the vector $(x_j - a_j,\ j = 1,\ldots, n)$ is orthogonal to the vectors $(\partial(p_j/p_0)/\partial u_i,\ j = 1,\ldots n)$. The equations in $\mathcal{T}$ are exactly the numerators of the dot products expressing this orthogonality. The common zeroes of the p_i are clearly zeroes of $\mathcal{T}$; if they are infinite in number (this may be the case only if $k > 2$), they have to be removed by localization by p_0.

ii. Let us consider, for the moment, that the a_j are unknowns; thus the elements of $\mathcal{S}$ lie in $\mathbb{R}[u_1, \ldots, u_k, a_1, \ldots, a_n]$ and define a variety $\mathcal{V}$ in $\mathbb{A}^k \times \mathbb{A}^n$. Let π be the projection of $\mathcal{V}$ on the $\mathbb{A}^n$. For a given point A we are looking for $\pi^{-1}(A)$. As $\pi^{-1}(A)$ is clearly not empty for almost all A, we know that the dimension of $\mathcal{V}$ is at least n and that, if it is equal to n, the fiber $\pi^{-1}(A)$ is finite for almost any A; in this case, the A for which $\pi^{-1}(A)$ is infinite lie on some subvariety of $\mathbb{A}^n$. Therefore we have to determine under which condition the dimension of $\mathcal{V}$ is higher than n.

For doing this, let us consider an irreducible component $\mathcal{W}$ of $\mathcal{V}$; let P be a generic point of the projection $\mathcal{W}'$ of $\mathcal{W}$ on $\mathbb{A}^k$ and let h be the rank at P of the matrix of the $p_j \partial p_0/\partial u_i - p_0 \partial p_j/\partial u_i$. The dimension of $\mathcal{W}$ is clearly the sum of $n - h$ and of the dimension of $\mathcal{W}'$. Thus, if $\dim(\mathcal{V}) > n$ there is some $\mathcal{W}$ such that $\dim(\mathcal{W}') > h$, and the inclusion $\mathcal{W}' \subset \mathcal{V}_h$ implies $\dim(\mathcal{V}_h) > h$. Conversely, if $\mathcal{W}'$ is a component of $\mathcal{V}_h$ of dimension $> h$, its inverse image under the projection of $\mathbb{A}^k \times \mathbb{A}^n$ on $\mathbb{A}^k$ is a subvariety of $\mathcal{V}$ of dimension $> n$.

iii. In this case, the projection by π of this subvariety has a generic fiber of positive dimension and $\mathcal{S}$ has infinitely many solutions for almost any A.

As the parameterization is valid, the rank of the matrix of the

$$p_j \partial p_0/\partial u_i - p_0 \partial p_j/\partial u_i$$

is k for almost all points of $\mathbb{A}^k$ and thus $\dim(\mathcal{V}_h) < k$ for $h < k$. It follows that a component of $\mathcal{V}$ of dimension higher than n comes from a $\mathcal{V}_h$ with $h \leqslant k - 2$, and such a $\mathcal{V}_h$ is contained in all the hypersurfaces defined by the minors of rank k or $k - 1$ of the matrix of the $p_j \partial p_0/\partial u_i - p_0 \partial p_j/\partial u_i$. Thus the localization by any of these hypersurfaces removes the components of $\mathcal{V}$ of dimension higher than n.

iv. If $k = 1$ the system $\mathcal{S}$ reduces to a single univariate polynomial which has only finitely many roots. If $k = 2$, the system $\mathcal{S}$ consists in 2 bivariate equations, and it becomes zero-dimensional by dividing them by their gcd.

Example 4. There may exist some points A such that the system $\mathcal{S}$ has infinitely many solutions. For example, if the variety of the parameterization is a torus, the set of "bad" points consists of the axis of the torus and in the circle of centers.

Example 5. An example where the situation of item *iii* occurs may be obtained by taking $p_i = c_i + g^2 q_i$ with c_i constant: all coefficients of the matrix

$$x_j \partial p_0/\partial u_i - \partial p_j/\partial u_i$$

vanish on the hypersurface $g = 0$.

Remark 6. If the conditions of item *ii* are satisfied, the system $\mathcal{S}$ may fail to output the closest point in two cases: if it has infinitely many solutions for the given point A, or if the closest point of the variety corresponds to infinite values of the parameters or to a common zero of the p_i (i.e. if the closest point belongs to the closure of the image but not to the image). All these situations imply that A belongs to some strict subvariety of $\mathbb{A}^n$, and the problem disappears (with probability 1) if A is slightly moved in a random direction. Therefore, *in practice, the system $\mathcal{S}$ always provides the closest point on the parameterized variety if the point A is approximately given*. On the other hand, there may be some problem if the point A is exactly given.

Practical computation and numerical stability

As the algorithms that we use (Gröbner bases) need exact computation, it is worthwhile to convert the approximate data (coordinates of A) to rational numbers. Thus, if there are numerical instabilities, they do not come from the computation, but from the problem itself; we will consider below this problem more precisely.

If $k \leqslant 2$, the system to be solved is a single univariate equation or a system of two bivariate equations, and there is no practical problem of computation. On the other hand, if $k \geqslant 3$, the localization by p_0 is frequently needed, and the computation may be rather difficult, even with the best available software. We describe below the most difficult example coming from applications that we have encountered.

When the critical points have been found, it is easy to deduce the corresponding coordinates of the points on the variety and their distances from A and therefore the closest point of the variety. However, among the various problems which may be solved by the knowledge of these critical points and their distance from A, one of them needs a special consideration, because it may generate instabilities.

This is when one knows that A is the approximation of a point of the surface, and the problem consists in finding the corresponding values of the parameters: if A is near from a point where the variety crosses itself we may find several points on the variety, corresponding to very different values of the parameters which are all at a distance from A which is lower than the error of the approximation. In such a case, it happens frequently that the statement of the problem implies inequalities on the parameters which allow the exclusion of all solutions but one. Otherwise, it is a case of unidentifiability or intrinsic instability, and the algorithm has to produce several possible solutions.

Example 7. We illustrate above the algorithm and the remarks by the following parametric variety of dimension 3, which comes from a problem of statistics [2].

$$W = w$$
$$X = -\frac{6u^3 + 2v^2 - 4vw - w^2}{3u}$$
$$Y = -\frac{4vu^3 + 20u^3w + 2v^3 - 5vw^2}{3u^2}$$
$$Z = \frac{144u^6 + 72u^3v^2 - 180u^3vw + 4v^4 - 90u^3w^2 - 30v^3w + 30v^2w^2 + 5vw^3}{9u^3}$$
$$u > 0$$

This example is a clearly a rather big one, with an implicit equation of degree 23, with 195 monomials and integer coefficients with up to 14 decimal digits. The above algorithm provides the critical points of the distance from A in less than 3 (resp. 5) minutes on a laptop if A is given with 5 (resp. 10) decimal digits. Again, an indication of the size of the example is given by the number of complex critical values, which is 176.

The problem of identifying the values of parameters for an approximation of a point of the image of the parameterization, is unstable in the neighborhood of a surface (dimension 2) of degree 151 where the variety crosses itself. Near this surface, there are two close points on the variety with very different values of the parameters. On the other hand, if one knows that the value of u corresponding to the point is positive, there is instability only in the neighborhood of the single plane $W = X = 0$ (which is a component of the above surface), which is included in the difference between the image of the parameterization and its closure. The computational method used to prove these results is sketched in Example 11 below.

These results show how important it is, to not eliminate the parameters for solving this kind of problems.

4.3 Intersection of parameterized varieties

The method described in this section may appear not to be very different from the usual method of computing an intersection. The main difference lies in the moment where a projection (elimination) is done and in the possible choices of this projection. We do not have enough examples to know the effect of this apparently slight difference on the efficiency of the computation. However, we are convinced that, on big problems, it is much more efficient to do the elimination as late as possible, because of the swell of the coefficients induced by an elimination.

To compute the intersection of a parameterized variety and another variety defined by its implicit equation(s), it is clear that the easiest way is to substitute the parameterization in the implicit equation(s), which gives the implicit equation(s) of a variety in the space of the parameters which is mapped by the parameterization to the intersection of the implicit variety with the image of the parameterization. If the implicit variety is a hypersurface (one equation), the resulting variety in the space of the parameters is also defined by one equation, which is very interesting in practice.

When dealing with two parametric varieties, it is usually asserted that the best way to compute the intersection is to implicitize one of the varieties and to apply the above method.

In this section, we propose another approach which is more flexible: it provides the intersection as a variety in the product of the two spaces of parameters. With it, the output of the previous method may be obtained by eliminating one of the set of parameters, but other ways of working with this variety are possible. Especially, if the intersection is zero-dimensional, one may apply mult_solve directly without any elimination.

We have not done enough experimentation to compare the efficiency of the two approaches. However, it often happens in this kind of problems that the most costly step is the elimination and that it is faster to do it at the end of the computation, when possible. Therefore, we guess that our approach is competitive.

Proposition 8. Let $x_i = p_i/p_0$ and $x_i = q_i/q_0$ ($i = 1, \ldots, n$) with $p_i \in \mathbb{R}[t_1, \ldots, t_k]$ and $q_i \in \mathbb{R}[u_1, \ldots, u_l]$ be two parameterizations. The localization of the set of polynomials $p_i q_0 - q_i p_0$ by $p_0 q_0$ defines a variety $\mathcal{V}$ in the product of the spaces of parameters. The intersection of the images of the two parameterizations is equal to the image by any of the parameterizations of the projection of $\mathcal{V}$ on the corresponding space of parameters.

The dimension of $\mathcal{V}$ is equal to the dimension of the intersection. In particular, if the intersection is zero-dimensional, the same is true for $\mathcal{V}$.

Proof. This follows immediately from the fact that the intersection of the images are defined by values of the parameters such that $p_0 q_0 \neq 0$ and $p_i/p_0 = q_i/q_0$.

The localization is necessary because the common zeroes of the p_i (resp. q_i) defines components whose projection on the space of the u_j (resp. t_j) is $\mathbb{A}^l$ (resp. $\mathbb{A}^k$).

The assertion on the dimension follows from the fact that both projections and parameterizations have finite fibers outside the zeroes of $p_0 q_0$, and thus preserve the dimension.

Remark 9. If the intersection is zero-dimensional ($k + l = n$), it may be computed by mult_solve without any elimination. If the intersection has positive dimension, one may project it on any of the spaces of the parameters. The result is similar to what would be obtained by implicitizing one of the varieties, except that, by our method, the components of the intersection which do not intersect with the image of the parameterization are removed; it depends on the application if this is an advantage or not.

With the system introduced by Proposition 8, one may compute the projection on the spaces of the parameters, but it may be better to project on other subspaces of the product of the parameters spaces, or to do other operations on the system. This depends on what one wants to do with the intersection (drawing, studying its topology, etc); it is outside of the scope of this paper to describe how to do all such post-processings.

5 Implicitization and singularities

It appears that the Gröbner basis computation which gives the implicit equation(s) of a parameterized variety gives also the singular locus without further computation. This means that the direct computation of the singular locus of a parameterized variety is always less costly than the implicitization followed by the computation of the singular locus of the implicit variety.

Before stating the exact result, we have to recall the notion of discriminant variety introduced in [3]: given an algebraic or a semi-algebraic set $\mathcal{V}$ in $\mathbb{A}^k \times \mathbb{A}^n$ and the projection π onto $\mathbb{A}^n$, a discriminant variety $\mathcal{D}$ is a sub-variety of the Zariski closure Π of $\pi(\mathcal{V})$ such that π is an analytic covering of $\pi^{-1}(\Pi \setminus \mathcal{D}) \cap \mathcal{V}$ over $\Pi \setminus \mathcal{D}$. It is shown in [3] that there exists a minimal discriminant variety which may be computed by classical algebraic algorithms (Gröbner bases and prime decomposition of the radical of an ideal). We will specialize this notion to the case where $\mathcal{V}$ is the graph of a parameterization. In this case, a single Gröbner basis computation gives a discriminant variety which is usually the minimal one (we do not believe that it is always the case but we are unable to provide a counter-example), and, if the parameterization is proper, the minimal discriminant variety is exactly the union of the singular locus of Π and of the limits of the points of the image of the parameterization when some parameters tend to the infinity.

Thus we start from the Gröbner basis of the graph of a parameterization, as defined in Sect. 4.1. The first thing to do is to compute its Gröbner basis for a block ordering eliminating the parameters:

`basis_elim(graph_eqns,[`$u_1, \ldots, u_k$`], [`$x_1, \ldots, x_n$`]):`

With this Gröbner basis at hand, most of the relevant properties of the parameterization may be extracted with almost no computation. These properties are the implicit equation(s) of the variety, the index of the parameterization (for simplicity we describe only how to test if the parameterization is proper) and the discriminant variety which is the union of the singular locus and of the points which are not in the image of the parameterization.

The following theorem states precisely how this information may be extracted from the Gröbner basis.

Theorem 10. *Let $\mathcal{G}$ be the reduced Gröbner basis, for a block ordering eliminating the parameters, of the graph of a parameterization. Let $\mathcal{E}_0$ be the set of elements of $\mathcal{G}$ which are independent from the parameters. For each parameter u_i, let $\mathcal{E}_i \subset \mathbb{R}[x_1, \ldots, x_n]$ be the set of coefficients of u_i in the polynomials in $\mathcal{G}$ whose leading term is linear in u_i and independent from the other u_j. Then*

i. $\mathcal{E}_0$ is a reduced Gröbner basis which is a set of implicit equations for the variety defined by the parameterization; in the case of a hypersurface ($k = n - 1$), $\mathcal{E}_0$ is reduced to a single polynomial.

ii. The parameterization is proper if and only if none of the $\mathcal{E}_i$ is empty.

iii. $\mathcal{E}_i \cup \mathcal{E}_0$ is a Gröbner basis of an ideal which defines a variety $\mathcal{V}_i$ included in the Zariski closure of the image of the parameterization. If this basis is not minimal or not reduced, then all superfluous or nonreduced elements in it belong to $\mathcal{E}_0$. Especially, if all leading terms of the elements of $\mathcal{E}_0$ are reducible by $\mathcal{E}_i$ (which arise frequently), then $\mathcal{E}_i$ is a reduced Gröbner basis of the same ideal.

iv. If the parameterization is proper, $\mathcal{D} = \cup_i \mathcal{V}_i$ is a discriminant variety for the projection of the graph of the parameterization.

v. $\mathcal{D}$ is usually the minimal discriminant variety. The points of the minimal discriminant variety divides in three subsets which are not necessarily disjoint: the points which are the image of several points of the graph; the critical values of the projection where the rank of its Jacobian matrix drops; the images of "infinite values" of the parameters. The singular locus is the union of the two first ones.

Proof. *i.* It is classical (see for example [1]) that $\mathcal{E}_0$ is a Gröbner basis of the intersection with $\mathbb{R}[x_1, \ldots, x_n]$ of the ideal generated by the equations of the graph, and that the variety of this intersection is the Zariski closure of the projection of the graph onto $\mathbb{A}^n$.

ii. By substituting in $\mathcal{G}$ the coordinates of a point of the variety, we get a system of equations for the corresponding values of the parameters, which is a non minimal Gröbner basis if the point is generic. This system has exactly one solution if there is at least one linear equation for each parameter, i.e. if none of the $\mathcal{E}_i$ are empty.

iii. The inclusion of $\mathcal{V}_i$ in the variety results immediately from *i* and from the inclusion of $\mathcal{E}_0$ in the equations. For proving the Gröbner basis properties, it is worthwhile to look on the polynomials as polynomials in the u_i with coefficients in $\mathbb{Q}[x_1, \ldots, x_n]$. In this context we will talk of U-leading term, U-leading coefficients, etc. Thus $\mathcal{E}_i$ is the set of U-leading coefficients of polynomials in $\mathcal{G}$ with x_i as U-leading term. Let $\mathcal{F}_i$ be the set of elements of the ideal generated by $\mathcal{G}$ with x_i as U-leading term. The S-polynomial of two elements of $\mathcal{F}_i$ (or of one polynomial in $\mathcal{F}_i$ and one in $\mathcal{E}_0$) followed by the leading term reduction by $\mathcal{G}$ implies only elements of $\mathcal{F}_i \cup \mathcal{E}_0$ until one get a polynomial independent from u_i. Looking at the coefficients of u_i, this shows that any proof that $\mathcal{G}$ is a Gröbner basis implies that the U-leading coefficients of x_i are also a Gröbner basis. Finally, the fact that $\mathcal{G}$ is a reduced Gröbner basis implies immediately that the elements of $\mathcal{E}_i$ are irreducible by the other elements of $\mathcal{E}_i \cup \mathcal{E}_0$, which is another formulation of the last assertion.

iv. Let us substitute in $\mathcal{G}$ the coordinates of a point P of the variety which lies outside of $\mathcal{D}$. By definition of the $\mathcal{E}_i$, one may extract k polynomials from $\mathcal{G}$ which define a linear system in the u_i which is regular at P and thus also in some neighbourhood of P. Thus the u_i are analytic functions of the x_j in this neighborhood, and this defines locally an inverse of the projection of the graph on the variety.

v. A point P lies in the difference between the discriminant variety and $\mathcal{D}$ if it is a zero of $\mathcal{E}_i \cup \mathcal{E}_0$ such that the substitution of the coordinates of P in $\mathcal{G}$ gives an ideal which contains a polynomial with U-leading term u_i. This is a very strong condition which makes it rare that $\mathcal{D}$ differs from the minimal discriminant variety.

The ideal of the graph being prime, it follows from [3] that the minimal discriminant variety is the union of the images of "infinite values of the parameters", of the critical values of the projection, and of the singular points of the variety. If P is a singular point which does not belong to the first two sets, we have to show that it is the projection of several points of the graph. Suppose the contrary: it is the projection of exactly one point Q of the graph. As P is not a critical value of the projection, the Jacobian matrix at Q with respect to the u_i has rank k, which implies that Q is regular on the graph and, by the implicit functions theorem, that the variety is locally isomorphic to the graph at P, which is a contradiction with the fact we have supposed P singular.

Example 11. We describe here how this theorem applies to Example 7. The equations of the graph are computed in less than a second, and about one minute is needed to compute $\mathcal{G}$. The implicit equation $\mathcal{E}_0$ has degree 23 and its 195 coefficients have up to 14 digits.

The variety $\mathcal{V}_w$ is clearly empty, while $\mathcal{V}_u$ and $\mathcal{V}_v$ depend on the monomial ordering: if $v > u$ then both have dimension 2, $\mathcal{V}_u$ is the union of the plane $W = X = 0$ and a surface of degree 150; $\mathcal{V}_v$ is the union of the same plane and a variety of dimension 1 and degree at least 318 (including the multiplicities of some components) which is included in the surface. On the other hand, if $v < u$ then $\mathcal{V}_v$ is the union of the same surface and of the double plane $W = X^2 = 0$, while $\mathcal{V}_u$ is a variety of dimension 1 and degree 398 included in $\mathcal{V}_v$. All these results may be deduced from the equations of $\mathcal{V}_u$ and $\mathcal{V}_v$ by simple Gröbner bases computations. Note that in all these cases, $\mathcal{E}_i$ is a reduced Gröbner basis of $\mathcal{E}_i \cup \mathcal{E}_0$.

By computing recursively the discriminant variety of the discriminant variety, one may decompose it into cells where the fiber of the projection of the graph on the variety is constant, and by choosing a sample point in each cell, one may show that we have $u \leqslant 0$ for the critical points, that above any point of the surface such that $W \neq 0$, there is at most one point of the graph with $u > 0$ and that the points of the graph with $u = 0$ project to the plane $W = X = 0$. This proves the stability results sketched in Example 7.

Remark 12. The above theorem may be rather easily generalized to improper parameterizations. We have not done this because it would be harder to state. Also, the higher is the index of the parameterization, the higher is the probability that the computed discriminant variety is much larger than the minimal one. Therefore, a study would be needed to decide at which index the algorithm in [3] becomes better than such a generalization.

References

1. Cox, D., Little, J., O'Shea, D.: *Using algebraic geometry*. Graduate Texts in Mathematics, 185. Springer-Verlag, New York (1998)
2. Lazard, D.: Injectivity of real rational mappings: the case of a mixture of two Gaussian laws. *Math. Comput. Simulation* **67** 67–84 (2004)
3. Lazard, D., Rouillier, F.: Solving Parametric Polynomial Systems. Research report INRIA RR-5322 (2004). Submitted to *J. Symb. Comp.*
4. Szilagyi, I.: Local Parameterization of Cubic Surfaces. Submitted to *J. Symb. Comp.*

Implicitization and Distance Bounds

Martin Aigner[1], Ibolya Szilágyi[2], Bert Jüttler[1], and Josef Schicho[3]

[1] Institute of Applied Geometry, Johannes Kepler University, A-4040 Linz, Austria
    martin.aigner@jku.at, Bert.Juettler@jku.at,
[2] RISC–Linz/RICAM, Johannes Kepler University, A-4040 Linz, Austria
    szibolya@risc.uni-linz.ac.at
[3] RICAM, Austrian Academy of Sciences, A-4040 Linz, Austria
    josef.schicho@oeaw.ac.at

Summary. We address the following problem: given a curve in parametric form, compute the implicit representation of another one that approximates the parametric curve on a certain domain of interest. We study this problem from the numerical point of view: what happens with the output curve if the input curve is slightly changed? It is shown that for any approximate parameterization of the given curve, the curve obtained by an approximate implicitization with a given precision is contained within a certain perturbation region.

1 Introduction

In geometric modeling and computer aided design, various different representations for curves and surfaces exist, such as implicitly defined curves and surfaces, parametric representations by (piecewise) rational functions, procedurally defined surfaces, or triangular meshes. The duality of implicit and parametric representations makes each of them especially well suited for certain applications.

In order to exploit the potential benefits of both representations, one has to be able to transform one representation into the other. Theoretically, the conversion from a parametric representation to an implicit one (implicitization) is always possible, whereas the reverse conversion (parameterization) is generally impossible. This paper focuses on implicitization.

Various symbolic-computation-based methods for implicitization have been introduced, based on Gröbner bases [2, 3], resultants [4, 15], moving curves and surfaces [21, 17], residue calculus [10] or on other methods [13, 6]. These techniques produce an exact implicit representation.

In applications, the input curve (or surface) is often not given exactly, but it may be contaminated by numerical errors. In this situation, using approximate techniques may be more appropriate [5, 7, 8, 9, 19]. Also, these techniques are able to deal with the more complicated data needed for industrial applications [18].

This paper studies the effects caused by using an approximate implicitization. More precisely, we address the following problem: Given a parametric representation $\mathbf{p} = \mathbf{p}(u)$, $u \in [0,1]$, of a planar curve segment with domain $[0,1]$. Let $\mathcal{V}(\mathbf{p}) = \mathbf{p}([0,1])$ be the point set defined by the curve. Consider the zero set $\mathcal{C}_f$ of an approximate implicitization f of the parametric curve, where the coefficients of the residuum $f \circ \mathbf{p}$ are bounded by a positive constant ϵ. How close is it to the given curve? As an answer, we derive an upper bound for the one-sided Hausdorff distance of $\mathcal{V}(\mathbf{p})$ and $\mathcal{C}_f$. The bound is valid for all approximate implicitizations f, where the residuum can be bounded by ϵ.

Our approach is based on earlier results of [16], who introduced a condition number that allows to estimate the distance of the two coefficient vectors of two approximate implicitizations. For curves (and similarly surfaces) with a high condition number, the computation of the coefficients of an approximate implicitization is not numerically stable, no matter which numerical method for implicitization is chosen.

Unfortunately, even if the coefficients can be computed in a numerically stable way, it is not guaranteed that the zero sets of two approximate implicitizations are close in a geometric sense. However, one can estimate the one-sided Hausdorff distance of zero sets of an exact and a perturbed equation, using a result by [1]. This leads to a constant expressing the robustness of an implicit representation.

Combining these two robustness results allows to examine the suitability of a given rational parametric curve for approximate implicitization. A curve could be said to be "well behaved", if

(1) the computation of the coefficients of an approximate implicitization is numerically stable, and
(2) the resulting implicit representation is geometrically robust with respect to small perturbations of its coefficients.

For the sake of simplicity, the results in this paper are presented in the case of planar curves. In principle, they can be generalized to hypersurfaces in any dimension (such surfaces in three-dimensional space), but the computations are especially simple in the planar situation.

The paper is organized as follows. After introducing some notations in the next section, we consider the stability of the implicitization in Section 3. In Section 4 we give a distance bound between a parametrically and an implicitly given curve. Section 5 collects the obtained results which enable us to show, that for any approximate parameterization $\mathbf{p}_\delta$ of $\mathbf{p}$, the curve obtained by any approximate implicitization with precision ϵ lies within a certain perturbation region. Finally we conclude the paper.

2 Preliminaries

Throughout this paper, we shall use three spaces P, I, and R of polynomials, where the letters stand for parameterization, implicitization, and residuals respectively.

Consider two positive integers n, d, and let P be the set of *triples of polynomials* of degree less or equal than n in the variable u over $\mathbb{R}$. These polynomials will be described by their coefficient vectors with respect to the Bernstein basis with respect to the parameter domain $[0,1]$.

The elements of P define rational parametric curves of degree less or equal than d in homogeneous coordinates,

$$\mathbf{p}(u) := (p_0(u), p_1(u), p_2(u)) = \sum_{i=0}^{n} \mathbf{b}_i \binom{n}{i} u^i (1-u)^{n-i}, \quad u \in [0,1], \quad (2.1)$$

where the corresponding Cartesian coordinates are $(\frac{p_1}{p_0}, \frac{p_2}{p_0})$.

Let I be the set of all *homogeneous polynomials* of degree d in the variables x_0, x_1, x_2 over $\mathbb{R}$. These functions serve to represent a (possibly approximate) implicitization of a given curve $\mathbf{p}$, by an algebraic curve segment $\mathcal{C}$ of order d

$$\mathcal{C} = \{\, (x_1, x_2) \,|\, (x_1, x_2) \in \Omega \subset \mathbb{R}^2 \land f(1, x_1, x_2) = 0 \,\}. \quad (2.2)$$

Such an algebraic curve is defined in a certain planar domain $\Omega \subset \mathbb{R}^2$ by the zero contour of a bivariate polynomial $f \in I$ of degree d. This polynomial is given by its homogeneous monomial representation

$$f(x_0, x_1, x_2) := \sum_{i,j,k \in \mathbb{N},\ i+j+k=d} b_{ijk}\, x_0^i x_1^j x_2^k \quad (2.3)$$

with certain coefficients b_{ijk}. Sometimes we will also use the inhomogeneous representation $f(1, x_1, x_2)$. If no confusion can arise, we shall write $f(x_1, x_2)$ instead. Alternatively, one can use a Bernstein-Bézier representation with respect to a suitable domain triangle, see [11].

Finally, we denote by R the set of *polynomials* of degree less or equal than nd in the variable u over $\mathbb{R}$. Again, these polynomials will be described by their coefficient vectors with respect to the Bernstein basis with respect to the parameter domain $[0,1]$.

Clearly, the sets P, I, R are linear spaces with a finite dimension, which can be identified with $\mathbb{R}^m$, where m is the corresponding dimension. Then, the usual inner product in $\mathbb{R}^m$ defines an inner product in P, I, and R respectively. The associated norm is defined by $\|x\| := \sqrt{\langle x, x\rangle}$. In order to simplify the notation, any element of R and I will be identified with its coefficient vector with respect to the corresponding basis.

Finally, we define the *evaluation map*

$$\mathrm{eval} : I \times P \to R \ \text{ by } \ (f, \mathbf{p}) \mapsto \mathrm{eval}(f, \mathbf{p}) = f \circ \mathbf{p}.$$

Note that the evaluation map is linear in its first argument, but non–linear in the second one.

3 Condition number of the implicitization

Following earlier results of [16], we define the condition number of the curve implicitization problem, and we give an algorithm for computing it.

3.1 Definition and computation of the condition number

Assume that $\mathbf{p} \in P$, $\|\mathbf{p}\| = 1$, $f \in I$, and $f \circ \mathbf{p} = 0$.
For any $h \in I$, $\mathbf{p} \in P$ we get

$$\text{eval}(h, \mathbf{p}) = M_\mathbf{p} \cdot h \tag{3.4}$$

where $M_\mathbf{p}$ is a matrix depending on the coefficients of $\mathbf{p}$ of size $\bar{d} \times \bar{n}$, where

$$\bar{d} = \binom{dn+2-1}{dn}, \quad \text{and} \quad \bar{n} = \binom{d+2}{d}. \tag{3.5}$$

Using a singular value decomposition, the matrix $M_\mathbf{p}$ can be factorized as

$$U \cdot \Sigma \cdot V^t, \tag{3.6}$$

where $\Sigma \in \mathbb{R}^{\bar{d} \times \bar{n}}$ is a diagonal matrix containing the singular values $\sigma_1 \geq \sigma_2 \geq \cdots \geq \sigma_{\bar{n}} \geq 0$, and $U \in \mathbb{R}^{\bar{d} \times \bar{d}}$, $V \in \mathbb{R}^{\bar{n} \times \bar{n}}$ are orthogonal matrices. We have the following result:

Proposition 1. *Let f and $\mathbf{p}$ as above, and assume additionally that $\|f\| = 1$. Then the following are true.*

1. *The smallest singular value of $M_\mathbf{p}$ is zero.*
2. *The coefficient vector of the polynomial f is a right singular vector to the smallest singular value $\sigma_{\bar{n}} = 0$.*
3. *The right singular vector belonging to the least but one singular value $\sigma_{\bar{n}-1}$, where $\bar{n} = \binom{d+2}{d}$, minimizes the function $g \mapsto \|\text{eval}(g, \mathbf{p})\|$ in the unit sphere of $f^\perp$, where $f^\perp := \{J \in I | \langle f, J \rangle_I = 0\}$.*

If $\sigma_{\bar{n}-1} = 0$, then there are several linearly independent equations h with $\text{eval}(h, \mathbf{p}) = 0$. If $\sigma_{\bar{n}-1}$ is small, we are close to such a case. Hence, in some sense, the reciprocal of $\sigma_{\bar{n}-1}$ is a numerical measurement of the uniqueness of the implicitization.

Let us now drop the assumption on $\mathbf{p}$ that there exists f such that $f \circ \mathbf{p} = 0$. We still keep the assumption $\|\mathbf{p}\| = 1$. We define the *condition number* K as the inverse of the *formally second smallest* singular value of $M_\mathbf{p}$,

$$K = \frac{1}{\sigma_{\bar{n}-1}}. \tag{3.7}$$

With "formally second smallest singular value", we mean that we take multiplicities into account. For instance, if 0 is a multiple singular value, then the condition number is infinity.

Note, that the condition number K depends not just on $\mathbf{p}$, but also on the (estimated) degree d of f. If $\|\mathbf{p}\| \neq 1$, then the condition number always refers the condition number of the normed equation.

Remark 2. In order to compute K, the implicit equation of the parametrically given curve is not needed. The computation of the formally second smallest singular value is easier than the computation of the implicit equation, at least numerically. Singular values are numerically stable, whereas the implicitization problem can be very badly conditioned.

Remark 3. If the last singular value of $M_\mathbf{p}$ is sufficiently small, then there exists an f of degree d such that $\text{eval}(f, \mathbf{p})$ is small. Namely, f is the right singular vector belonging to the smallest singular value.

Here is an algorithm for computing the condition number.

Algorithm 1 ("Condition Number").
Input: A triple of polynomials $\mathbf{p} = (p_0, p_1, p_2)$ of total degree n in the parameter u, such that $\|\mathbf{p}\| = 1$, and an $d \in \mathbb{Z}$.
Output: Condition number K of the implicitization problem.

1. Initialize $M_\mathbf{p}$ by a zero matrix.
   for each b_i in the basis B_I of I, $i = 1, \ldots, \bar{n}$
   a) substitute $\mathbf{p}$ into b_i,
   b) expand the result in the basis B_R of R
   c) append the column to $M_\mathbf{p}$
   (Now we have constructed the matrix $M_\mathbf{p}$)
2. Compute the singular value decomposition of the matrix $M_\mathbf{p}$.
3. The condition number is $K = 1/\sigma_{\bar{n}-1}$, where $\bar{n} = \binom{d+2}{d}$

3.2 Examples

The planar algebraic curves shown in Table 1 will serve as test examples. We considered segments of four well known curves. First we computed an approximate parametric representation and then an approximative implicitization. Both representations are given in the Table. In addition, the domain triangle has been specified, and it is also shown in the figures. Although the computations are done in Bernstein-Bézier-representation, the parametric representation is given with respect to the monomial basis, since many coefficients vanish and a basis transformation is relatively simple [12].

Using Algorithm 1 above we compute the condition number of the four curves. The condition number of the implicitization problem depends not only on the parametric form, but also on the estimate of the degree of the implicit form. From classical algebraic geometry it is known that any degree n polynomial or rational parametric curve can be represented using a degree n algebraic equation. However, if the implicitization is only done approximatively, a lower degree than n may be sufficient in order to gain a result of a desired accuracy.

The Cardioid

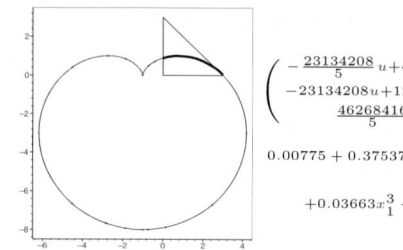

$$\begin{pmatrix} -\frac{23134208}{5}u+4194304+\frac{63799808}{25}u^2-\frac{87973954}{125}u^3+\frac{15527402881}{160000}u^4 \\ -23134208u+12582912+\frac{63799808}{5}u^2-\frac{263921862}{125}u^3-\frac{15527402881}{160000}u^4 \\ \frac{46268416}{5}u-\frac{63799808}{5}u^2+\frac{703791632}{125}u^3-\frac{15527402881}{20000}u^4 \end{pmatrix}$$

$$0.00775 + 0.37537x_1 + 0.19263x_2 + 0.00762x_1^2 + 0.061820x_1x_2 - 0.41813x_2^2$$
$$+0.03663x_1^3 + 0.35689x_1^2x_2 + 0.42381x_1x_2^2 - 0.68875x_2^3$$

$(0,0), (3,0), (0,3)$

The Tacnode

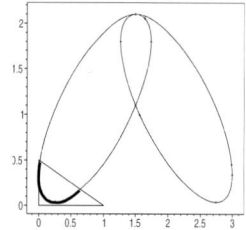

$$\begin{pmatrix} 0.95233\cdot 10^{-2}u^3 + 0.43723u^2 - 0.5478u + 0.89 \\ 0.15\cdot 10^{-1} - 0.1848u + 0.1132\cdot 10^{-1}u^3 + 0.6836u^2 + 0.8523\cdot 10^{-4}u^4 \\ 0.11043\cdot 10^{-3}u^4 + 0.20358\cdot 10^{-1}u^3 + 0.93016u^2 - 1.2295u + 0.429 \end{pmatrix} 10^{-2}$$

$$0.5449 - 0.64792x_1 - 0.25435x_2 + 0.3853x_1^2 - 0.26083x_1x_2 + 0.04559x_2^2$$

$(0,0), (1,0), (0,0.5)$

The Bicorn

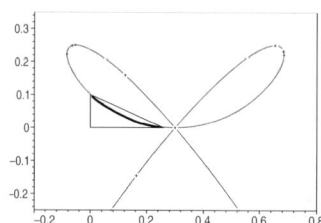

$$\begin{pmatrix} 1 \\ 0.399 + 0.5335u - 0.16637u^3 - 0.0.9075u^2 \\ 0.28435u^2 + 0.1078u - 0.0915u^4 - .06655u^3 + 0.0099 \end{pmatrix}$$

$$-0.35901 - 0.34195x_1 + 0.33064x_2 - 0.22577x_1^2 + 0.57652x_1x_2 - 0.10127x_2^2 - 0.11083x_1^3 + 0.46583x_1^2x_2$$
$$-0.14776x_1x_2^2 + 0.01032x_2^3$$

$(0,0), (0.25,0), (0,0.1)$

The Trifolium

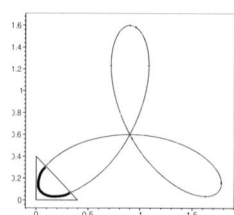

$$\begin{pmatrix} 1.86624u^4 + 3.59424u^3 + 4.58784u^2 - 1.86336u + 0.19629 \\ -1.34784u^2 - 1.08864u + 0.45096 + 3.31776u^4 + 3.31776u^3 \\ 1.1881 + 3.6576u^2 + 1.5696u + 2.0736u^4 + 2.0736u^3; \end{pmatrix}$$

$$-0.1034x_1^4 + 0.37226x_1^3 - 0.020681x_1^2x_2^2 - 0.62044x_1^2x_2$$
$$-0.39088x_1^2 - 0.37226x_1x_2^2 + 0.11168x_1x_2 + 0.10051x_1 - 0.1034x_2^4$$
$$+0.35158x_2^3 - 0.57701x_2^2 + 0.15076x_2 - 0.01312$$

$(0,0), (0.4,0), (0,0.4)$

Table 1. The four examples. From top to bottom, the table shows name, approximative parametric representation, approximative implicit representation and domain triangle.

curve	degree	$\sigma_{\tilde{n}}$	K
cardioid	3	$0.22976 \cdot 10^{-5}$	$0.43523 \cdot 10^6$
tacnode	2	0.07907	12.687
trifolium	4	$0.40345 \cdot 10^{-3}$	2478.62
bicorn	4	$0.50253 \cdot 10^{-8}$	$0.19899 \cdot 10^9$

Table 2. Condition numbers of the implicitization.

Example 4. Table 2 contains the singular value $\sigma_{\tilde{n}}$, the condition number K and the degree of the implicit representation of each curve. In the case of the tacnode and the trifolium we can say that the implicitization problem is well-conditioned.

4 Distance between implicit and parametric curves

This Section is dedicated to the computation of the distance between a parametric and an implicitly given curve. For quantifying the distance between the two curves we use the concept of the Hausdorff distance.

Since we use a triangular Bernstein-Bézier-representation, the implicit curve is tied to its domain triangle. In the remainder of this paper we assume that this triangle is chosen in such way, that the whole parametric curve is contained in it.

In order to avoid some technical difficulties which may arise if the parametric curve hits the boundary of the triangular domain $\triangle$, we consider the distance between $\mathcal{V}(\mathbf{p})$ and

$$\mathcal{C}^* := \mathcal{C} \cup \partial\triangle. \tag{4.8}$$

More precisely, we consider the distance

$$\mathrm{HD}_\triangle(\mathcal{V}(\mathbf{p}), \mathcal{C}^*) = \sup_{\mathbf{y} \in \mathcal{V}(\mathbf{p}) \cap \triangle} \inf_{\mathbf{x} \in \mathcal{C}^* \cap \triangle} ||\mathbf{x} - \mathbf{y}||. \tag{4.9}$$

We call this distance the *one–sided Hausdorff distance* [4] of $\mathcal{V}(\mathbf{p})$ and $\mathcal{C}^*$ with respect to the domain triangle $\triangle$. For the effect of replacing $\mathcal{C}$ by $\mathcal{C}^*$, see Figure 3.

Lemma 5. *Let $\mathcal{C} \subset \triangle \subset \mathbb{R}^2$ be an algebraic curve which is defined by the homogeneous polynomial f of degree d and $\mathcal{C}^* := \mathcal{C} \cup \partial\triangle$. We assume that the gradient field $\nabla f(1,.,.)$ of the inhomogeneous polynomial does not vanish in $\triangle$. Furthermore a curve $\mathcal{V}(\mathbf{p}) \subset \triangle$ is given by its parametric representation $\mathbf{p} = \mathbf{p}(u)$, $u \in [0,1]$. If*

---
[4] The symmetric version is $\max(\mathrm{HD}_\triangle(\mathcal{V}(\mathbf{p}), \mathcal{C}^*), \mathrm{HD}_\triangle(\mathcal{C}^*, \mathcal{V}(\mathbf{p})))$.

$$c \leq \|\nabla f(1,.,.)\|_{(x_1,x_2)}\|_{\mathbb{R}^2} \ \forall \ (x_1, x_2) \in \triangle \ \text{ and } \ \|\text{eval}(f, \mathbf{p})\| \leq \epsilon$$

and

$$p_0(u) \geq N \quad \forall u \in [0, 1]$$

hold, where c, ϵ and N are certain positive constants, then the one–sided Hausdorff distance can be bounded by

$$\text{HD}_\triangle(\mathcal{V}(\mathbf{p}), \mathcal{C}^*) \leq \frac{\epsilon}{c \, N^d}. \tag{4.10}$$

Proof. The proof is a consequence of the mean value theorem, which is applied to the integral curves of the gradient field emanating from the parametric curve and hitting the implicit curve or the boundary of the triangle. More precisely, we consider the integral curve $\gamma(t) := (x_1(t), x_2(t))$ of the normalized gradient field $V = \nabla \bar{f} / \|\nabla \bar{f}\|$ of $\bar{f}(x_1, x_2) = f(1, x_1, x_2)$. We choose the starting point $\gamma(0)$ of the integral curve $\gamma(t)$ to lie on the parametric curve $\mathbf{p}$. In the absence of points with a vanishing gradient in the domain of interest, $\gamma(t)$ hits the implicit curve or the boundary of the triangle in $\gamma(s)$ for some $s \in \mathbb{R}$. Since the curve is parameterized by its arc length,

$$\|\gamma(s) - \gamma(0)\| \leq s.$$

Let $F(t)$ be the restriction of $\bar{f}$ to $\gamma(t)$:

$$F(t) := \bar{f}(x_1(t), x_2(t)).$$

Applying the mean value theorem we obtain

$$\exists \xi \in]0, s[: \ \frac{F(s) - F(0)}{s - 0} = F'(\xi)$$

Due to $F'(t) = \dot{\gamma}(t) \cdot \nabla \bar{f}|_{\gamma(t)} = \|\nabla \bar{f}|_{\gamma(t)}\|$ we obtain

$$\frac{|F(s) - F(0)|}{|s - 0|} = |F'(\xi)| = \|\nabla \bar{f}|_{\gamma(\xi)}\|, \text{ hence } s = \frac{|F(s)|}{\|\nabla \bar{f}(\gamma(\xi))\|} = \frac{|\bar{f}(\gamma(s))|}{\|\nabla \bar{f}(\gamma(\xi))\|}$$

Finally, we observe that the values

$$f\left(1, \frac{p_1(u)}{p_0(u)}, \frac{p_2(u)}{p_0(u)}\right), \quad u \in [0, 1], \tag{4.11}$$

of the values of the inhomogeneous implicit representation along the parametric curve are bounded by ϵ/N^d, which completes the proof. $\square$

The next step is the computation of the constants c and N which are needed in Lemma 5. One can see that for curves that have singular points in the domain of interest, (4.10) is not defined, since the gradient vanishes. Consequently, such cases have to be excluded. More precisely, vanishing gradients correspond

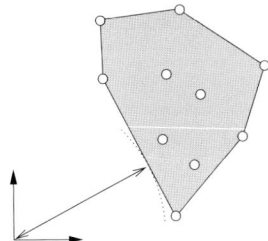

Fig. 1. Bounding the minimal norm of the gradient.

to singular points (including isolated points) of the original curve or of other iso-value curves ("algebraic offsets").

In the regular case, a lower bound for the minimal gradient can be computed. The essential ingredient of this algorithm is the convex hull property of Bernstein–Bézier representations. Clearly, this algorithm gives only a conservative lower bound on $||\nabla f(1,.,.)||$. The result can be made more accurate by splitting the domain into smaller triangles.

Algorithm 2.
<u>Input</u>: control points c_{ijk} of a bivariate polynomial.
<u>Output</u>: lower bound for the minimal gradient

1. Compute the partial derivatives of $f(1, x_1, x_2)$ with respect to x_1 and x_2.
2. Describe them in Bernstein–Bézier form with respect to the domain triangle.
3. Combine the corresponding coefficients of the derivative patches together to vector-valued control points $d_{ijk} \in \mathbb{R}^2$
4. Compute the minimal distance from the origin to the convex hull of the d_{ijk}, see Figure 1.
5. This distance serves as constant c in Lemma 5.

Remark 6. The procedure can be generalized to the surface case. While the algorithms for the three-dimensional convex hull computations are more involved and need special data structures for the storage of the data points [14], the time complexity is still the same as in the planar case.

Remark 7. The constant N can be chosen as the minimum value of the 0–th components of the control points in (2.1), provided that all of them are positive. More precisely, the so–called 'weights' of the rational curve have to be positive, and N can be chosen as the minimum weight.

5 Distance bound for approximate implicitization

Based on the previous result, we derive an upper bound on the one–sided Hausdorff distance between the parametric curve and any approximate implicitization which has a certain accuracy.

5.1 Estimating the implicitization error

In order to estimate the Haussdorff distance of a parametric and an approximate implicit curve we use a known result [16] that allows to estimate the error in the coefficient vector in terms of the condition number K.

Lemma 8. *Let* $\mathbf{p}$ *be a triple of polynomials of parametric degree* n, *with* $\|\mathbf{p}\| = 1$. *Furthermore, let* $f_1, f_2 \in I$ *be polynomials of degree* d *with* $\|f_1\| = \|f_2\| = 1$ *such that* $\|\mathrm{eval}(f_1, \mathbf{p})\| \leq \epsilon$ *and* $\|\mathrm{eval}(f_2, \mathbf{p})\| \leq \epsilon$. *Then we have one of the following:*

$$\|f_1 - f_2\| \leq K \cdot 4 \cdot \epsilon, \quad \|f_1 + f_2\| \leq K \cdot 4 \cdot \epsilon,$$

where K *is the condition number of* $\mathbf{p}$.

Proof. The complete proof (of a more general result) is given in [16]. Here we restrict ourselves to a sketch of the proof. Let f_3 be such that $\|f_3\| = 1$ and $\|\mathrm{eval}(f_3, \mathbf{p})\|$ is minimal. It follows that $\|\mathrm{eval}(f_3, \mathbf{p})\| \leq \|\mathrm{eval}(f_1, \mathbf{p})\| \leq \epsilon$. In first order approximation we have, that

$$r_1 := f_1 - f_3, \quad \text{and} \quad r_2 := f_3 - f_2,$$

are in $f_3^\perp$. In order to estimate $\|r_i\|$, $i = 1, 2$, we get

$$\|r_i\| = K \cdot \|\mathrm{eval}(r_i, \mathbf{p})\| \leq K \cdot \epsilon.$$

It follows, that

$$\|f_1 - f_2\| \leq \|f_1 - f_3\| + \|f_3 - f_2\| \leq K \cdot 2 \cdot \epsilon.$$

If we take terms of higher order into account, then we get the inequality $\|f_1 - f_2\| \leq 4 \cdot K \cdot \epsilon$, see [16]. □

5.2 Bounding the minimal gradient

From the previous Section we can conclude that the output of the implicitization process is no longer an exact polynomial, but that its coefficients can only be specified up to a certain tolerance. This means that the possible outputs of the implicitization process form a whole set of curves.

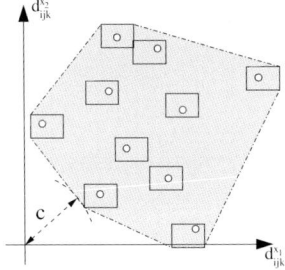

Fig. 2. Bounding the minimal norm of the gradient of a perturbed polynomial. The rectangles represent the areas where all possible control points lie in.

We denote the set of defining polynomials of the implicitized curves by

$$F_\epsilon := \{f \mid \|\mathrm{eval}(f, \mathbf{p})\| \leq \epsilon, \|f\| = 1\}.$$

The corresponding coefficients of all $f \in F_\epsilon$ lie in certain intervals. The length of these intervals can be bounded using Lemma 8.

On the other hand, Lemma 5 allows us to bound the Hausdorff distance between a parametric and an algebraic curve. In order to get a distance bound between a parametric and an arbitrary $f \in F_\epsilon$ we need to compute a lower bound for the minimal gradient for all possible $f \in F_\epsilon$. This bound is given by

$$G := \min_{f \in F_\epsilon} \min_{(x_1,x_2) \in \Omega} \| \nabla f(1, \cdot, \cdot)|_{(x_1,x_2)} \|.$$

In order to compute this bound the same technique as in Section 4 can be applied. One has to replace the exact control points by intervals. Hence, standard techniques for interval arithmetics have to be applied for computing the d_{ijk}. These are no longer points in $\mathbb{R}^2$, but rectangles containing all possible positions of the control points. Consequently, one has to compute the convex hull of these rectangles in order to gain a lower bound for the minimal gradient, cf. Figure 2. For further informations on interval arithmetics and related techniques see [20] and the references cited therein.

Algorithm 3 ("Minimal Gradient").
Input: parametric representation $\mathbf{p}$ of a curve, $\epsilon > 0$
Output: G

1. Compute K using Algorithm "Condition Number".
2. Compute the bound given in Lemma 8 and an approximate implicitization f for $\mathbf{p}$.
3. Generate intervals using the coefficients of f as center and adding/subtracting the bound derived in the previous step.
4. Determine the derivative patches of $f(1, x_1, x_2)$ in x_1 and x_2 direction and describe them in Bernstein–Bézier (with interval coefficients !) form with respect to the domain triangle.

curve	K	G	position err.
cardioid	$0.43523 \cdot 10^6$	0	∞
bicorn	$0.19899 \cdot 10^9$	0	∞
trifolium	2478.62	0	∞
tacnode	12.687	0.17537	$0.98718 \cdot 10^{-5}$

Table 3. Geometric robustness and position bound

5. For all pairs of corresponding coefficients of the derivative patches, generate the Cartesian product of the intervals.
6. Collect the vertices of all these rectangles and determine their convex hull.
7. The shortest distance from the origin to this convex hull serves as G. (If the convex hull contains the origin, then $G = 0$)

5.3 Perturbation regions of parametric curves

In this Section we combine the results of the previous parts and determine an upper bound for the Hausdorff distance between an exact parametric and approximatively computed implicit curve.

Theorem 9. *Let* $\mathbf{p}$ *with* $\|\mathbf{p}\|$ *be a triple of polynomials of degree less or equal than* n, *and let* $\mathcal{V}(\mathbf{p}) \subset \triangle$ *be the curve defined by* $\mathbf{p}$. *If the bound* G *computed by Algorithm 3 is nonzero, then for any algebraic curve segment* $\mathcal{C} \subset \triangle$ *defined by an* $f \in F_\epsilon$ *of degree* d,

$$\mathrm{HD}_\triangle(\mathcal{V}(\mathbf{p}), \mathcal{C}^*) \leq \frac{\epsilon}{GN^d},$$

where the constant N *is defined as in Lemma 5.*

Proof. The proof is an immediate consequence of the previous two lemmas. □

The bound given in Theorem 9 defines two offset curves to $\mathbf{p}$ that enclose a perturbation region within the triangle. For any approximate parameterization $\mathbf{p}$ of a curve and for any approximate implicitization with a given precision ϵ the obtained curve $\mathcal{C}$ lies within this perturbation region.

Clearly, the result of this Theorem is only meaningful for points that are further away from the boundary than $\frac{\epsilon}{GN^d}$. This is due to the fact, that we do not only measure the distance to the implicitly defined curve, but also to the boundary of the triangle, see Figure 3.

Example 10. In Table 3 we determine for each of the four examples the bound provided in Theorem 9. Using Algorithm 3 we compute a lower bound G for the minimal gradient. In the examples we set $\epsilon = 10^{-6}$.

In the first two cases the high condition number K reflects the fact that the implicitization process is very unstable. Consequently, the coefficient bounds

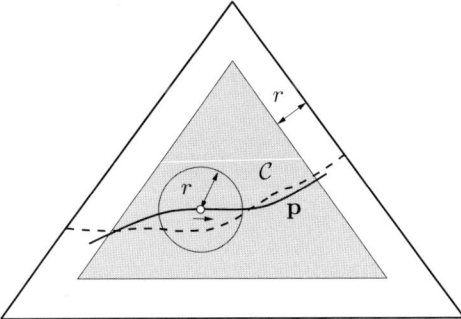

Fig. 3. For each point that lies on **p** and in the grey shaded triangle exists within a certain bound $r := \epsilon/(GN^d)$ a point on the implicit curve $\mathcal{C}$.

are very poor and the bound for the minimal gradient yields zero. No prediction for the position error of the obtained implicit curve can be made.

For the trifolium the implicitization is robust but the obtained implicit representation is unstable with respect to small errors in the coefficients. Again the geometric robustness is poor and the position bound is infinity.

In the last example the implicitization as well as the obtained implicit representation are stable under numerical perturbations; the geometrical robustness is good. Knowing the precision of the implicitization process we are able to predict the maximal displacement of the implicit curve.

6 Conclusion

Schicho and Szilágyi [16] have analyzed the robustness of approximate implicitizations with respect to the resulting coefficients. On the other hand, Aigner et al. [1] treated the geometric robustness of implicit representations if some error in the coefficient vector is allowed. In the present paper, we combined these results, in order to predict the geometric stability of approximate implicitization. More precisely, given a parametric representation, we may derive an error estimate for the obtained coefficients of the implicit curve. With this result we can compute a bound such that any approximate implicitization lies within a certain neighborhood of the original parametric curve. The width of this vicinity is determined by the bound. While we have presented the results for curves, they can be generalized immediately to the case of general hypersurfaces.

Acknowledgments

The first and the third author have been supported by the European Commission through RTD project IST 2001-35512, entitled "Intersection algorithms for geometry based IT-applications using approximate algebraic methods". The second author has been supported by Austrian Science Fund (FWF) in the frame of the Special Research Area SFB 013 "Numerical and Symbolic Scientific Computing", subproject 03.

References

1. M. Aigner, B. Jüttler, and M.-S. Kim. Analyzing and enhancing the robustness of implicit representations. In *Geometric modeling and Processing*, pages 131–142. IEEE Press, 2004.
2. C. Alonso, J. Gutierrez, and T. Recio. An implicitization algorithm with fewer variables. *Comp. Aided Geom. Design*, 12:251–258, 1995.
3. B. Buchberger. Applications of Gröbner bases in nonlinear computational geometry. In *Trends in computer algebra*, pages 52–80. Springer, Berlin, 1988.
4. L. Busé. Residual resultant over the projective plane and the implicitization problem. In *Proc. ISSAC*, pages 48–55, New York, 2001. ACM.
5. F. Chen. Approximate implicitization of rational surfaces. In *Computational geometry (Beijing, 1998)*, volume 34, pages 57–65. Amer. Math. Soc., Providence, RI, 2003.
6. F. Chen and L. Deng. Interval implicitization of rational curves. *Comput. Aided Geom. Design*, 21:401–415, 2004.
7. R.M. Corless, M.W. Giesbrecht, I.S. Kotsireas, and S.M. Watt. Numerical implicitization of parametric hypersurfaces with linear algebra. In *AISC 2000*, LNCS, pages 174–183. Springer, Berlin, 2001.
8. T. Dokken. Approximate implicitization. In *Mathematical methods for curves and surfaces*, pages 81–102. Vanderbilt Univ. Press, Nashville, TN, 2001.
9. T. Dokken and J. Thomassen. Overview of approximate implicitization. In *Topics in algebraic geometry and geometric modeling*, volume 334, pages 169–184. Amer. Math. Soc., Providence, RI, 2003.
10. M. Elkadi and B. Mourrain. Residue and implicitization problem for rational surfaces. *Appl. Algebra Engrg. Comm. Comp.*, 14:361–379, 2004.
11. G. Farin, J. Hoschek, and M.-S. Kim, editors. *Handbook of Computer Aided Geometric Design*. Elsevier, 2002.
12. R.T. Farouki. On the stability of transformations between power and Bernstein polynomial forms. *Comp. Aided Geom. Design*, 8:29–36, 1991.
13. L. Gonzalez-Vega. Implicitization of parametric curves and surfaces by using multidimensional Newton formulae. *J. Symb. Comp.*, 23:137–151, 1997.
14. J.E. Goodman and J. O'Rourke, editors. *Handbook of discrete and computational geometry*. Chapman & Hall, Boca Raton, FL, 2004.
15. A. Marco and J.J. Martínez. Implicitization of rational surfaces by means of polynomial interpolation. *Comput. Aided Geom. Design*, 19:327–344, 2002.
16. J. Schicho and I. Szilágyi. Numerical Stability of Surface Implicitization. *Journal of Symbolic Computation*, page to appear, 2004.
17. T.W. Sederberg, T. Saito, D.X. Qi, and K.S. Klimaszewski. Curve implicitization using moving lines. *Comp. Aided Geom. Des.*, 11:687–706, 1994.
18. M. Shalaby et al. Piecewise approximate implicitization: Experiments using industrial data. In *Algebraic Geometry and Geometric Modeling (this volume)*. Springer, 2006, 37–51.

19. M. Shalaby, B. Jüttler, and J. Schicho. C^1 spline implicitization of planar curves. In *Automated deduction in geometry*, volume 2930 of *LNCS*, pages 161–177. Springer, Berlin, 2004.
20. H. Shou, H. Lin, R.R. Martin, and G. Wang. Modified affine arithmetic is more accurate than centred interval arithmetic or affine arithmetic. In *The Mathematics of Surfaces X*, volume 2768 of *LNCS*, pages 355–365. Springer, Heidelberg, 2003.
21. J. Zheng, T.W. Sederberg, E.-W. Chionh, and David A. Cox. Implicitizing rational surfaces with base points using the method of moving surfaces. In *Topics in algebraic geometry and geometric modeling*, pages 151–168. AMS, Providence, RI, 2003.

Singularities and their deformations: how they change the shape and view of objects

Alexandru Dimca

Laboratoire J.A. Dieudonné, UMR du CNRS 6621,
Université de Nice-Sophia-Antipolis,
Parc Valrose, 06108 Nice Cedex 02, FRANCE.
dimca@math.unice.fr

Summary. We show how the presence of singularities affects the geometry of complex projective hypersurfaces and of their complements. We illustrate the general principles and the main results by a lot of explicit examples involving curves and surfaces.

1 The setting and the problem

Let $\mathbb{P}^{n+1} = \mathbb{P}^{n+1}(\mathbb{C})$ denote the complex projective $(n+1)$-dimensional space. It can be regarded as the set of complex lines passing through the origin of $\mathbb{C}^{n+2}$ or, alternatively, as the simplest compactification of the affine space $\mathbb{C}^{n+1}$. The homogeneous coordinates of a point $x \in \mathbb{P}^{n+1}$ are denoted by

$$x = (x_0 : x_1 : \ldots : x_{n+1}).$$

Let $\mathbb{C}[X_0, X_1, \ldots, X_{n+1}]$ be the ring of polynomials in $X_0, X_1, \ldots, X_{n+1}$ with complex coefficients. For a homogeneous polynomial $f \in \mathbb{C}[X_0, X_1, \ldots, X_{n+1}]$ we define the corresponding projective hypersurface by

$$V(f) = \{x \in \mathbb{P}^{n+1}; f(x) = 0\}$$

i.e. $V(f)$ is the zero set of the polynomial f in the complex projective $(n+1)$-dimensional projective space. We consider $\mathbb{P}^{n+1}$ endowed with the strong complex topology (coming from the metric topology on $\mathbb{C}^{n+1}$) and all subsets in $\mathbb{P}^{n+1}$ are topological spaces with the induced topology. Note that this topology is quite different from the Zariski topology used in Algebraic Geometry over an arbitrary algebraically closed field.

A point $x \in V(f)$ is a singular point if the tangent space of $V(f)$ at x is not defined. Formally the set of such singular points of $V(f)$ is called the *singular locus* of $V(f)$ and is given by

$$Sing(V(f)) = \{x \in \mathbb{P}^{n+1}; f_0(x) = \ldots = f_{n+1}(x) = 0\}$$

where f_j denotes the partial derivative of f with respect to X_j. We assume in the sequel that the hypersurface $V(f)$ is reduced (i.e. we have chosen a simple equation for $V(f)$, without multiple factors) and then $\dim Sing(V(f)) < \dim V(f) = n$.

In this survey we will investigate an algebraic view of the *shape* of the hypersurface $V(f)$, expressed by various invariants from Algebraic Topology such as the homology groups, cohomology groups, fundamental groups. For the definition of these invariants we refer to [13, 29]. This will give a precise idea about the intrinsic geometry of the hypersurface and helps a lot in understanding the possible deformation of that object.

To understand the topology of a space A it is usual to give its homology group with integer coefficients $H_j(A, \mathbb{Z})$ (which measures how many j-dimensional holes there are in the the space A) or at least the corresponding Betti numbers

$$b_j(A) = \text{rank} H_j(A, \mathbb{Z})$$

defined when the prop of this $\mathbb{Z}$-module is finite. To give the Betti numbers of a space A is the same as giving its rational homology groups $H_j(A, \mathbb{Q})$. Indeed, one has

$$b_j(A) = \dim_{\mathbb{Q}} H_j(A, \mathbb{Q}).$$

A weaker invariant is the Euler characteristic of the space A given by

$$\chi(A) = \sum_j (-1)^j b_j(A)$$

when these Betti numbers exist and are all trivial except finitely many. For algebraic varieties these numerical invariants are always defined since a quasi-projective n-dimensional complex algebraic variety has the homotopy type of a finite CW-complex of (real) dimension $2n$.

In order to understand the position of $V(f)$ inside the complex $(n+1)$-dimensional projective space, in other words *its view from outside*, we have to study the topology of the complement

$$M(f) = \mathbb{P}^{n+1} \setminus V(f).$$

This will tell us how much freedom we have to move around the hypersurface $V(f)$. This idea was very fruitful in Knot Theory. Here one studies various embeddings of the circle S^1 into the sphere S^3. The image of such an embedding is a knot K and the fundamental group of the complement $\pi_1(S^3 \setminus K)$ is called *the group of the knot K*.

For any knot K one has

$$H_1(S^3 \setminus K, \mathbb{Z}) = \mathbb{Z}$$

and $S^3 \setminus K$ is a $K(\pi, 1)$-space, i. e. all the topological information about it is contained in its fundamental group. Refer to [29] for a formal definition.

Note that the homology says nothing about the view of our knot K. A key result due to Papakyriakopoulos says that

$$\pi_1(S^3 \setminus K, \mathbb{Z}) = \mathbb{Z}$$

if and only if the knot K is trivial, i.e. isotopic to a linear embedding of the circle. For all these results concerning Knot Theory we refer to [26].

This trip into the realm of knot theory is related to the above discussion through the following construction. Let $n = 1$ and O be any point on the curve $V(f)$ such that $V(f)$ has just one branch at O. A small closed ball B in $\mathbb{P}^2$ centered at O (defined using any local parametrization at O) has a boundary ∂B homeomorphic to the sphere S^3. Moreover, we have that the intersection $V(f) \cap \partial B$ is homeomorphic to the circle S^1. The corresponding knot is trivial if and only if O is not a singular point on the curve $V(f)$.

The main message of our paper is that the larger the dimension of the singular locus of $V(f)$, the more difficult it is to give accurate answers to the above problems concerning the shape and the view of the hypersurface $V(f)$.

We warn the reader that the setting discussed here is the simplest possible one. The answers to the above questions become much more complicated in either of the following three apparently simpler settings.

(RS) The *real setting* consists of replacing all the objects above by the corresponding real objects. This study is clearly more interesting for applications than the complex setting (CS) considered above. However, usually, a real problem is first solved in the complex setting and then we try to get as much real information out of the complex solution. For more on this see [2, 4, 20, 24, 28].

(AS) The *affine setting* consists of working in an affine (or numerical) space $\mathbb{C}^{n+1}$. The objects are easier to define but the behavior at infinity causes many technical problems. For more on this see [7, Chapter 6, section 3], [12, 11, 8].

(RB) The *real bounded setting* consists of studying bounded pieces of real algebraic varieties, e.g. the intersections of real affine algebraic varieties with balls or cubes.

2 The smooth case

In this section we consider only smooth hypersurfaces $V(f)$, i.e. hypersurfaces with an empty singular locus

$$\mathrm{Sing}(V(f)) = \emptyset.$$

The first result says that in this case the coefficients of the polynomial f play no role in determining the shape and the view of the smooth hypersurface $V(f)$, see [6, p. 15]. In terms of deformations, we can say that a small deformation of a smooth hypersurface is smooth and its shape and view are unchanged.

Theorem 1. *Let f and g be two homogeneous polynomials in $\mathbb{C}[X_0, X_1, \ldots, X_{n+1}]$ of the same degree d such that the corresponding hypersurfaces $V(f)$ and $V(g)$ are smooth. Then the following hold.*

(i) The hypersurfaces $V(f)$ and $V(g)$ are diffeomorphic. In particular they have exactly the same invariants coming from Algebraic Topology.
(ii) The complements $M(f)$ and $M(g)$ are diffeomorphic.

Example 2. (i) Consider first the case of complex projective plane curves, i.e. $n = 1$. Such a curve C is the same as an oriented Riemann surface, so topologically it is obtained from the 2-dimensional sphere by adding a number of handles. This number is called the genus $g(C)$ of the curve C. In the case of a plane curve $C = V(f)$ one can easily show using the above theorem and taking $f = X_0^d + X_1^d + X_2^d$ (a Fermat type equation) that there is the following celebrated *genus-degree formula*

$$g(V(f)) = \frac{(d-1)(d-2)}{2}.$$

Hence for $d = 1$ and $d = 2$ we get the sphere $S^2 = \mathbb{P}^1$, for $d = 3$ we get an elliptic curve which is diffeomorphic to a torus $S^1 \times S^1$. One can also show that

$$H_0(V(f)) = H_2(V(f)) = \mathbb{Z} \text{ and } H_1(V(f)) = \mathbb{Z}^{2g}.$$

(ii) Consider now the case of real projective plane curves. The example of $f = X_0^2 + X_1^2 + X_2^2$ and $g = X_0^2 - X_1^2 + X_2^2$ shows that the above theorem is false in the real setting. A smooth real curve $V(f)$ is a collection of circles, but their exact number and relative position depends heavily on the coefficients of f and this is an area of active research, see [4].

(iii) Consider now the affine setting, i.e. complex curves in $\mathbb{C}^2$. The example of $f = X^3 + Y^3 - 1$ and $g = X + X^2Y - 1$ shows that the above theorem is false in this setting. Indeed, topologically $V(f)$ is a torus with 3 deleted points, while $V(g)$ is a punctured plane. Hence

$$b_1(V(f)) = 4 \neq 1 = b_1(V(g)).$$

There is a similar description of the homology of a smooth hypersurface $V(f)$ in general, see for instance [6, p. 152].

Proposition 3. *Let V be an n-dimensional smooth hypersurface of degree n. Then the integral homology of V is torsion free and the corresponding Betti numbers are as follows.*

(i) $b_j(V) = 0$ for $j \neq n$ odd or $j \notin [0, 2n]$;
(ii) $b_j(V) = 1$ for $j \neq n$ even and $j \in [0, 2n]$;
(ii) $\chi(V) = \frac{(1-d)^{n+2}-1}{d} + n + 2$.

Example 4. Let S_3 be a smooth cubic surface in $\mathbb{P}^3$. Then the corresponding sequence of integral homology groups $H_j(S_3, \mathbb{Z})$ for $0 \leq j \leq 4$ is the following: $\mathbb{Z}, 0, \mathbb{Z}^7, 0, \mathbb{Z}$.

Now we turn to the study of the complement $M(f)$ in this case. One way to study it is to consider the *Milnor fiber* $F(f)$ associated to the polynomial f. This is the following *affine* hypersurface

$$F(f) = \{x \in \mathbb{C}^{n+2}; f(x) = 1\}.$$

If d is the degree of f as above, then there is a monodromy automorphism of $F(f)$ given by

$$h : F(f) \to F(f), x = (x_0, \ldots, x_{n+1}) \mapsto (\lambda x_0, \ldots, \lambda x_{n+1})$$

with $\lambda = \exp(2\pi i/d)$. Let G be the cyclic group of order d spanned by h. Then the quotient $F(f)/G$ can be identified to the complement $M(f)$. This gives the second part of the following.

Theorem 5. *Assume that $n > 0$. Then the following hold.*

(i) *The Milnor fiber $F(f)$ is homotopy equivalent to a bouquet of $(n+1)$-dimensional spheres. In particular, $F(f)$ is simply-connected and the reduced integral homology groups of $F(f)$ vanish in degrees up-to n.*
(ii) *The complement $M(f)$ has $\pi_1(M(f)) = \mathbb{Z}/d\mathbb{Z}$ and the reduced rational homology groups of $F(f)$ vanish in degrees up-to n.*

Using the homotopy exact sequence of the fibration $G \to F(f) \to M(f)$, refer to [29] for a definition, one gets information on the higher homotopy groups, namely $\pi_j(M(f)) = \pi_j(F(f)) = 0$ for $1 < j < n+1$ and $\pi_j(M(f)) = \pi_j(F(f)) = \mathbb{Z}^\mu$ for $j = n+1$ where

$$\mu = (d-1)^{n+2}$$

is the *Milnor number* of f. In particular, even in the simplest case, the complement $M(f)$ is not a $K(\pi, 1)$-space.

In conclusion, in the case of smooth hypersurfaces, the spaces $V(f)$ and $M(f)$ are not very complicated. They depend only on the dimension n and the degree d, and their topological invariants can be computed to a large extent.

3 The isolated singularities case

In this section we consider hypersurfaces $V(f)$ having at most isolated singularities, i.e. hypersurfaces with a finite singular locus

$$\mathrm{Sing}(V(f)) = \{a_1, \ldots, a_m\}.$$

In order to study the topology of such an object we have to study first the local situation, i.e. the topology of an isolated hypersurface singularity $(V,0)$ defined at the origin of $\mathbb{C}^{n+1}$ by a reduced analytic function germ

$$g : (\mathbb{C}^{n+1}, 0) \to (\mathbb{C}, 0).$$

The topological study was essentially done by Milnor in [21] where the following facts are obtained.

Theorem 6. *Let B_ϵ be a closed ball of radius $\epsilon > 0$, centered at the origin of $\mathbb{C}^{n+1}$ with boundary the sphere S_ϵ.*

(i) For all $\epsilon > 0$ small enough, the intersection $V \cap B_\epsilon$ is the cone over the link $K = V \cap S_\epsilon$ of the singularity $(V, 0)$. This link is an $(n-2)$-connected submanifold of the sphere S_ϵ and $\dim K = 2n - 1$.

(ii) For all $\epsilon \gg \delta > 0$ small enough, the Milnor fiber of the singularity $(V, 0)$, defined as $F = B_\epsilon \cap g^{-1}(\delta)$ is a smooth manifold, homotopy equivalent to a bouquet of n-spheres. The number of spheres in this bouquet is the Milnor number $\mu(V, 0)$ of the singularity $(V, 0)$ and is given by

$$\mu(V, 0) = \dim_\mathbb{C} \frac{\mathcal{O}_{n+1}}{J(g)}$$

where $\mathcal{O}_{n+1}$ is the ring of germs of analytic function germs at the origin of $\mathbb{C}^{n+1}$ and $J(g)$ is the Jacobian ideal of g, i.e., the ideal spanned by all the partial derivatives of g. Alternatively, the Milnor number $\mu(V, 0)$ is given by the degree of the gradient mapping germ

$$\mathrm{grad}(g) : (\mathbb{C}^{n+1}, 0) \to (\mathbb{C}^{n+1}, 0).$$

The Milnor fiber should be regarded as a smooth deformation of the singular fiber $g^{-1}(0)$ of g over the origin.

By the above theorem, it follows that the only interesting homology group of the Milnor fiber F is the group

$$L(V, 0) = H_n(F, \mathbb{Z}) = \mathbb{Z}^{\mu(V,0)}.$$

This group is endowed with a $(-1)^n$-symmetric bilinear form $<,>$ coming from the intersection of cycles. Regarded with this additional structure, the free abelian group $L(V, 0)$ is called the *Milnor lattice* of the singularity $(V, 0)$. It is known that this intersection form $<,>$ is non-degenerate exactly when the link K is a $\mathbb{Q}$-homology sphere, i.e. $H_*(K,\mathbb{Q}) = H_*(S^{2n-1},\mathbb{Q})$, see for details [6, p. 93].

Similarly, the Milnor lattice is unimodular (i.e. the corresponding bilinear form has as determinant $+1$ or -1) if and only if the link has the same integral homology as the sphere S^{2n-1}.

The Milnor lattice tells a lot about the possible deformations of an isolated hypersurface singularity. Indeed, the singularity $(V,0)$ can be deformed into the singularity $(W,0)$ only if there is an embedding of lattices

$$L(W,0) \to L(V,0).$$

The Milnor number $\mu(V,0)$ is also called *the number of vanishing cycles* at the singularity $(V,0)$. Ample justification for this name is given below.

One may ask which singularities among the isolated ones are the simplest. The answer depends on our interests, but in a lot of questions the class of *simple singularities* introduced by Arnold, see for details [1], are very useful. These singularities are by definition the singularities which can be deformed only into finitely many isomorphism classes of singularities. Their classification, up to isomorphism, is given in dimension $n = 2$ by the following list of local equations at the origin, see [5] where the possible deformations are discussed in detail.

$$A_k : x^{k+1} + y^2 + z^2, \quad \text{for} \quad k > 0;$$

$$D_k : x^2 y + y^{k-1} + z^2, \quad \text{for} \quad k > 3;$$

$$E_6 : x^3 + y^4 + z^2, \quad E_7 : x^3 + xy^3 + z^2 \quad \text{and} \quad E_8 : x^3 + y^5 + z^2.$$

To get the corresponding equations for the curve singularities, i.e. $n = 1$, we have just to discard the last term z^2 from the above equations. Note that for curves A_1 is just a *node*, while A_2 is just a *cusp*. These names are used for higher dimensional singularities A_k as well.

In the above list of simple surface singularities, all the associated Milnor lattices are non-degenerate and only the lattice $L(E_8)$ is unimodular.

Remark 7. Using the real parts of some of the above expressions defining the simple singularities, one can obtain *simple real equations for hypersurfaces in the real projective space* $\mathbb{P}^3(\mathbb{R})$ which represent up to diffeomorphism all the surfaces, i.e., all the compact, connected 2-manifolds. For example, any compact, connected 2-manifold M which is orientable can be constructed up to diffeomorphism from the sphere S^2 by attaching g handles, where $g \geq 0$ is the genus of M exactly as in Example 2. This integer g is completely determined by the equality

$$\chi(M) = 2 - 2g.$$

Let X, Y, Z and W be the homogeneous coordinates on the real projective space $\mathbb{P}^3(\mathbb{R})$. Then the equation

$$Re(X+iY)^{2g} + (X^2+Y^2+Z^2+W^2)^{g-1}Re(Z+iW)^2 = 0,$$

which is essentially the real part of the simple singularity A_{2g-1}, defines a compact, connected 2-manifold of genus g. For more details and a similar formula for non-orientable surfaces based on the real part of the simple singularity D_k, see [10]. It is quite natural to look for equations of the non-orientable surfaces in the real projective space $\mathbb{P}^3(\mathbb{R})$ since they are not embeddable in the usual affine space $\mathbb{R}^3$. See [14, p. 181] for a topological immersion of the projective plane $\mathbb{P}^2(\mathbb{R})$ in $\mathbb{R}^3$ having as singularity a cross cap (also called a Whitney umbrella) described as the image of the mapping

$$\phi: \mathbb{R}^2 \to \mathbb{R}^3, (x,y) \mapsto (x^2, y, xy).$$

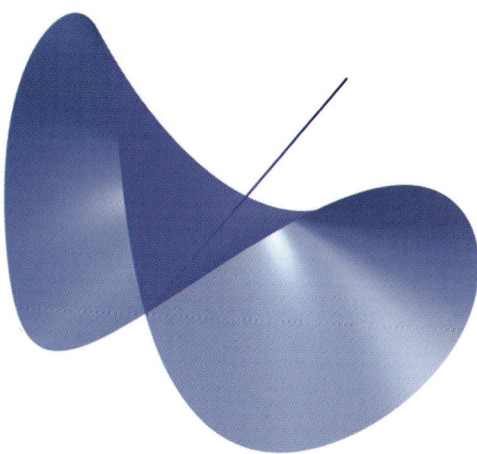

Fig. 1. A cross cap given by $XY^2 - Z^2 = 0$.

The next result compares the topology of the hypersurface $V(f)$ having at most isolated singularities to the known topology of a smooth hypersurface $V(f)_{\text{smooth}}$ having the same dimension n and degree d as $V(f)$. For a proof we refer to [6, p. 162].

Theorem 8. *(i)* $H_j(V(f), \mathbb{Z}) = H_j(V(f)_{\text{smooth}}, \mathbb{Z})$ *for all* $j \notin \{n, n+1\}$. *In addition,* $H_{n+1}(V(f), \mathbb{Z})$ *is torsion free.*
(ii) $\chi(V(f)) = \chi(V(f)_{\text{smooth}}) + (-1)^{n-1} \sum_{k=1}^{m} \mu(V, a_k)$.

Example 9. For a plane curve C, the above result coupled with the following easy facts gives a complete description of the integral homology.

(a) $b_2(C)$ is equal to the number of irreducible components of C;
(b) the first homology group is torsion free.

As an explicit example, consider a 3-cuspidal quartic curve C_4. Any such curve is *projectively equivalent* to the curve defined by the equation

$$X^2Y^2 + Y^2Z^2 + Z^2X^2 - 2XYZ(X+Y+Z) = 0.$$

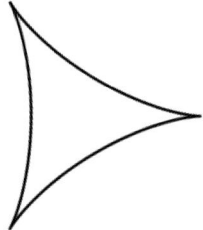

Fig. 2. A symmetric quartic with 3 cusps, of equation $x^4 + 2x^2y^2 - 8x^3 + 18x^2 + y^4 + 24y^2x + 18y^2 - 27 = 0$.

The corresponding smooth curve has genus $g = 3$ and hence $b_1 = 6$. The singular curve C_4 is irreducible (since the only singularities are cusps, hence locally irreducible!) and has 3 cusps located at the points $(1:0:0), (0:1:0), (0:0:1)$. Hence one would expect a loss of $6 = 3 \times \mu(A_2)$ cycles due to the presence of singularities. Using Theorem 8 and the above remarks, it follows that indeed $b_1(C_4) = 0$. This result is confirmed by the known fact that the normalization of C_4 is the projective line $\mathbb{P}^1$ and the normalization morphism is a homeomorphism in this situation.

For any curve C, its homology is determined by its degree, the list of singularities on C and the number of irreducible components of C.

Beyond the curve case, new phenoma may occur. First of all, torsion can appear in the homology, see for details [6, p. 161].

Theorem 10. *(i) If all the Milnor lattices $L(V, a_k)$ for $k = 1, \ldots, m$ are unimodular, then $H_j(V(f), \mathbb{Z}) = H_j(V(f)_{\text{smooth}}, \mathbb{Z})$ for all $j \neq n$ and $H_n(V(f), \mathbb{Z})$ is torsion free of rank $b_n(V(f)_{\text{smooth}}) - \sum_{k=1}^{m} \mu(V, a_k)$. In this situation, $V(f)$ is an integral homology manifold and in particular the Poincaré Duality holds over $\mathbb{Z}$.*

(ii) *If all the Milnor lattices $L(V, a_k)$ for $k = 1, \ldots, m$ are nondegenerated, then $H_j(V(f), \mathbb{Z}) = H_j(V(f)_{\text{smooth}}, \mathbb{Z})$ for all $j \neq n$, $b_n(V(f)) = b_n(V(f)_{\text{smooth}}) - \sum_{k=1}^{m} \mu(V, a_k)$ and the torsion part of $H_n(V(f), \mathbb{Z})$ is determined by a lattice morphism defined on the direct sum of the lattices $L(V, a_k)$.*

In the second case, the hypersurface $V(f)$ is a rational homology manifold and in particular the Poincaré Duality holds over $\mathbb{Q}$. For precise information on the determinant of the cup-product in this case, see [6, p. 171]. For general facts on Poincaré Duality and cup-product, see [13].

Example 11. The list of cubic surfaces with isolated singularities can be found in [3]. We list some of the cases below.

(a) A cubic surface S can have s nodes A_1, for $s = 1, 2, 3, 4$. The only case which produces torsion is $s = 4$ and then the torsion part of the second homology group of S is given by $\text{Tors}H_2(S) = \mathbb{Z}/2\mathbb{Z}$.

(b) A cubic surface S can have s cusps A_2, for $s = 1, 2, 3$. The only case which produces torsion is $s = 3$ and then $\text{Tors}H_2(S) = \mathbb{Z}/3\mathbb{Z}$.

For a complete discussion and proofs we refer to [6, p. 165].

Note also that the determinant of the cup-product can be used to distinguish hypersurfaces having the same integral homology. For instance, the three cubic surfaces with singularity type $3A_1$, A_1A_2 and A_3 have all the same integral homology, but they are not homotopy equivalent since the cup-products are different, see [6, p. 171].

A second major phenomenon is the dependence of the Betti numbers of the hypersurface $V(f)$ on the position of singularities.

Example 12. The classical example here, going back to Zariski in the early '30's, is that of sextic surfaces

$$S_6 : f(X, Y, Z) + T^6 = 0$$

where $f(X, Y, Z) = 0$ is a plane sextic curve C_6 having 6 cusps. Two situations are possible here.

(a) The six cusps of the sextic curve C_6 are all situated on a conic. This is the case for instance for

$$f(X, Y, Z) = (X^2 + Y^2)^3 + (Y^3 + Z^3)^2.$$

Then it can be shown that $b_2(S_6) = 2$, see for instance [6, p. 210].

(b) The six cusps of the sextic curve C_6 are not situated on a conic. Then it can be shown that $b_2(S_6) = 0$, see loc.cit.

The explanation of this difference is that the two types of sextic curves cannot be deformed one into the other even though they are homeomorphic.

A good way to understand this strange behaviour of the Betti numbers is to use *algebraic differential forms defined on $M(f)$ and with poles along $V(f)$* to describe the topology of $V(f)$ and of the complement $M(f)$. See [6, Chapter 6], for details on this approach and many more examples.

This remark has brought into discussion the complement $M(f)$. For $n = 1$, the main topological invariant is the fundamental group $\pi_1(M(f))$. Usually this group is highly non-commutative.

Example 13. (a) For the 3-cuspidal quartic C_4 considered in Example 9, the fundamental group $\pi_1(M(f))$ is the metacyclic group of order 12 which can be described by generators and relations as the group

$$G = \{u, v; u^2 = v^3 = (uv)^2\}.$$

(b) For the two types of 6-cuspidal sextic curves discussed in Example 12, one has

$$\pi_1(M(f)) = \mathbb{Z}/2\mathbb{Z} * \mathbb{Z}/3\mathbb{Z},$$

a free product, for C_6 of the first type, and

$$\pi_1(M(f)) = \mathbb{Z}/2\mathbb{Z} \times \mathbb{Z}/3\mathbb{Z} = \mathbb{Z}/6\mathbb{Z},$$

a direct product, for C_6 of the second type, see [6, p. 134].

This example shows that the fundamental group $\pi_1(M(f))$ depends on the position of singularities even for plane curves.

In higher dimension, i.e. for $n > 1$, the complement $M(f)$ has a commutative fundamental group, which is cyclic of order d, and the object of study is the homology $H_*(M(f)^c, \mathbb{Q})$ of the infinite cyclic covering $M(f)^c$ of the space $M(f) \setminus H$, where H is a generic hyperplane. Then the groups $H_*(M(f)^c, \mathbb{Q})$ can be regarded in a natural way as Λ-modules of finite type, where $\Lambda = \mathbb{Q}[T, T^{-1}]$ and, as such, they are called the Alexander invariants of the hypersurface $V(f)$. See [12, 15, 16, 19] for more on this beautiful subject.

In conclusion, when the number or the type of singularities on the hypersurface $V(f)$ is small compared to the degree d, then the list of singularities is enough to determine the topology of $V(f)$, even the embedded topology, see [6, pp. 17–19]. In such a case there is usually no torsion in homology.

On the other hand, when the number or the type of singularities on the hypersurface $V(f)$ is large compared to the degree d, then torsion is likely to occur in homology and the position of singularities may influence the Betti numbers of the hypersurface $V(f)$.

4 The general case

In this section we consider hypersurfaces $V(f)$ having an s-dimensional singular locus, for $0 \leq s \leq n-1$. Note that $s < n-1$ implies that the hypersurfaces $V(f)$ is irreducible. Little is known in general about the homology of such a hypersurface; for the following result see [6, p. 144].

Theorem 14. *With the above notation, there are isomorphisms*

$$H_j(V(f)) = H_j(\mathbb{P}^{n+1})$$

for all $j < n$ and $j > n + s + 1$. The complement $M(f)$ has a commutative fundamental group, which is cyclic of order d, if $s < n - 1$.

Remark 15. For an arbitrary hypersurface $V(f)$, Parusiński has defined a global Milnor number $\mu(V(f))$ such that one has the following generalization of Theorem 8, (ii).

$$\chi(V(f)) = \chi(V(f)_{\text{smooth}}) - \mu(V(f)).$$

For more details and applications see [23]. An alternative approach via vanishing cycles is described in [7, pp. 179–183].

For special classes of hypersurfaces the information we have is complete. This is the case for instance when $V(f)$ is a *hyperplane arrangement*, i.e., $V(f)$ is a finite union of hyperplanes in $\mathbb{P}^{n+1}$. Then not only the homology is known, but also the cohomology algebra, see [22].

In general, one can adopt various approaches which we briefly describe below.

4.1 Using algebraic differential forms

There is a spectral sequence whose E_2-term consists of various homogeneous components of the homology of the Koszul complex of the partial derivatives of f and converging to the cohomology of $M(f)$, see [9]. Recall that the Koszul complex describes the linear relations involving the partial derivatives of f, then the relations among the relations, and so on. Hence it can be successfully handled by the computer algebra packages.

As an example, using this approach one can determine the Betti numbers of the cubic surface

$$S_3' : X^2Z + Y^3 + XYT = 0.$$

The singular locus here is 1-dimensional (i.e. the line $X = Y = 0$) and it turns out that the surface S_3' has the same rational cohomology as the projective plane $\mathbb{P}^2$, see [9] for very explicit computations.

4.2 Building the hypersurface inductively out of successive hyperplane sections

We have seen in Theorem 6 that a key role is played by the fact that the Milnor fiber F in that case has a very simple topology, i.e. it is homotopically equivalent to a bouquet of spheres. In the case of projective hypersurfaces we have a similar result, see [11].

Theorem 16. *For any hypersurface $V(f)$ and any transversal hyperplane H, the complement $V_a(f) = V(f) \setminus H$, which can be regarded as the affine part of $V(f)$ with respect to the hyperplane at infinity H, is homotopically equivalent to a bouquet of n-spheres. The number of spheres in this bouquet is given by the global Milnor number of f, defined as the degree of the gradient mapping*

$$\mathrm{grad}(f) : M(f) \to \mathbb{P}^{n+1}.$$

The remaining difficult problem is to glue the information we have on $V_a(f)$ and on the hyperplane section $V(f) \cap H$ in order to get information on the hypersurface $V(f)$.

Another possibility is to look for the Alexander invariants in this setting, and this was recently done by Maxim, see [19].

4.3 Cyclic coverings of a projective space

Let $p : X \to \mathbb{P}^{n+1}$ be a cyclic covering ramified along the hypersurface $V(f)$.

For simplicity, we assume below that the degree of p coincides with the degree d of the hypersurface $V(f)$. The general case can be treated similarly, using weighted projective spaces instead of the usual projective spaces, see [6, Appendix B].

Under our assumption, it follows that X can be identified to the hypersurface $V(\widetilde{f})$ given in the projective space $\mathbb{P}^{n+2}$ by the equation

$$\widetilde{f}(X_0, \ldots, X_{n+1}, T) = f(X_0, \ldots, X_{n+1}) + T^d = 0.$$

We have already seen this construction in Example 12.

If $M(\widetilde{f})$ denotes the corresponding complement, then we have the following isomorphism for the cohomology with rational coefficients coming from Alexander Duality

$$H_0^{n+1+j}(V(\widetilde{f})) = H^{n+2-j}(M(\widetilde{f}))$$

where H_0^* denotes the primitive cohomology as defined in [6, p. 146]. Let now F (resp. $\widetilde{F}$) be the Milnor fibers associated to the homogeneous polynomials f (resp. $\widetilde{f}$) as in Theorem 5. It follows from the Thom–Sebastiani Theorem, see [6, p. 196], that one has the following isomorphisms.

$$H^{n+2-j}(M(\widetilde{f})) = H^{n+2-j}(\widetilde{F})_1 = H^{n+1-j}(F)_{\neq 1}.$$

Here $H^{n+2-j}(\widetilde{F})_1$ is the eigenspace of the monodromy corresponding to the eigenvalue 1, and $H^{n+1-j}(F)_{\neq 1}$ has a similar meaning.

Now several results in [7, Section 6.4], obtained via the theory of *perverse sheaves* give sufficient conditions for the vanishing of the groups $H^{n+1-j}(F)_{\neq 1}$ for all $j > 0$, see Example 6.4.14, Corollary 6.4.15, Theorem 6.4.18. As a sample result, we give the following.

Proposition 17. *Assume that the hypersurface $V(f)$ is a normal crossing divisor along one of its irreducible components. Then the associated cyclic covering $X = V(\widetilde{f})$ satisfies $H_0^{n+1+j}(V(\widetilde{f})) = 0$ for all $j > 0$. In other words, the Betti numbers of the associated cyclic covering $X = V(\widetilde{f})$ are known once the Euler characteristic $\chi(X)$ is known.*

In conclusion, in the general case we encounter two difficult problems, whose solution is far from being complete even from the theoretical viewpoint. The first one is to classify the simplest non-isolated singularities and to understand their local topology. For recent progress in this area we refer to [27]. The second one is to glue the local information in order to obtain information on the global topology. Here the most powerful approach is to use the theory of constructible sheaves, [7]. In this theory, the topology of a singularity is reflected in a number of invariants related to the Euler obstruction and the polar cycles, see [25] for details on this classical and beautiful approach.

References

1. V.I. Arnold, S.M. Gusein-Zade, A.N. Varchenko: Singularities of Differentiable Maps, Vol. 1 and 2, Monographs in Math., vol. 82 and 83, Birkhäuser 1985, 1988.
2. J. Bochnak, M. Coste, M.-F. Roy: Géométrie algébrique réelle, Erg. der Mathematik, vol. 12, Springer Verlag 1987.
3. J.W. Bruce and C.T.C. Wall: On the classification of cubic surfaces, J. London Math. Soc. 19, 1979, 245–256.
4. A. Degtyarev and V. Kharlamov: Topological properties of real algebraic varieties: Rokhlin's way. Russ. Math. Surveys 55, 2000, 735–814.
5. A. Dimca: Topics on Real and Complex Singularities, Vieweg Advanced Lectures in Math.,Vieweg 1987.
6. A. Dimca: Singularities and Topology of Hypersurfaces, Universitext, Springer, 1992.
7. A. Dimca: Sheaves in Topology, Universitext, Springer-Verlag, 2004.
8. A. Dimca: Hyperplane arrangements, M-tame polynomials and twisted cohomology, Commutative Algebra, Singularities and Computer Algebra, Eds. J. Herzog, V. Vuletescu, NATO Science Series, Vol. 115, Kluwer, 2003, pp. 113–126.
9. A. Dimca: On the Milnor fibrations of weighted homogeneous polynomials, Compositio Math. 76, 1990, 19–47.
10. A. Dimca, J. Hillman, L. Paunescu: On hypersurfaces in real projective spaces, Far East J. Math. Sciences 5(1997),159–168.
11. A. Dimca and S. Papadima: Hypersurface complements, Milnor fibers and higher homotopy groups of arrangements, Annals of Math. 158, 2003, 473–507.
12. A. Dimca and A. Némethi: Hypersurface complements, Alexander modules and monodromy, 'Proceedings of the 7th Workshop on Real and Complex Singularities, Sao Carlos, 2002, M. Ruas and T. Gaffney Eds, Contemp. Math. AMS. 354, 2004, 19–43.

13. W. Fulton: Algebraic Topology, a first course, GTM 153, Springer, 1995.
14. M. Golubitsky and V. Guillemin: Stable Mappings and their Singularities, GTM 14, Springer, 1973.
15. A. Libgober: Alexander invariants of plane algebraic curves, Proc. Symp. Pure Math. 40, Part 2, AMS, 1983, 135–144.
16. A. Libgober: Homotopy groups of the complements to singular hypersurfaces II, Annals of Math. 139, 1994, 117–144.
17. A. Libgober: Eigenvalues for the monodromy of the Milnor fibers of arrangements. In: Libgober, A., Tibăr, M. (eds) Trends in Mathematics: Trends in Singularities. Birkhäuser, Basel, 2002.
18. A. Libgober: Isolated non-normal crossing, Proceedings of the 7th Workshop on Real and Complex Singularities, Sao Carlos, 2002, M. Ruas and T.Gaffney Eds, Contemp. Math. AMS. 354, 2004, 145–160.
19. L. Maxim: Intersection homology and Alexander modules of hypersurface complements. arXiv:math.AT/0409412.
20. C. McCrory, A. Parusiński: Complex monodromy and the topology of real algebraic sets, Compositio Math. 106, 1997, 211–233.
21. J. Milnor: Singular Points of Complex Hypersurfaces, Annals Math. Studies 61, Princeton, 1968.
22. P. Orlik and H. Terao: Arrangements of Hyperplanes, Grundlehren 300, Springer-Verlag, 1992.
23. A. Parusiński, P. Pragacz: A formula for the Euler characteristic of singular hypersurfaces, J. Algebraic Geometry, 4, 1995, 337–351.
24. L. Paunescu: The topology of the real part of a holomorphic function, Math. Nachrichten, 174, 1995, 265–272.
25. R. Piene: Cycles polaires et classes de Chern pour les variétés projectives singulières, in: Introduction à la théorie des singularités II, Travaux en cours 37, Hermann Paris, 1988.
26. D. Rolfsen: Knots and Links, Publish or Perish, 1976.
27. D. Siersma: The vanishing topology of non-isolated singularities, in: D. Siersma, C.T.C. Wall, V.Zakalyukin (Eds.), New Developments in Singularity Theory, NATO Science Series 21, Kluwer, 2001, pp. 447–472.
28. R. Silhol: Real Algebraic Surfaces, Lecture Notes in Math. 1392, Springer-Verlag, 1989.
29. E. Spanier: Algebraic Topology, McGraw-Hill, 1966.

Overview of topological properties of real algebraic surfaces

Viatcheslav Kharlamov

Institut de Recherche Mathématique Avancée
Université Louis Pasteur et CNRS
7 rue René-Descartes
67084, Strasbourg, France
kharlam@math.u-strasbg.fr

Summary. These notes reproduce the content of a short, 50-minutes, survey talk given at the Nice University in September 2004. We added a few topics that have not been touched on in the lecture by lack of time.

1 Introduction

Topology of real algebraic varieties is a broad subject. Thus, it is reasonable to specify the "level" of objects and the goal of study. In what concerns the level, one may distinguish between affine, projective and abstract varieties, and, from a certain point of view, it is natural to start with abstract varieties, and then descend to projective and affine ones. As to the goal, I'll give preference to those "real results" that require "complex proofs" (even though in what follows the proofs will almost always stay behind the scenes) and, moreover, admit "complex statements."

Another major, and traditional, simplification is to consider *nonsingular* varieties, at least at the first stage. Certainly, a complete separation from the singular world is never possible. However, de facto, in all the cases when a complete understanding was achieved in the nonsingular case, it turned out that the singular case could be treated, at least in principle, by similar methods.

In such a setting, it is natural to consider not only algebraic, but arbitrary Kähler compact complex manifolds, and to call a complex manifold X *real*, if it can be equipped with a *real structure*, that is an anti-holomorphic involution $c: X \to X$. Real points are then, by definition, the fixed points of the real structure. We denote by $X_{\mathbb{R}}$ the set of real points, or the *real part*, of X. For the sake of symmetry, we often denote X by $X_{\mathbb{C}}$.

Of course, the principal source of examples is given by nonsingular varieties defined by systems of real polynomial equations; in these c is the complex conjugation. Note that by our convention a real variety is nonsingular if it does not have singular points, be they real or imaginary.

We consider real algebraic varieties up to the following equivalence relations: diffeomorphism of real part, diffeomorphism of real structure, and deformation of variety together with real structure. As usual, by an *elementary real deformation* of a real variety we mean a smooth, differentially locally trivial, family of real varieties (say, an equivariant deformation in the sense of Kodaira-Spencer). Two real varieties are called real deformation equivalent if there exists a chain of elementary deformations connecting them. *Topology of the real part and that of the real structure are preserved under deformation.* This phenomenon is one of the main motivations for the study of topological properties of real varieties. Note in advance that in many important cases the topology of real structure (which includes, in fact, the topology of real part) determines the deformation class.

A fundamental example of deformation is a small variation of a nonsingular system of polynomial equations (that is, a system whose Jacobian has maximal rank at each solution of the system) or, more generally, any variation which is represented in the total space of systems of a given number of equations (equal to the codimension of variety) of a given degree in a given number of variables by a smooth path in the complement of the discriminant locus, that is a smooth path avoiding singular systems. However, other deformations may exist as well; thus, surfaces of degree ≥ 5 in $\mathbb{P}^3$ have big deformations failing to be embedded in $\mathbb{P}^3$ while surfaces of degree 4 have small Kähler deformations failing to be projective. By contrast, any deformation of a surface of degree 3 is realized in $\mathbb{P}^3$.

Our choice of surfaces as the topic of this talk is motivated by the fact that this is the first nontrivial case with respect to the above equivalence relations. In fact, in dimension zero the topology is determined by the natural number $b_0(X_\mathbb{R})$ satisfying the following relations

$$b_0(X_\mathbb{R}) \leq b_0(X_\mathbb{C}), \quad b_0(X_\mathbb{R}) = b_0(X_\mathbb{C}) \mod 2 \tag{1.1}$$

(here and in what follows $b_i(\cdot)$ denotes the rank of $H_i(\cdot\,;\mathbb{Z}/2\mathbb{Z})$, so that b_0 is nothing but the number of connected components). Similarly, in dimension one everything is determined by the number $b_0(X_\mathbb{R})$ (or, equivalently, by $b_1(X_\mathbb{R}) = b_0(X_\mathbb{R})$), and the only relations linking this number with the invariants of $X_\mathbb{C}$ are as follows:

$$b_0(X_\mathbb{R}) \leq g(X_\mathbb{C}) + 1, \quad \text{and} \quad b_0(X_\mathbb{R}) = g(X_\mathbb{C}) + 1 \mod 2 \text{ if } X_\mathbb{R} \text{ divides } X_\mathbb{C}. \tag{1.2}$$

Here, g is the usual genus, $g(X_\mathbb{C}) = \frac{1}{2}b_1(X_\mathbb{C})$ if X is irreducible; otherwise, $g(X_\mathbb{C}) + 1$ is the sum of $g + 1$ over all irreducible components. The condition that $X_\mathbb{R}$ divides $X_\mathbb{C}$ is equivalent to the orientability of the quotient $X_\mathbb{C}/c$. The above relations demonstrate a general phenomenon: $\mathrm{Top}(X_\mathbb{R}) \leq \mathrm{Top}(X_\mathbb{C})$ (the topology of a real variety is bounded by that of its complexification).

Moreover, in dimensions less than two topology, and even homology determines the deformation equivalence classes: $\text{Top}(X_\mathbb{C}) = \text{Def}(X_\mathbb{C})$ over $\mathbb{C}$ and $\text{Top}(X_\mathbb{C}, c) = \text{Top}(X_\mathbb{C}, X_\mathbb{R}) = \text{Def}(X_\mathbb{C}, c)$ over $\mathbb{R}$. For example, two real irreducible curves are deformation equivalent, if and only if they have the same number of real components, $b_0(X_\mathbb{R})$, and they both either divide $X_\mathbb{C}$ or not (in other words, either, in both cases, $X_\mathbb{R}$ is homologous to 0 in $X_\mathbb{C}$, or not).

An advantage of dimension two is that algebraic topology still suffices to determine $\text{Top } X_\mathbb{R}$, which is no more the case in dimensions ≥ 3 (that is why the dimensions ≥ 3 are still very far from being well understood; cf., however, J. Kollár's papers [13, 14] and references in [11] for some nontrivial partial results).

Many of the tools used in dimension ≤ 2 can be extended to higher dimensions, and when consideration of higher dimensions does not lead to complications we present our results in full generality.

For lack of time, we do not discuss arrangements of curves on surfaces, construction of surfaces (and curves on surfaces) with prescribed topology, or enumerative results. We also will not have time to discuss history of the subject, but it is worth mentioning that many of the results presented below emerged due to collective efforts of many insiders (in particular, a group of Russian mathematicians inspired by I. G. Petrovsky and V. I. Arnol'd in Moscow, D. A. Gudkov in the former Gor'kii, and V. A. Rokhlin in the former Leningrad; the author gratefully dedicates these notes to the memory of the latter). A reader interested to know better who, when and how discovered these results is invited to consult [5] and the references therein.

2 Topology of real varieties versus topology of their complexifications

2.1 Smith theory

Smith theory provides the following relations valid for all dimensions:

$$\sum b_i(X_\mathbb{R}) \leq \sum b_i(X_\mathbb{C}), \quad \sum b_i(X_\mathbb{R}) = \sum b_i(X_\mathbb{C}) \mod 2, \qquad (2.1)$$

where $b_i(\cdot) = \dim H_i(\cdot\,; \mathbb{Z}/2\mathbb{Z})$. Behind these relations there are such useful tools as Smith's exact sequence and Kalinin's spectral sequence, see, for example, [5]. The latter starts with

$$E^1_* = H_*(X_\mathbb{C}; \mathbb{Z}/2\mathbb{Z}), \quad E^2_* = \text{Ker}(1 + c_*)/\text{Im}(1 + c_*),$$

and converges to $H_*(X_\mathbb{R}; \mathbb{Z}/2\mathbb{Z})$ (here and in what follows c_* stands for the homomorphism $H_*(X_\mathbb{C}; \mathbb{Z}/2\mathbb{Z}) \to H_*(X_\mathbb{C}; \mathbb{Z}/2\mathbb{Z})$ induced by the real structure $c : X_\mathbb{C} \to X_\mathbb{C}$). Already the existence of such a spectral sequence implies (2.1). According to the above formula for E^2_*, there is a "stronger" inequality

$$\sum b_i(X_\mathbb{R}) \leq \dim H_1(\mathbb{Z}/2\,\mathbb{Z}, H_*(X_\mathbb{C}; \mathbb{Z}/2\,\mathbb{Z})), \tag{2.2}$$

where $H_1(\mathbb{Z}/2\,\mathbb{Z}, H_*(X_\mathbb{C}; \mathbb{Z}/2\,\mathbb{Z})) = \mathrm{Ker}(1+c_*)/\mathrm{Im}(1+c_*)$.

There are two important classes of real varieties enjoying special features, viz.

- *M-varieties*, i.e. varieties for which $E_*^1 = H_*(X_\mathbb{R}; \mathbb{Z}/2\,\mathbb{Z})$ (which is equivalent to $\sum b_i(X_\mathbb{R}) = \sum b_i(X_\mathbb{C})$, the extremal case of (2.1), or to the existence of c-invariant cycles in any homology class in $H_*(X_\mathbb{C}; \mathbb{Z}/2\,\mathbb{Z})$); here M stands for "maximal";
- and *GM-varieties*, i.e. varieties for which $E_*^2 = H_*(X_\mathbb{R}; \mathbb{Z}/2\,\mathbb{Z})$ (the extremal case of (2.2), which is equivalent to the existence of c-invariant cycles in any of the classes in $H_*(X_\mathbb{C}; \mathbb{Z}/2\,\mathbb{Z})$ fixed by c_*; note that all real surfaces with $\pi_1(X_\mathbb{C}) = 1$ are GM); here GM stands for "Galois maximal."

In dimensions 0 and 1 the relations (2.1) yield (1.1) and (1.2), except for the congruence in (1.2) which is a mod 4 relation for $\sum b_i$. The latter can be generalized in the following way: *if X is a GM-variety of odd dimension, $\dim X = 2k+1$, and $X_\mathbb{R}$ is homologous to 0 in $X_\mathbb{C}$, then*

$$\frac{1}{2}\sum b_i(X_\mathbb{R}) = \frac{1}{2}\sum b_i(X_\mathbb{C}) \mod 2.$$

Note that both parts of the above congruence are integers. Since $X_\mathbb{R} \sim 0$, for any $x \in H_{2k+1}(X_\mathbb{C}; \mathbb{Z}/2)$ one has

$$(x, c_*x) = (x, [X_\mathbb{R}]) = 0 \mod 2,$$

which implies that, besides the summands with constant action of c, any irreducible orthogonal decomposition of $c_* : H_*(X_\mathbb{C}; \mathbb{Z}/2) \to H_*(X_\mathbb{C}; \mathbb{Z}/2)$ contains only irreducible components of rank 4 (each such component is a permutation of two $\mathbb{Z}/2\,\mathbb{Z}$-planes). Thus,

$$\sum b_i(X_\mathbb{R}) = \dim \mathrm{Ker}(1+c_*)/\mathrm{Im}(1+c_*) = \sum b_i(X_\mathbb{C}) - 4p,$$

where p is the number of the above irreducible components of rank 4.

Smith theory helps to answer the following questions: *Is $X_\mathbb{R}$ non empty (existence of real solutions)? How many connected components does it have?* and, more generally, *What are the possible values of $b_i(X_\mathbb{R})$?*

Already the congruence part of (2.1) gives a simple, often useful, sufficient condition: $X_\mathbb{R}$ is non-empty, if $\sum b_*(X_\mathbb{C})$ is odd. Note that this condition does not depend on the choice of real structure.

Smith theory is not well adapted to work with individual Betti numbers, and so in dimensions > 2 it is hard to get more information and to answer the other questions using Smith theory alone. Hopefully, in dimension 2 it is sufficient to perform homological calculations in Kalinin's spectral sequence (or in Smith's exact sequence) and to add the information coming from the Lefschetz trace formula. In particular, in the case of real surfaces with $\pi_1(X_\mathbb{C}) = 1$ we get the following formulae:

$$b_0(X_\mathbb{R}) = \frac{1}{2}(\sum b_i(X_\mathbb{C}) - b_2^{-1} - a),$$

$$b_1(X_\mathbb{R}) = b_2 - b_2^{+1} - a,$$

where $b_2^{\pm 1}$ are the dimensions of the eigenspaces of the involution

$$c_* : H_2(X_\mathbb{C}; \mathbb{R}) \to H_2(X_\mathbb{C}; \mathbb{R}),$$

and a is the number of nontrivial components in an irreducible $\mathbb{Z}/2\mathbb{Z}$-vector space decomposition of $c_* : H_*(X_\mathbb{C}; \mathbb{Z}/2) \to H_*(X_\mathbb{C}; \mathbb{Z}/2)$ (each such component is generated by two elements permuted by c_*).

We recall that the above definition of GM-variety is equivalent to the existence of an equivariant cycle in each invariant $\mathbb{Z}/2\mathbb{Z}$-homology class, and the definition of M-variety is equivalent to the existence of an equivariant cycle in each $\mathbb{Z}/2\mathbb{Z}$-homology class. The latter happens, for example, if all $\mathbb{Z}/2\mathbb{Z}$-homology classes of $X_\mathbb{C}$ are algebraic and have a real representative. This is the case for projective spaces and Grassmann varieties equipped with their tautological real structures. Many other special varieties also have this maximality property. An important class of varieties for which the maximality question is open is provided by the discriminants of polynomials in three or more (homogeneous) variables. For three variables, this reduces to the question of maximality of the space of all singular plane curves of a given degree. In degrees 1, 2, and 3 the discriminant is indeed an M-variety, as can be verified by a more or less straightforward calculation using the Alexander–Pontryagin duality (one should first compute the Betti numbers of the space of nonsingular curves). But already in degree 4, the question is open.

The relations (2.1) apply, in fact, to any finite-dimensional space with involution, and, in particular, to singular varieties. In many cases, for example in the case of projective hypersurfaces, the corresponding upper bounds are the best known one. These relations can also be applied to semi-algebraic sets; to do this, it suffices to replace such a set by its tubular neighborhood and then apply (2.1) to its boundary, which is a hypersurface. Y. Laszlo and C. Viterbo [15] recently addressed the following question: how to bound the total Betti number of a nonsingular real projective variety X in terms of its degree d and dimension n. Combining (2.1) with some inequalities due to Demailly–Peternell–Schneider they proved that

$$\sum b_i(X_\mathbb{R}) \leq 2^{n^2+n+3} d^{n+1}, \tag{2.3}$$

while all the previously known estimates (like those of R. Thom [20] and J. Milnor [17]) were of the type

$$\sum b_i(X_\mathbb{R}) \leq C d^{2n+1}. \tag{2.4}$$

Using Lefschetz pencil, resp. Morse function type arguments applied to $X_\mathbb{C}$, resp. $X_\mathbb{R}$, one can improve the leading coefficient in the Laszlo-Viterbo bound and get

$$\sum b_i(X_\mathbb{R}) \leq d^{n+1} + O(d^n), \quad \sum b_i(X_\mathbb{C}) \leq d^{n+1} + O(d^n) \quad (2.5)$$

(without appealing to Demailly-Peternell-Schneider inequalities). What is an optimal choice of $O(d^n)$? I do not know. At least,

$$\sum b_i(X_\mathbb{R}) \leq \sum b_i(X_\mathbb{C}) \leq d^{n+1} \quad \text{if} \quad d > 1, \quad (2.6)$$

and it seems reasonable to expect that $d^{n+1} + O(d^n)$ in (2.5) can be replaced by the polynomial $d^{n+1} + \sum_{k=0}^{n} a_k d^k$ representing the total Betti number of a degree d nonsingular hypersurface in $\mathbb{CP}^{n+1}$ (cf. (2.10) for the case $n = 2$). The key point in the proof of (2.5) and (2.6) is a similar bound,

$$d^* \leq d(d-1)^n = d^{n+1} + O(d^n), \quad (2.7)$$

for the degree d^* of the variety X^* projectively dual to X (I am grateful to F. Zak who explained to me how such a general bound for d^* is deduced from the, classical, computation of d^* in the special case of hypersurfaces; he also proposed to replace $O(d^n)$ by 0 in (2.5)). Indeed, the number of singular fibers of the pencil of hyperplane sections of $X_\mathbb{C}$ and, respectively, the number of singular values of the linear Morse function on an affine part of $X_\mathbb{R}$ is bounded by d^*, so that an induction on the dimension n of $X_\mathbb{R}$ gives a sequence $x_n = \sum b_i(X_\mathbb{R})$ (resp. $y_n = \sum b_i(X_\mathbb{C})$) with the property

$$x_n - 2x_{n-1} + x_{n-2} \leq d^* \quad (\text{respectively} \quad y_n - 2y_{n-1} + y_{n-2} \leq d^*)$$

(here $X = X_n, X_{n-1}, \ldots, X_0$ is the sequence formed by X and its consecutive hyperplane sections). Combined with (2.7), this yields (2.5). For $d > 2$ and $n > 2$ the bound (2.6) follows from the inequalities $y_n \leq d^* + 2y_{n-1} \leq d(d-1)^n + 2d^n \leq d^{n+1}$, and if $d = 2$ or $n = 1, 2$, then it is easy to prove the bound by an ad hoc argument.

Returning to the Smith bound, let us forewarn that it is impossible in general to replace $b_i(\cdot) = \dim H_*(\cdot\,; \mathbb{Z}/2\,\mathbb{Z})$ in the inequality (2.1) by the ordinary Betti numbers $\beta_i(\cdot) = \dim H_*(\cdot\,; \mathbb{Q})$. For example, there exist real Enriques surfaces X with the real part consisting of two real components, one homeomorphic to a torus and another to a connected sum of 10 real projective planes, while for such real surface $\dim H_*(X_\mathbb{C}; \mathbb{Q}) = 12 < 14 = \dim H_*(X_\mathbb{R}; \mathbb{Q})$.

2.2 Higher order congruences

Higher order congruences can be found based on Smith theory and arithmetic of integral quadratic forms. Here is a typical example: *if* $\dim X = 2k$ *and* $X_\mathbb{R}$ *is* $\mathbb{Z}/2$-*homologous to the middle dimensional Wu class of* $X_\mathbb{C}$, *then*

$$\chi(X_\mathbb{R}) = \sigma(X_\mathbb{C}) \mod 8$$

(here χ is the Euler characteristic and σ is the signature). Its one-line proof given below is a model for finding other higher order congruences. It is based on the Lagrangian property of the real part and the Lefschetz–Hirzebruch signature formula:

$$\chi(X_\mathbb{R}) = (-1)^k \sigma(c) =_8 (-1)^k (\operatorname{Wu}_c, c \operatorname{Wu}_c) =$$
$$(-1)^k (\operatorname{Wu}_X, c \operatorname{Wu}_X) = (\operatorname{Wu}_X, \operatorname{Wu}_X) =_8 \sigma(X_\mathbb{C})$$

(here Wu_c is the Wu integral characteristic class of the quadratic form (x, cy) on $H_{2k}(X_\mathbb{C}, \mathbb{Z})/\operatorname{Tors}$, $\sigma(c)$ is its signature, Wu_X is an integral algebraic representative of the Wu-class of $X_\mathbb{C}$, and $=_8$ stands for congruence modulo 8). It should be mentioned that in the case when $X_\mathbb{C}$ is a complete intersection there are various methods for explicit computation of $\sigma(X_\mathbb{C})$ (see, for example, [5] and references therein).

There is a series of congruences refining the above one (see, for example, [5] and references therein); the two simplest of them are as follows:

if $\dim X = 2k$ and $X_\mathbb{R}$ is a M-variety, then

$$\chi(X_\mathbb{R}) = \sigma(X_\mathbb{C}) \mod 16;$$

if $\dim X = 2k$ and $X_\mathbb{R}$ is a $(M-1)$-variety, then

$$\chi(X_\mathbb{R}) = \sigma(X_\mathbb{C}) \pm 2 \mod 16.$$

Note that $\sigma(X_\mathbb{C}) = (-1)^k \chi(X_\mathbb{C}) \mod 4$ (which most easily can be seen from the Hodge decomposition), which yields an analog of congruence (1.2) in even dimensions:
$\chi(X_\mathbb{R}) = (-1)^k \chi(X_\mathbb{C}) \mod 4$ if $X_\mathbb{R}$ is $\mathbb{Z}/2$-homologous to the middle dimensional Wu class of $X_\mathbb{C}$.

2.3 An application of Hodge theory and some other inequalities

As is shown in [7], from the Hodge decomposition and the Lefschetz formula it follows that

$$|\chi(X_\mathbb{R}) - 1| \leq h^{k,k}(X_\mathbb{C}) - 1 \quad \text{if} \quad \dim X = 2k \qquad (2.8)$$

(where $h^{k,k}$ is the Hodge number of bidegree (k, k); various explicit computations of the Hodge numbers are found in [5]; an expression for $h^{1,1}$ in the case of surfaces in 3-space is given below in (2.11)). See [8, 9, 5] and references therein for an odd-dimensional version of this Comessatti–Petrovskii type inequality and for generalizations to varieties with singularities (naturally, in the singular case pure Hodge structure is to be replaced by mixed one). It would certainly be nice to find other applications of Hodge theory giving more detailed information than (2.8). Especially challenging is to somehow relate Hodge theory with Smith theory.

Let me indicate here only a very special amusing application of (2.8) to the case of odd dimension. It concerns plane curves, and, more specifically, line arrangements. We consider a generic configuration of $2k$ real lines in the projective plane. The number of connected components, called cells, of the complement of the arrangement is equal then to $2k^2 - k + 1$. Since the number of lines is even, the cells can be chess-board colored, and an application of (2.8) shows that an upper bound for the number of projective cells of one color is $\frac{3}{2}k(k-1) + 1$, so that a bound from below for the other color is given by $\frac{1}{2}k(k+1)$.

More special inequalities, not directly related to Hodge theory, can be obtained using the Lagrangian property of $X_\mathbb{R}$. Thus, in the case of surfaces one can easily show that *the number p_- of orientable components of $X_\mathbb{R}$ with $\chi < 0$ has an upper bound*

$$p_- \leq \frac{1}{2}(\sigma^+(X_\mathbb{C}) - 1)$$

(*where σ^+ is the positive index of inertia of the intersection form*).

Note also that in the case of surfaces the inequality (2.8) can be extended to non-Kähler surfaces in the form

$$2 - \dim H^2(X_\mathbb{C}; \mathbb{R}) + 2h^{2,0}(X_\mathbb{C}) \leq \chi(X_\mathbb{R}) \leq 4 + \dim H^2(X_\mathbb{C}; \mathbb{R}) - 2h^{2,0}(X_\mathbb{C}),$$

which is weaker than (2.8), but differs from it only by 4 in the left- and right-hand parts. This difference is due to the absence of a Kähler class in H^2 and the asymmetry $H^{1,0} = H^{0,1} - 1$ in H^1.

It would be interesting to find analogs of (2.8) for the signature $\sigma(X_\mathbb{R})$ instead of $\chi(X_\mathbb{R})$, of course under the hypothesis that $X_\mathbb{R}$ is orientable. The best bound known to me does not involve Hodge theory. It says that

$$|\sigma(X_\mathbb{R})| \leq \frac{1}{3}c_2^2(X_\mathbb{C}), \tag{2.9}$$

and follows directly from evaluating an algebraic representative of the second Chern class of $X_\mathbb{C}$ on $X_\mathbb{R}$, viz. $\operatorname{inj}_* c_2(X_\mathbb{C}) \cap [X_\mathbb{R}] = p_1(X_\mathbb{R}) \cap [X_\mathbb{R}] = 3\sigma(X_\mathbb{R})$. This bound holds under an additional hypothesis that the tangent, or cotangent, or some other vector bundle of $X_\mathbb{C}$ with the same c_2 is generated by its sections (this moving condition allows to put an algebraic representative of the second Chern class in a general position with respect to $X_\mathbb{R}$ and thus to get (2.9)). More general and considerably more subtle bounds for arbitrary Pontryagin numbers can be found in a recent paper by Y. Laszlo and C. Viterbo [15].

2.4 Special surfaces

The above tools allow to understand thoroughly the topology of $X_\mathbb{R}$ for many special types of surfaces. For example, they lead to a complete topological classification of $X_\mathbb{R}$, and even of $(X_\mathbb{C}, c)$, for cubic and quartic surfaces in $\mathbb{P}^3$. We describe it in terms of generators: each topological type generates a list of its Morse simplifications, that is the topological types obtained from the initial one by a series of Morse surgeries decreasing the total Betti number (removing a spherical component or contracting a handle).

There are 5 classes of nonsingular cubics generated by $\#_7 \mathbb{P}^2(\mathbb{R})$ and $\mathbb{P}^2(\mathbb{R}) \sqcup S^2$ (here and in what follows $\#$ stands for a connected and $\sqcup$ for a disjoint sum), and 66 classes of nonsingular quartics generated by three M-surfaces $\#_{10}(S^1 \times S^1) \sqcup S^2$, $\#_6(S^1 \times S^1) \sqcup 5S^2$, $\#_2(S^1 \times S^1) \sqcup 9S^2$, two $(M-2)$-surfaces $\#_7(S^1 \times S^1) \sqcup 2S^2$, $\#_3(S^1 \times S^1) \sqcup 6S^2$, and a pair of tori $2(S^1 \times S^1)$.

Surfaces in $\mathbb{P}^3(\mathbb{R})$ can also be studied up to different equivalence relations, such as: *ambient isotopy in $\mathbb{P}^3(\mathbb{R})$*, *rigid isotopy* (i.e., isotopy in the class of nonsingular or, more generally, equisingular in some appropriate sense surfaces of the same degree), and *rough projective equivalence* (i.e., projective transformation and rigid isotopy). The difference between the last two relations is due to the fact that the group $PGL(4; \mathbb{R})$ of projective transformations of $\mathbb{P}^3(\mathbb{R})$ has two connected components. Of course, the transformations in the component of unity transform a surface into a rigidly isotopic one. To what extent the classifications up to rigid isotopies and up to rough projective equivalence are topological is an open question for surfaces of degree 5 and higher, cf. the discussion in 3.1.

Topologically, the non-spherical component of the real part of a nonsingular cubic surface is embedded in $\mathbb{P}^3(\mathbb{R})$ as the standard $\mathbb{P}^2(\mathbb{R})$ with unlinked and unknotted handles. Moreover, for cubic surfaces not only the isotopy equivalence relation, but all the other relations mentioned above coincide with the purely topological one.

The embedding of a quartic surface in $\mathbb{P}^3(\mathbb{R})$ is also simple: it is isotopic to a union of ellipsoids and hyperboloids with unknotted and unlinked handles. With one exception, the components are outside each other; in the exceptional case the real part consists of two nested spheres. In all other cases the isotopy type of the real part $X_\mathbb{R}$ of a real quartic surface in $\mathbb{P}^3(\mathbb{R})$ is determined by its topological type and contractibility or noncontractibility of $X_\mathbb{R}$ in $\mathbb{P}^3(\mathbb{R})$. It turns out that in the case of degree 4 surfaces all the four classifications (topological, isotopic, rough projective, and rigid) are different. Note that the only difference between rough projective equivalence and rigid isotopy is in chirality which tells whether or not a surface is rigidly isotopic to its mirror image. Rough projective equivalence is discussed in [19] and chirality in [10].

Any nonsingular degree 4 surface in $\mathbb{P}^3$ is a *K3-surface*, that is a compact complex surface with $\pi_1 = 1$ and $c_1 = 0$. Other examples of $K3$-surfaces are given by double coverings of a nonsingular quadric in $\mathbb{P}^3$ branched in a transverse section by a quartic, by double coverings of a nonsingular cubic in $\mathbb{P}^3$ branched in a transverse section by a quadric, by transverse intersections of three quadrics in $\mathbb{P}^5$, etc. Classification of all the real projective $K3$-surfaces up to rough projective equivalence can also be found in [19].

The methods used in the study of real $K3$-surfaces are based on the above tools, including Hodge theory, as well as the Torelli theorem which plays a key role. In what concerns rigid isotopies and rough projective equivalence, using the surjectivity of the period map, one can reduce the study of real structures to a study of arithmetic properties of integral lattices. Similar methods can be used to study $K3$-surfaces with simple singularities, but this problem has never been treated systematically.

Starting with degree 5, our knowledge is much more limited. It is not even known what are the extremal values of the Betti numbers of nonsingular quintics. We only know that the maximal number of connected components is somewhere in between 23 and 25 and that the maximal first $\mathbb{Z}/2\mathbb{Z}$-Betti number is either 45 or 47 (for the surfaces in the same deformation class one has $\max b_1 = 47$). The best known general bounds for the Betti numbers are those given by the inequalities described in the previous sections. For a surface in $\mathbb{P}^3$ and, more generally, for a transversal complete intersection X in $\mathbb{P}^q$, the complex ingredients of these bounds can easily be found. To wit, if X is a complete intersection in $\mathbb{P}^q$ of polydegree $(m_1, \ldots, m_{q-2})$ then

$$b_1(X) = 0, \qquad b_2(X) = \chi(X) - 2,$$

$$h^{1,1}(X) = \frac{1}{2}[b_2(X) - \sigma(X)] + 1$$

$$\chi(X) = \mu_{q-2}\left(\mu_1^2 - \mu_2 - (q+1)\mu_1 + \tfrac{1}{2}q(q+1)\right),$$

$$\sigma(X) = -\tfrac{1}{3}\mu_{q-2}(\mu_1^2 - 2\mu_2 - q - 1),$$

where μ_i is the i-th elementary symmetric polynomial in $(m_1, \ldots, m_{q-2})$. In particular, for a surface of degree m in $\mathbb{P}^3$

$$\sum b_i(X_\mathbb{C}) = \chi(X_\mathbb{C}) = m^3 - 4m^2 + 6m, \tag{2.10}$$

$$h^{1,1}(X_\mathbb{C}) - 1 = (m-1)^3 - \frac{1}{3}m(m-1)(m-2). \tag{2.11}$$

As it was already noticed, the same tools can be applied to singular objects as well (see, for example, [9] and [21]). For instance, one can use them to bound the number of real double points in the following very simple way. In the case of surfaces there are two types of such points, viz. solitary points and nodes (in local coordinates their equations are $x^2 + y^2 + z^2 = 0$ and $x^2 + y^2 = z^2$ respectively). One can resolve the nodes (which is differentially equivalent to replacing a neighborhood of a node by its perturbation $x^2 + y^2 = z^2 + \epsilon^2$) and replace the solitary points by spheres (which means replacing a neighborhood of a solitary point in $X_\mathbb{C}$ by its perturbation $x^2 + y^2 + z^2 = \epsilon^2$). As a result, we obtain a 4-manifold diffeomorphic to the minimal desingularization $\tilde{X}_\mathbb{C}$ of $X_\mathbb{C}$ and an involution on it such that the fixed point set is diffeomorphic to a disjoined sum of the minimal desingularization $\tilde{X}_\mathbb{R}$ of $X_\mathbb{R}$ and S spheres, where S is the number of solitary points of $X_\mathbb{R}$. Now, applying the Smith inequality, one gets

$$2S + \sum b_i(\tilde{X}_\mathbb{R}) \le \sum b_i(\tilde{X}_\mathbb{C}).$$

Thus in the case of surfaces of degree m in $\mathbb{P}^3$

$$2S \le 2S + \sum b_i(\tilde{X}_\mathbb{R}) \le m^3 - 4m^2 + 6m.$$

This implies, in particular, that $S \le \frac{1}{2}(m^3 - 4m^2 + 6m)$. Using the congruences mod 16 described in 2.2, this can be improved to a sharp bound: *the number of solitary points of a real quartic and, more generally, of any real singular $K3$-surface, is ≤ 10.* (This may be worth comparing with the upper bound 16 for the number of complex nodes of a complex quartic. This bound, which is probably due to R.W.H.T. Hudson, was extended to any singular $K3$-surface by V. Nikulin [18] who used arithmetic of integral quadratic forms. As is well known, probably since Fresnel and Kummer, real quartics with 16 real nodes do exist.)

Let me notice that the frontier of our knowledge of surfaces in $\mathbb{P}^3$ is similar to the frontier between special surfaces and surfaces of general type in the Enriques-Kodaira classification of compact complex surfaces: surfaces of degree ≥ 5 are of general type while surfaces of degree 4 are $K3$-surfaces and surfaces of degree 3 are rational. This gives additional motivation to turn to real structures on complex surfaces in various Enriques-Kodaira classes.

3 Deformation classes

Even the above very sketchy discussion shows that a thorough topological study of surfaces leads unavoidably to their study up to variation of equations and then to their study up to deformation (see Introduction for the definition; recall that we have chosen to work with Kähler surfaces).

3.1 Quasi-simplicity

As is pointed in 2.4, two nonsingular real cubic surfaces are real deformation equivalent if and only if their real point sets are homeomorphic. Furthermore, the real structures of two real nonsingular cubic surfaces are diffeomorphic if (and only if) the real point sets of the surfaces are homeomorphic. This is a manifestation of what we call the quasi-simplicity property: a real surface X is called *quasi-simple* if it is real deformation equivalent to any other real surface X' such that, first, X' is deformation equivalent to X as a complex surface, and, second, the real structure of X' is diffeomorphic to the real structure of X.

In fact, all rational real surfaces are quasi-simple. For $\mathbb{R}$-minimal (i.e., minimal over $\mathbb{R}$) rational surfaces this result is essentially due to Comessatti, Manin, and Iskovskikh (see e.g. the survey [16]). In full generality this is proved in [6], where, in addition, it is shown that the real deformation type of a real rational surface is determined by certain homological data.

Ruled $\mathbb{C}$-minimal surfaces of any genus are also quasi-simple, see [22]. Another class of surfaces whose real deformation theory is well understood is formed by minimal surfaces of Kodaira dimension 0. This class consists of Abelian, hyperelliptic, $K3$-, and Enriques surfaces. They are all quasi-simple (see [2] and [1]; recall that, by definition, hyperelliptic and Enriques surfaces are respectively quotients of Abelian and $K3$-surfaces by free involutions). Furthermore, quasi-simplicity of hyperelliptic and Enriques surfaces extends to quasi-simplicity of the quotients of Abelian and $K3$-surfaces by certain finite group actions, see [3].

Whether elliptic surfaces and irrational ruled non $\mathbb{C}$-minimal surfaces are quasi-simple is, as far as I know, an open question.

It is natural to expect that for surfaces of general type there is no quasi-simplicity: there should exist examples of real deformation distinct real surfaces with diffeomorphic real structures. A challenging problem is to find convenient deformation invariants which are not covered by the differential topology of $(X_{\mathbb{C}}, c)$.

Existence of non quasi-simple families of surfaces of general type does not prevent certain particular classes of surfaces of general type from being quasi-simple. And examples of quasi-simple real surfaces of general type do exist. One such example is given by real Bogomolov–Miyaoka–Yau surfaces, that is, surfaces covered by a ball in $\mathbb{C}^2$, see [12]. (Note in passing that in [12] it is also shown that there exist diffeomorphic, in fact complex conjugated, Bogomolov–Miyaoka–Yau surfaces which are not real and thus, being rigid, are not deformation equivalent as complex surfaces. These surfaces are counter-examples to the so called Diff = Deff problem in complex geometry, see [12] for precise definitions and references to counter-examples not related to the complex conjugation. This problem is a kind of substitute of quasi-simplicity for complex varieties. The existence of Diff $\neq$ Deff examples explains why we need to fix complex deformation class in the definition of quasi-simplicity of real varieties.)[1]

3.2 Finiteness

While the problem of quasi-simplicity is solved for rational surfaces and is essentially open and very difficult for surfaces of general type, the situation with finiteness is an opposite one: finiteness holds both for each complex deformation class of surfaces of general type (deformation finiteness) and for any fixed surface of general type (individual finiteness).

---

[1] *Added in proof.* When this paper had been already finished, we with Vik. Kulikov have constructed examples of non quasi-simple real surfaces of general type.

To wit, since the composition of two real structures is a biholomorphic automorphism and since the group of automorphisms of any variety of general type is finite, there are only finitely many real structures on a variety of general type (the same argument works for nonsingular hypersurfaces of degree ≥ 3 in projective space of dimension $n \geq 3$ with the exception of $n = 3, d = 4$). This is what we call *individual finiteness*, which we understand as finiteness of the number of conjugacy classes of real structures on a given variety (note that individual finiteness understood in this way extends to hypersurfaces of degree 4 in projective spaces of dimension 3, see [2]).

On the other hand, the Hilbert scheme of varieties of general type with given characteristic numbers is quasi-projective, which implies *deformation finiteness*: real structures on the varieties which, as complex varieties, are deformation equivalent to a given variety of general type split into a finite number of real deformation classes (where, according to our definitions, both variety and real structure are subject to deformation).

Unlike surfaces of general type, a rational surface may have a huge automorphism group, and, as far as I know, the problem of individual finiteness for rational surfaces is open. The situation is different with regard to deformation finiteness of rational surfaces which is an easy consequence of their quasi-simplicity.

In fact, *the deformation finiteness holds for any type of surfaces*. Indeed, the only birational classes of surfaces for which such a result is not contained in the literature, either explicitly or implicitly, are elliptic surfaces and irrational ruled surfaces, but for these classes the proof is more or less straightforward. It would be useful to find a conceptual proof dealing with all types of surfaces in a unified way.

Some finiteness results are also known for Klein actions of finite groups on $K3$- and Abelian surfaces. In particular, *the number of equivariant deformation classes of $K3$- and Abelian surfaces with faithful Klein actions of finite groups is finite*, see [3].

Another, higher-dimensional, generalization of finiteness of real structures on $K3$-surfaces extends it to so called holomorphic symplectic (hence hyperkähler) manifolds: *the number of equivariant deformation classes of real structures in a given deformation class of compact holomorphic symplectic manifolds is finite*, see [4].

The differential topology of $(X_\mathbb{C}, c)$ is preserved under deformation, and therefore deformation finiteness implies topological finiteness. Another, more direct, approach to topological finiteness was recently developed by Y. Laszlo and C. Viterbo. They proposed to study finiteness of diffeomorphism types of real forms on complex projective varieties of a given degree. Here one should distinguish between the real and complex degree. For example, there exists a sequence X_n of complex $K3$-surfaces of degree 4 (quartics in $\mathbb{P}^3$) such that, for appropriate real structures c_n on X_n, their real degrees (the minimal degree of a real projective embedding $X_n \to \mathbb{P}^{q_n}$) converge to infinity (so that these real structures are not induced from $\mathbb{P}^3$ and, moreover, can not be induced from $\mathbb{P}^q$ with bounded q). Of course, varieties of a given real degree split into a finite number of families. Whether the same is true for real varieties of a given complex degree is still an open question of utmost importance. But, thanks to Y. Laszlo and C. Viterbo [15], we now have some explicit bounds for the Pontryagin numbers of varieties of a given real degree and, as a consequence, some explicit bounds for the number of cobordism classes of such varieties.

References

1. F. Catanese, P. Frediani: Real hyperelliptic surfaces and the orbifold fundamental group. *J. Inst. Math. Jussieu*, **2** (2003), 163–223.
2. A. Degtyarev, I. Itenberg, V. Kharlamov: Real Enriques Surfaces. *Lecture Notes in Math., Springer, Berlin*, **1746** (2000).
3. A. Degtyarev, I. Itenberg, V. Kharlamov: Finiteness and quasi-simplicity for symmetric $K3$-surfaces. *Duke Math. J.*, **122** (2004), no. 1, 1–49.
4. A. Degtyarev, I. Itenberg, V. Kharlamov: Finiteness for real hyperkähler manifolds. *in preparation*.
5. A. Degtyarev, V. Kharlamov: Topological properties of real algebraic varieties: Rokhlin's way. *Russ. Math. Surveys*, **55** (2000), no. 4, 735–814.
6. A. Degtyarev, V. Kharlamov: Real rational surfaces are quasi-simple. *J. Reine. Angew. Math.*, **551** (2002), 87–99.
7. V. Kharlamov: Generalized Petrovskii inequality. *Funkz. Anal. i Priloz.*, **9** (1974), 50–56.
8. V. Kharlamov: Generalized Petrovskii inequality II. *Funkz. Anal. i Priloz.*, **10** (1975), 93–94.
9. V. Kharlamov: Topology of real algebraic varieties. *in Collected Papers by Petrovskii, Nauka*, 1986, 546–598.
10. V. Kharlamov: On non-amphicheiral surfaces of degree 4 in $\mathbb{R}P^3$. *Lecture Notes in Math.* **1346** (1988), 349–356.
11. V. Kharlamov: Variétés de Fano réelles (d'après C. Viterbo). *Astérisque, Séminaire Bourbaki*, **276** (2002), 189–206.
12. V. Kharlamov, V. Kulikov: On real structures of rigid surfaces. *Izv. Math.*, **66** (2003), no. 1, 133–150.
13. J. Kollár: The topology of real and complex algebraic varieties. *Mathematical Society of Japan. Adv. Stud. Pure Math.* **31** (2001), 127–145.
14. J. Kollár: The Nash conjecture for nonprojective threefolds. *Contemp. Math.* **312** (2002), 137–152.
15. Y. Laszlo, C. Viterbo: Estimates of characteristic numbers of real algebraic varieties. *Topology* **45** (2006), no. 2, 261–280.
16. Y. I. Manin. M. A. Tsfasman: Rational varieties: Algebra, geometry and arithmetic. *Russ. Math. Surv.*, **41** (1986), no. 2, 51–116.
17. J. Milnor: On the Betti numbers of real varieties. *Proc. Amer. Math. Soc.* **15** (1964), 275–280.
18. V. Nikulin: Kummer surfaces. *Math. USSR - Izv.*, **9** (1975), no. 2, 261–275.
19. V. Nikulin: Integer symmetric bilinear forms and some of their geometric applications. *Math. USSR - Izv.*, **14** (1979), no. 1, 103–167.
20. R. Thom: Sur l'homologie des variétés algébriques réelles. In *Differential and Combinatorial Topology. Symp. Marston Morse* (1965), 255–265.

21. A. N. Varchenko: On a local residue and the intersection form in vanishing cohomologies. *in Izv. Akad. Nauk SSSR, Ser. Mat.* **49**, (1985), 32–54.
22. J. Y. Welschinger: Real structures on minimal ruled sufaces. *Comment. Math. Helv.*, **78** (2003), 418–466.

Illustrating the classification of real cubic surfaces

Stephan Holzer[1] and Oliver Labs[2]

[1] Johannes Gutenberg Universität, Mainz, Germany,
   StHolzer@Students.Uni-Mainz.de
[2] Johannes Gutenberg Universität, Mainz, Germany,
   Labs@Mathematik.Uni-Mainz.de, mail@OliverLabs.net

Summary. Knörrer and Miller classified the real projective cubic surfaces in $\mathbb{P}^3(\mathbb{R})$ with respect to their topological type. For each of their 45 types containing only rational double points we give an affine equation, s.t. none of the singularities and none of the lines are at infinity. These equations were found using classical methods together with our new visualization tool SURFEX. This tool also enables us to give one image for each of the topological types showing all the singularities and lines.

1 Introduction

A *projective real cubic surface in real projective three-space* $\mathbb{P}^3(\mathbb{R})$ is a homogeneous polynomial f of degree 3 in four variables x, y, z, w with real coefficients:

$$f = \sum_{i,j,k,l \in \mathbb{N}_0 \mid i+j+k+l=3} a_{i,j,k,l} x^i y^j z^k w^l,$$

where $a_{i,j,k,l} \in \mathbb{R}$. In 1987, Knörrer and Miller [13] classified all such surfaces with respect to their topological type. A similar classification had already been given by Schläfli in the 19th century [19], but Knörrer and Miller obtain more precise and more complete results. Some of these are based on ideas of Bruce and Wall [2] who gave a modern treatment of the complex case.

Here, we restrict ourselves to cubic surfaces with only rational double points which is the most interesting part of the classification. We summarize briefly Knörrer/Miller's main results on these surfaces and give an explicit real affine equation for each class in their list (see table 2 on page 125). These allow us to draw images for each class showing all singularities and lines (see fig. 2, 3, 4) using our software SURFEX [10].

In the already cited article, Schläfli also gave equations for each of his types and described their construction in a very geometric way. In many cases, it is easy to find real affine equations from these with the help of our tool SURFEX. But in the other cases, there are too many free parameters and we have to use other methods such as the deformation techniques described by Klein [11].

To perform these deformations explicitly, it is useful to have a visualization software at hand. We explain how to use our software SURFEX for such purposes. SURFEX can be used directly on our webpage [14]. It can also produce high quality raytraced images for publications in color or in black/white. All the images in the present paper are produced using SURFEX in connection with SINGULAR [9]. This computer algebra program was used to compute a primary decomposition of the ideal (f, F_9) describing the 27 lines of f with multiplicities which allowed us to draw the lines on the surfaces using SURFEX. Here, F_9 denotes Clebsch's covariant of degree 9 (see, e.g., [16, appendix 4.1] for a determinantal formula for this covariant).

The webpage www.CubicSurface.net [15] contains some movies and more images. SURFEX [10] uses S. Endraß's SURF [7] to produce the high quality raytraced images of the surfaces and R. Morris's LSMP [17] and K. Polthier's JAVAVIEW [18] to allow rotation and scaling of a triangulated preview.

Several mathematicians have already given real affine equations for particularly interesting cubic surfaces such as the Clebsch Diagonal Surface or the four-nodal cubic surface. Recently, the architect J. Chertok collected equations for Rodenberg's 100-year-old series of plaster-models. These equations were communicated to him by different people, mainly S. Endraßand the second author. With these the architect recreated Rodenberg's series using $3d$-printers. But also Rodenberg's series is restricted to some types of cubic surfaces, and several of Rodenberg's models do not show all the projective real lines because some are at infinity. In fact, this was Rodenberg's intention. His aim was to give an overview of the possible singularities on cubic surfaces and the possible affine views of the projective surfaces. Here instead, we do not show different affine views of the same surfaces. We choose real affine equations that allow us to show all singularities and lines in a single image.

The second author thanks the organizers of the AGGM 2004 workshop at Nice for their hospitality. He also thanks S. Endraßand D. van Straten, without whom our tool SURFEX would never have existed, for many valuable discussions and motivation. Furthermore, we thank R. Morris, who is a co-author of our tool SURFEX and with whom the second author had several discussions concerning the visualization of algebraic surfaces. The first author was supported by the E-Learning Förderprogramm 2004 of the Johannes Gutenberg Universität Mainz.

2 The main results of Knörrer/Miller on cubic surfaces with only rational double points

We briefly review some results of Knörrer and Miller. As already mentioned we restrict ourselves to those concerning only rational double points. In [13], the authors say that two cubic surfaces have the same topological type if they can be transformed continuously into each other without changing the shape. The precise definition uses the finer notion of equisingularity:

Definition 1 (4.1 in [13]).

1. A *differentiable family* $(Y_t)_{t \in [0,1]}$ *of cubic surfaces in* $\mathbb{P}^3(\mathbb{R})$ *with equations* f_t *is called* equisingular *if in the neighborhood of each point it can be extended to a family of diffeomorphisms of the surrounding space. I.e., if for each* $t_0 \in [0,1]$ *and each* $p \in Y_{t_0}$ *there exists a neighborhood* I *of* t_0 *in* $[0,1]$, *a neighborhood* U *of* p *in* $\mathbb{P}^3(\mathbb{R})$ *and a diffeomorphism* $\Phi : U \times I \to U \times I$, *s.t. the following diagram commutes:*

$$\begin{array}{ccc} U \times I & \stackrel{\Phi}{\to} & U \times I \\ \cup & & \cup \\ \{(x,t) \in U \times I \mid f_{t_0}(x) = 0\} & \stackrel{\Phi}{\to} & \{(x,t) \in U \times I \mid f_t(x) = 0\} \\ & \searrow \text{pr}_2 \quad \swarrow \text{pr}_2 & \\ & I & \end{array}$$

Two surfaces that can be transformed into each other by an equisingular family are called equisingular isotopic.

2. *Two cubic surfaces* $Y_0, Y_1 \in \mathbb{P}^3(\mathbb{R})$ *have the same* topological type *if there are projectively equivalent surfaces* Y_0', Y_1' *which are equisingular isotopic.*

There exist at most two different equisingular isotopy classes of cubic surfaces of the same topological type. Equisingular families are characterized by their configuration of singularities:

Proposition 2 (4.2 in [13]).

1. *If two cubic surfaces in* $\mathbb{P}^3(\mathbb{R})$ *with only isolated singularities have the same topological type then suitable neighborhoods of their singular sets are analytically isomorphic.*
2. *Any differentiable family of cubics in* $\mathbb{P}^3(\mathbb{R})$ *with only isolated singularities for which the configuration of singularities is constant is equisingular.*

Table 1 on the next page gives an overview of the rational double points occuring on cubic surfaces (see also [5] or [1]). The classical geometers associated to each surface f a *class* which is the number of tangency points f has with a generic pencil of hyperplanes (for computing the class see [2, sect. 3]). The subscript of the old names for the singularities is the number by which the class drops when a cubic surface possesses such a singularity (see table 1).

For the following definition we assume familiarity with some concepts from algebraic geometry, in particular with the blowup. A reader who is not familiar with this should simply read the numbers from table 1 on the following page.

Here, we just want to mention that it is well-known that the blowup $\widetilde{\mathbb{P}^2(\mathbb{C})}$ of the projective plane $\mathbb{P}^2(\mathbb{C})$ in a set of six points Σ (*basepoints*) which are in general position (i.e. no three on a line, no six on a conic) is a smooth complex cubic surface and that all smooth complex cubic surfaces can be obtained in this way. This blowup is a birational map which is a bijection away from the basepoints, i.e. for all points in $\mathbb{P}^2(\mathbb{C}) \backslash \Sigma$. In the real case, we have to be more careful: the cubic F_5 with two components is not the result of such a blowup:

Definition 3 (p. 54/55 in [13]).

1. $\mu_\mathbb{R}$ denotes the number of (-2)-curves defined over $\mathbb{R}$ in the dual resolution graph of a rational double point that is defined over $\mathbb{R}$. ν denotes the number of pairs of non-intersecting complex conjugate (-2)-curves in this graph.

Name	Old Name	Normal Form	Coxeter Diagram	$\mu_\mathbb{R}$	ν	
A_{2k}^-	B_{2k+1}	$x^{2k+1}+y^2-z^2$		$2k$	0	$k=1,2$
A_{2k}^+	B_{2k+1}	$x^{2k+1}+y^2+z^2$		0	$k-1$	$k=1$
A_{2k-1}^-	B_{2k}	$x^{2k}+y^2-z^2$		$2k-1$	0	$k=2,3$
A_{2k-1}^+	B_{2k}	$x^{2k}-y^2-z^2$		1	$k-1$	$k=2$
A_1^-	C_2	$x^2+y^2-z^2$		1	0	
A_1^+	C_2	$x^2+y^2+z^2$		1	0	
D_4^-	U_6	$x^2y-y^3-z^2$		4	0	
D_4^+	U_6	$x^2y+y^3+z^2$		2	1	
D_5^-	U_7	$x^2y+y^4-z^2$		5	0	
E_6^-	U_8	$x^3+y^4-z^2$		6	0	

Table 1. The types of singularities occuring on real cubic surfaces, their normal forms, and the numbers $\mu_\mathbb{R}$ and ν. For later use, we also give their Coxeter Diagrams.

2. Let Σ be a sequence of six points defined over $\mathbb{R}$ in almost general position in $\mathbb{P}^2(\mathbb{C})$ in the sense of [4, p. 39]. Then there exists $r(\Sigma) \in \mathbb{N}_0$, s.t. Σ consists of $2r$ points that are invariant under complex conjugation and $6-2r$ pairwise compl. conj. points. We call $r(\Sigma)$ the reality index of Σ.

3. Let X be a cubic surface in $\mathbb{P}^3(\mathbb{C})$ defined over $\mathbb{R}$ with only rational double points. The reality index $r(X)$ of X is defined as follows: Let $\widetilde{X}$ denote the desingularization of X and $\overline{X}(\Sigma)$ the blowup of $\mathbb{P}^2(\mathbb{C})$ along Σ. Then, $r(X) = r(\Sigma)$, if $\widetilde{X} \cong \overline{X}(\Sigma)$ for a sequence Σ of six points in almost general position in $\mathbb{P}^2(\mathbb{C})$. Otherwise, $r(X) = -1$.

Theorem 4 (Satz 2.8 in [13]). Let $X \subset \mathbb{P}^3(\mathbb{C})$ be a cubic surface defined over $\mathbb{R}$ with only rational double points as singularities. Suppose that the real part $X_\mathbb{R} \subset \mathbb{P}^3(\mathbb{R})$ of X has k singular points. Denote by $\mu_\mathbb{R}(X)$ the sum of the $\mu_\mathbb{R}$ for these singular points and by $\nu(X)$ the sum of the ν of all singularities on X. Then the real part $X_\mathbb{R}$ contains exactly $l(X_\mathbb{R})$ lines, where

$$l(X_\mathbb{R}) = \frac{(2+2r(X)-\mu_\mathbb{R}(X))(1+2r(X)-\mu_\mathbb{R}(X))}{2} - (r(X)-2) + k - \nu(X).$$

For a cubic surface $X \subset \mathbb{P}^3(\mathbb{C})$ we can read the topology of its real part $X_\mathbb{R} \subset \mathbb{P}^3(\mathbb{R})$ from the reality index. E.g., the five smooth cubic surfaces, classically denoted by $F_1, F_2, \ldots, F_5$ (see [20]), are classified by the reality index, e.g., $r(F_5) = -1$. Here is another result of Knörrer/Miller of this kind:

Lemma 5 (3.2, 3.3, 4.3 in [13]).

1. If $X_\mathbb{R}$ does not contain any singularity of type $A^\cdot$ and $r(X) \geq 0$ then $X_\mathbb{R}$ is connected and $\chi(X_\mathbb{R}) = 1 - 2r(X) + \mu_\mathbb{R}(X)$. If $r(X) = -1$ then $X_\mathbb{R}$ is diffeomorphic to the disjoint union $\mathbb{P}^2(\mathbb{R}) \sqcup S^2$.
2. If $p \in X_\mathbb{R}$ is a singular point of type $A^\cdot$ then $X_\mathbb{R}$ does not contain any other singularity and $X_\mathbb{R}$ is diffeomorphic to $\mathbb{P}^2(\mathbb{R}) \sqcup \{p\}$.
3. Cubic surfaces of the same topological type are homeomorphic.
4. Two cubic surfaces of the same topological type with only rational singularities have the same reality index.

The following is Knörrer/Miller's main result on cubic surfaces with only rational double points:

Theorem 6 (Classification, Liste 4 in [13]). *Let $X \subset \mathbb{P}^3(\mathbb{C})$ be a cubic surface defined over $\mathbb{R}$ with only rational double points and let $X_\mathbb{R} = X \cap \mathbb{P}^3(\mathbb{R})$ be its real part. Then the topological type of $X_\mathbb{R}$ is one of the 45 types given in table 2 on page 125. If X has exactly $3A_1^-$ singularities and X contains exactly 12 lines (no. 18/19 in the table) then its topological type can be determined by prop. 8 below. Otherwise, the topological type of X is determined by its singularities, its number of lines, and the reality index $r(X)$.*

To explain how to distinguish between the types 18 and 19, we need Knörrer/Miller's notion of a *configuration type of an A_1^- singularity*. We only give a sloppy definition and illustrate it using SURFEX, see [13, p. 63] for details. For this local study we have to work in affine space:

Recall that the *tangent cone* $tc(f)$ of a singularity f at the origin is the lowest non-zero homogeneous part of f. For an A_1^- singularity, it is a cone of the form $x^2 + y^2 - z^2$. The tangent cone intersects the cubic surface X in a curve of degree $2 \cdot 3 = 6$, which consists in fact of six lines, counted with multiplicities. Knörrer/Miller describe such a configuration by a small circle together with six points (counted with multiplicities), because a small real sphere around the singularity intersects X in two small real "circles" (fig. 1 on the following page). On each of these circles there lies one point of each of the real lines. Therefore, Knörrer/Miller denote a pair of complex conjugated lines by a point in the center of the circle, the real points are drawn on the circle in the correct order. Different such configurations correspond to cubic surfaces of different topological types.

Example 7. Example (a) is a configuration with one real point of multiplicity 2, two real ones of multiplicity 1, and two complex conjugated ones. The other two examples show two doubled and two simple points (see fig. 1):

(a) $\left(\begin{array}{c}\bullet\ 2\end{array}\right)$, (b) $\left(\begin{array}{c}2\ \ 2\end{array}\right)$ (KM$_{18}$ in fig. 1), (c) $\left(\begin{array}{c}2\\ \bullet\ \ 2\end{array}\right)$ (KM$_{19}$ in fig. 1). □

Proposition 8 (Topological Types 18/19, p. 63 in [13]). *If a cubic surface X has exactly $3A_1^-$ singularities and contains 12 lines then X has*

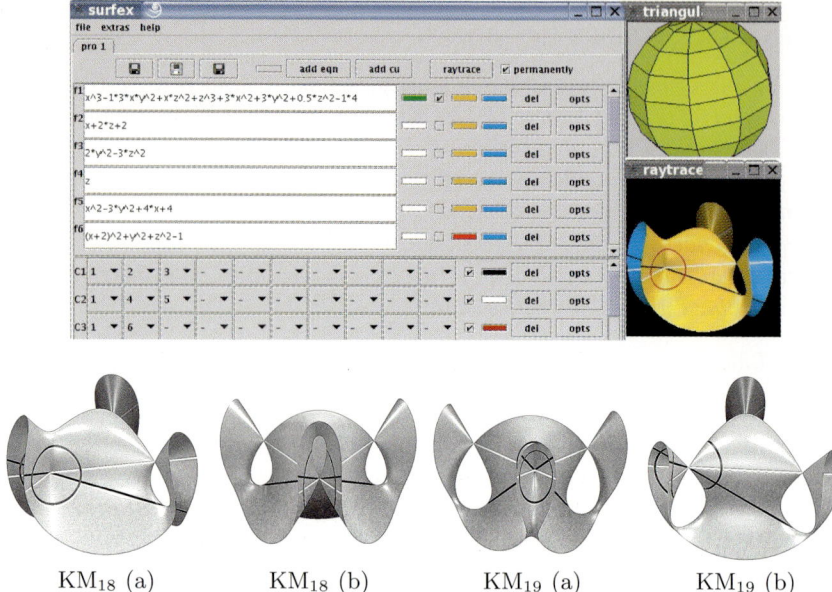

Fig. 1. The configuration of the lines cut out by the tangent cone at one of the three A_1^- singularities of our surfaces with topological types no. 18 and 19. For each of the surfaces, we show two views (a), (b) from different angles. The white lines have multiplicity two, the black ones have multiplicity one. The figure above illustrates how SURFEX can draw curves on surfaces using the corresponding feature of SURF. To draw the two doubled white lines, we computed the equations f4, f5 cutting these out on the surface using SINGULAR. Then we chose the numbers of the equations from the drop down menu in the row called C2 and selected the color white.

the topological type 18 *if the singular points have a configuration of type* ⬡ *(example 7 (b)). Otherwise, the* A_1^- *singularities of* X *have a configuration of type* ⬡ *(example 7 (c)) and* X *has the topological type* 19.

3 Constructing nice real affine equations

3.1 Nice equations

By a *nice* real affine equation f for a given topological type t we mean an equation, s.t. its projective closure $\overline{f}$ has the required topological type and s.t. the plane at infinity neither contains a singularity nor a line of $\overline{f}$. It has also to be possible to see all its singularities and lines in a single picture (modulo guessing using symmetries). This is not a precise definition. Nevertheless, we formulate our main result in the form of a theorem:

Illustrating the classification of real cubic surfaces   125

Name	Sp.	Cl.	Sing.	r	l	Equation
KM_1	I	12	$\emptyset$	3	27	$KM_{27} + \frac{3}{2}(x^2 + y^2 - z^3)$
KM_2	I	12	$\emptyset$	2	15	$KM_{27} + \frac{8}{5}((z+1)^2 - z^2)$
KM_3	I	12	$\emptyset$	1	7	$KM_{27} + \frac{2}{3}((z+1)^2 + (x-1)^2) - 4y^2$
KM_4	I	12	$\emptyset$	0	3	$KM_2 - 4$
KM_5	I	12	$\emptyset$	−1	3	$KM_{27} - \frac{2}{3}((z+1)^2 + z^2)$
KM_6	II	10	A_1^-	3	21	$KM_{27} + 2(x^2 + y^2)$
KM_7	II	10	A_1^-	2	11	$KM_{27} + z^3 + y^2$
KM_8	II	10	A_1^-	1	5	$KM_6 - 4y^2$
KM_9	II	10	A_1^-	0	3	$KM_6 - 3(x^2 + y^2)$
KM_{10}	II	10	$A_1^\bullet$	0	3	$pc + (z+1) \cdot z^2$
KM_{11}	IV	8	$2A_1^-$	3	16	$KM_{27} + y^2$
KM_{12}	IV	8	$2A_1^-$	2	8	$KM_{27} + z^2 - \frac{1}{5}(x + \frac{1}{2})^2$
KM_{13}	IV	8	$2A_1^-$	1	4	$KM_{27} - y^2$
KM_{14}	III	9	A_2^-	3	15	$KM_{21} + \frac{1}{10}(y-1)^2$
KM_{15}	III	9	A_2^-	2	7	$pl + z^3 - z^2(x-1) - \frac{1}{5}(x-y)^2$
KM_{16}	III	9	A_2^-	1	3	$KM_{43} - y^2$
KM_{17}	III	9	A_2^+	0	3	$pc + z^3$
KM_{18}	VIII	6	$3A_1^-$	3	12	$KM_{43} + z^2(x + \frac{1}{2})$
KM_{19}	VIII	6	$3A_1^-$	3	12	$KM_{43} + 2z^2$
KM_{20}	VIII	6	$3A_1^-$	2	6	$KM_{27} - z^2$
KM_{21}	VI	7	$A_2^- A_1^-$	3	11	$pl + z^3 + z^2(x + y - 2) + \frac{1}{10}(x-1)^2$
KM_{22}	VI	7	$A_2^- A_1^-$	2	5	$pl + z^3 + z^2(x + y) + \frac{1}{5}(x-1)^2$
KM_{23}	V	8	A_3^-	3	10	$wxy + (x+z)(y^2 - (\frac{2}{3}x)^2 - (\frac{3}{5}z)^2),\ w = 1 - x$
KM_{24}	V	8	A_3^-	2	4	$KM_{32} - \frac{1}{100}z^2(x-z)$
KM_{25}	V	8	A_3^-	1	2	$KM_{32} + \frac{1}{100}z^2(x-z)$
KM_{26}	V	8	A_3^+	1	4	$2(x^2 + y^2)w + 2xz(z^2 - 2x^2 - 4y^2),\ w = 1 - y$
KM_{27}	XVI	4	$4A_1^-$	3	9	$4(pc + \frac{1}{2}) + 3(x^2 + y^2)(z - 6) - z(3 + 4z + 7z^2)$
KM_{28}	XIII	5	$A_2^- 2A_1^-$	3	8	$KM_{43} + z^2(x + 2)$
KM_{29}	IX	6	$2A_2^-$	3	7	$KM_{43} + (x-1)z$
KM_{30}	IX	6	$2A_2^-$	2	3	$KM_{43} - \frac{3}{10}(x-1)^2$
KM_{31}	X	6	$A_3^- A_1^-$	3	7	$wxz - (x+z)(x^2 - y^2),\ w = 1 - z$
KM_{32}	X	6	$A_3^- A_1^-$	2	3	$wxy - (x+z)(x^2 + y^2),\ w = 1 - z$
KM_{33}	VII	7	A_4^-	3	6	$wxy + y^2z + yx^2 - z^3,\ w = 1 - x - y - z$
KM_{34}	VII	7	A_4^-	2	2	$wxy - y^2z + yx^2 - z^3,\ w = 1 - x - y - z$
KM_{35}	XII	6	D_4^-	3	6	$(x + y + z)^2 w + xyz,\ w = \frac{1}{2}(1 - x - y - z)$
KM_{36}	XII	6	D_4^+	1	2	$(x + y + z)^2 w + (x^2 + y^2)z,$ $w = \frac{1}{2}(1 - x - y - z)$
KM_{37}	XVII	4	$2A_2^- A_1^-$	3	5	$KM_{43} + (x-1)z^2$
KM_{38}	XVIII	4	$A_3^- 2A_1^-$	3	5	$wxz + y^2(x+z),\ w = 2(1 + x - y + z)$
KM_{39}	XIV	5	$A_4^- A_1^-$	3	4	$wxz - y^2z + \frac{1}{2}x^2y,\ w = \frac{1}{8}(1 - y - z)$
KM_{40}	XI	6	A_5^-	3	3	$wxz + y^2z + x^3 - z^3,\ w = 1 - x$
KM_{41}	XI	6	A_5^-	2	1	$wxz + y^2z + x^3 + z^3,\ w = 1$
KM_{42}	XV	5	D_5^-	3	3	$wx^2 + y^2z + xz^2,\ w = 1 + x$
KM_{43}	XXI	3	$3A_2^-$	3	3	$tl + z^3$
KM_{44}	XIX	4	$A_5^- A_1^-$	3	2	$wxz - y^2z - x^3,\ w = 1 - z$
KM_{45}	XX	4	E_6^-	3	1	$x^2w - xz^2 + y^3,\ w = 1 - x - y$

Table 2. Our nice real affine equations for Knörrer/Miller's 45 topological types. The abreviation Sp. denotes Schläfli's *species* of the surface, Cl. its class, Sing. its singularities. r denotes the reality index and l the number of real lines on the surface.

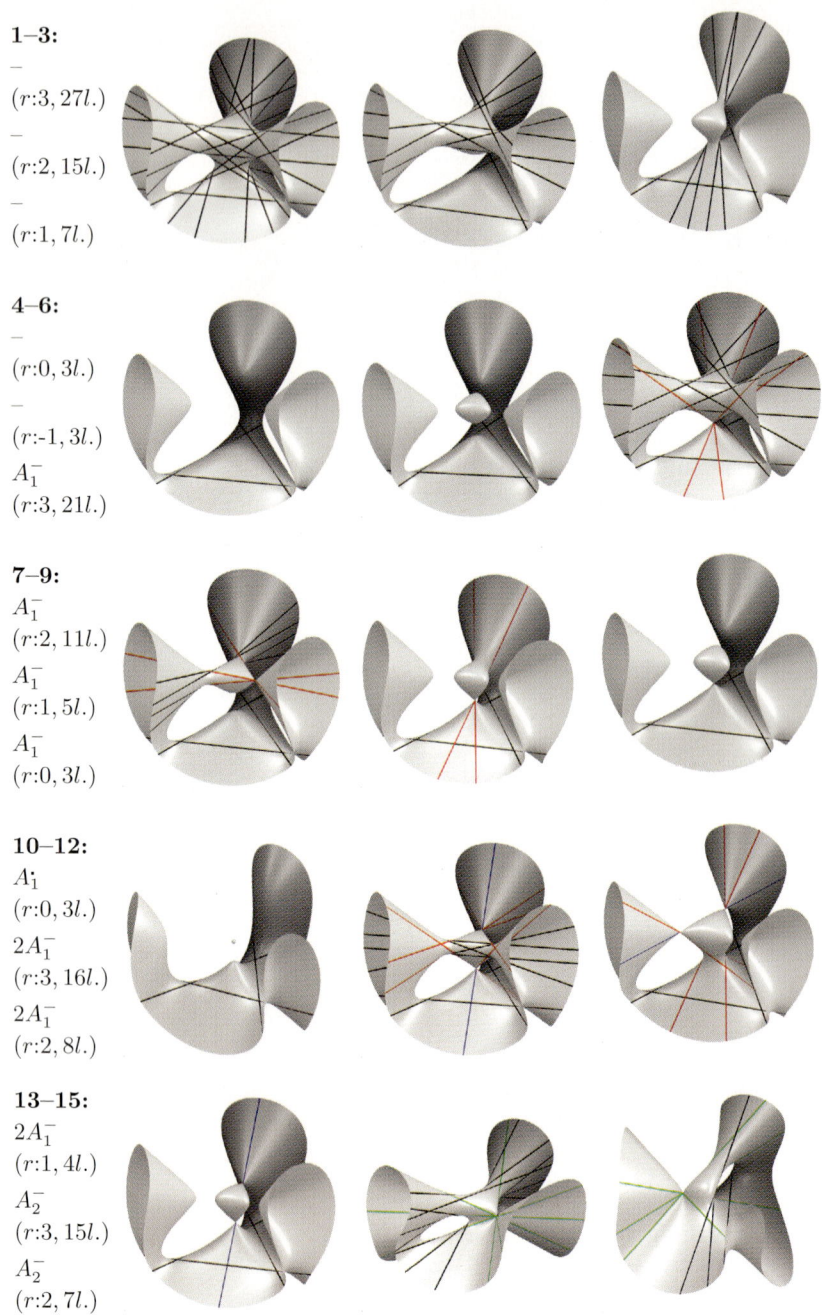

Fig. 2. The surfaces $KM_1, \ldots, KM_{15}$. The colors of the lines indicate their multiplicities: ■ 1, ■ 2, ■ 3, ■ 4, ■ 5, ■ 6, ■ 8, ■ 9, ■ 10, ■ 12, ■ 15, ■ 16, ■ 27.

16–18:
A_2^-
($r{:}1, 3l.$)
A_2^+
($r{:}0, 3l.$)
$3A_1^-$
($r{:}3, 12l.$)

19–21:
$3A_1^-$
($r{:}3, 12l.$)
$3A_1^-$
($r{:}2, 6l.$)
A_2^-, A_1^-
($r{:}3, 11l.$)

22–24:
A_2^-, A_1^-
($r{:}2, 6l.$)
A_3^-
($r{:}3, 10l.$)
A_3^-
($r{:}2, 4l.$)

25–27:
A_3^-
($r{:}1, 2l.$)
A_3^+
($r{:}1, 4l.$)
$4A_1^-$
($r{:}3, 9l.$)

28–30:
$A_2^-, 2A_1^-$
($r{:}3, 8l.$)
$2A_2^-$
($r{:}2, 7l.$)
$2A_2^-$
($r{:}2, 3l.$)

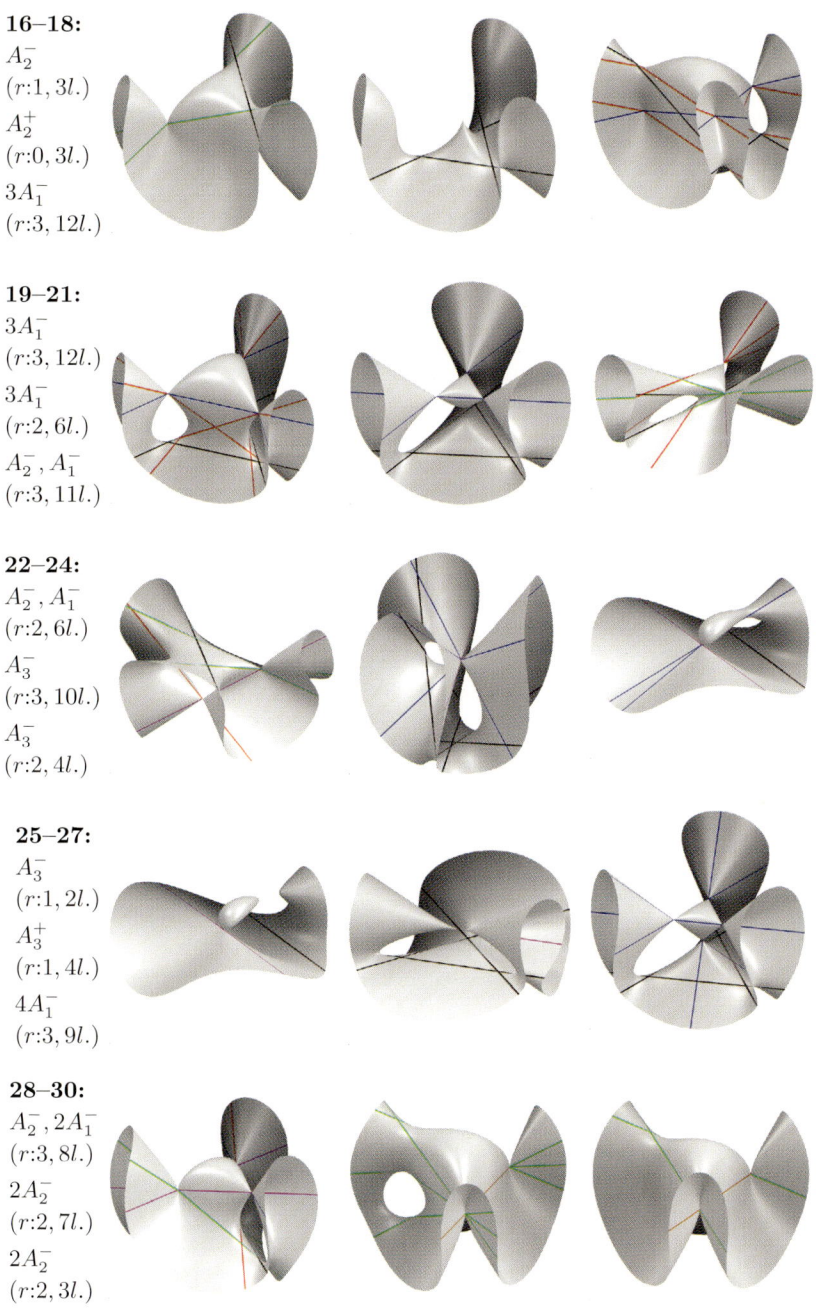

Fig. 3. The surfaces $KM_1, \ldots, KM_{15}$. The colors of the lines indicate their multiplicities: ■ 1, ■ 2, ■ 3, ■ 4, ■ 5, ■ 6, ■ 8, ■ 9, ■ 10, ■ 12, ■ 15, ■ 16, ■ 27.

31–33:
A_3^-, A_1^-
$(r{:}3, 7l.)$
A_3^-, A_1^-
$(r{:}2, 3l.)$
A_4^-
$(r{:}3, 6l.)$

34–36:
A_4^-
$(r{:}2, 2l.)$
D_4^-
$(r{:}3, 6l.)$
D_4^+
$(r{:}1, 2l.)$

37–39:
$2A_2^-, A_1^-$
$(r{:}3, 5l.)$
$A_3^-, 2A_1^-$
$(r{:}3, 5l.)$
A_4^-, A_1^-
$(r{:}3, 4l.)$

40–42:
A_5^-
$(r{:}3, 3l.)$
A_5^-
$(r{:}2, 1l.)$
D_5^-
$(r{:}3, 3l.)$

43–45:
$3A_2^-$
$(r{:}3, 3l.)$
A_5^-, A_1^-
$(r{:}3, 2l.)$
E_6^-
$(r{:}2, 1l.)$

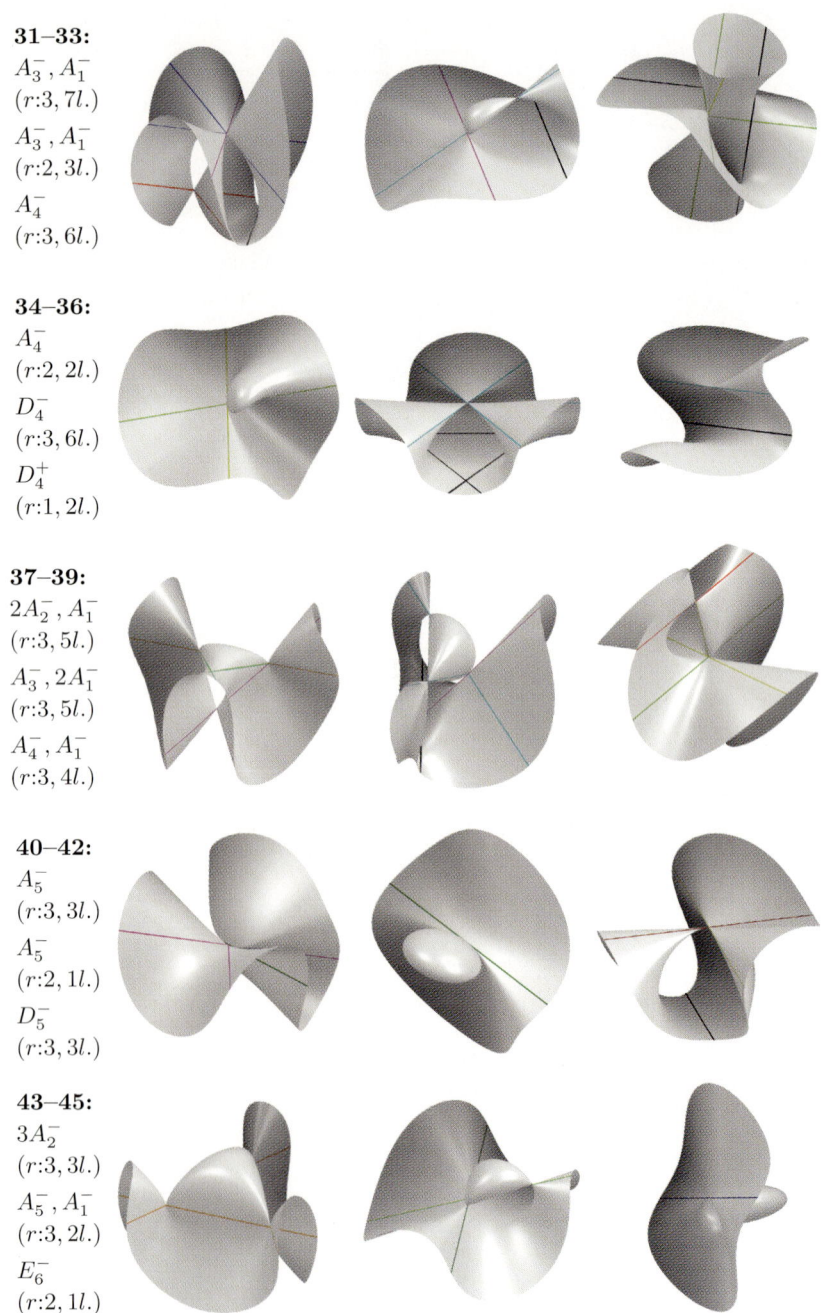

Fig. 4. The surfaces $KM_1, \ldots, KM_{15}$. The colors of the lines indicate their multiplicities: ■ 1, ■ 2, ■ 3, ■ 4, ■ 5, ■ 6, ■ 8, ■ 9, ■ 10, ■ 12, ■ 15, ■ 16, ■ 27.

Theorem 9. *For each topological type $t \in \{1, 2, \ldots, 45\}$ of cubic surfaces with only rational double points there is a nice affine equation KM_t in the sense of the preceding paragraph. The equations KM_t are given in table 2 on page 125 and the corresponding pictures are shown in the figures 2, 3, 4.*

Remark 10. For a nice equation for a given topological we do not require the greatest possible symmetry because we want the equations to be generic in the sense that the configuration of the lines on the surface should not be too special. E.g., the Clebsch Cubic Surface has 10 so-called *Eckardt Points* in which three of its 27 real lines meet, but a generic cubic surface with 27 lines does not have any such point.

Remark 11. Schläfli orders the cubic surfaces first by their class and then by the worst singularity occuring. This differs from Knörrer/Miller's order which is first by the sum of the Milnor numbers of the singularities and then by the worst singularity occuring.

In the following subsections we describe how to construct such surfaces.

3.2 Via projective equations

For the projective case, Schläfli already gave equations in [19]. He describes in a very geometric way how to construct them. In [3], Cayley gives the same equations again and computes a lot of additional data connected to the surfaces.[3]

To obtain a nice real affine equation from one of Schläfli's equations is an easy task for most topological types with higher singularities (A_3 or higher): We just have to choose a good hyperplane at infinity and maybe some constants which is not difficult using our tool SURFEX:

Example 12. Let us take the equation $wxz + y^2z + x^3 = 0$ given by Schläfli [19, p. 357] for a projective cubic surface with an A_1 and A_5 singularity. The choice $w = 1 - z$ gives our affine equation KM_{44}.

For those surfaces with only A_1 and A_2 singularities, this method does not work well because of the great number of free parameters. In this case, we can either write down the equation directly (section 3.3), or we can use a deformation process (section 3.4) already described by F. Klein in [11].

3.3 Direct construction

In some cases, it is easy to write down a nice real affine equation for a topological type directly using symmetry. For this purpose, we will use the three plane curves shown in figure 5.

---

[3] Attention, Cayley's list on p. 321 contains some typos.

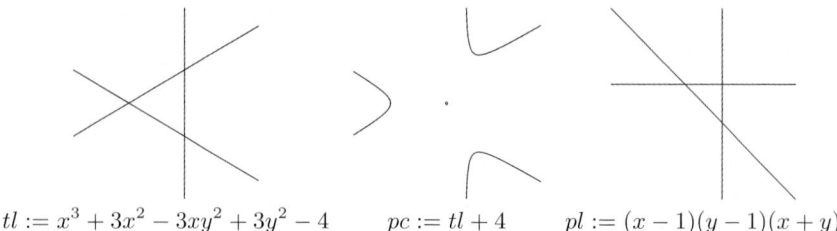

$tl := x^3 + 3x^2 - 3xy^2 + 3y^2 - 4$ $pc := tl + 4$ $pl := (x-1)(y-1)(x+y)$

Fig. 5. Three plane curves, useful for constructing nice equations for cubic surfaces.

Example 13 (Constructing KM_{43} *with three* A_2^- *Singularities).* We take the polynomial tl defining three triangle-symmetric lines (fig. 5) in the x,y-plane and add the term z^3: $\mathrm{KM}_{43} = tl + z^3$. At each intersection point of the lines tl, this gives a singularity of type A_2^- with z-coordinate 0, see fig. 8(a).

The four-nodal surface KM_{27} can be constructed in a similar way. This and a lot more information on nodal surfaces with dihedral symmetry can be found in S. Endraß's Ph.D. thesis [6]. The following example uses a plane curve with a solitary point. In the same way we obtain the surface KM_{26} with an A_3^+ singularity.

Example 14 (Constructing KM_{10} *with an* A_1^- *Singularity).* To construct a surface with an A_1^- Singularity which has the normal form $x^2 + y^2 + z^2$ we start with the triangle-symmetric plane cubic pc (fig. 5). The origin is a solitary point (i.e., a singularity with normal form x^2+y^2). Thus the surface $pc+z^2$ has an A_1^- singularity with normal form $x^2+y^2+z^2$ and is triangle-symmetric. To obtain the desired affine topology we require a third root on the $\{x=y=0\}$ axes at $z=-1$: $\mathrm{KM}_{10} = pc + (z+1) \cdot z^2$.

3.4 The deformation process

Klein's strategy for obtaining surfaces with fewer singularities from surfaces with many singularities is based on the fact that any singularity on a cubic surface can be deformed separately.

In the case of complex projective cubic surfaces, this fact can be formulated in the following way (see Knörrer/Barth's article in [8] for an overview on this and other visible properties of cubic surfaces): The configurations of rational double points occuring on cubic surfaces are exactly those for which the disjoint union of their Coxeter Diagrams is a subgraph of the Coxeter Diagram of $\widetilde{E}_6$, see fig. 6. A surface can be specialized into another one if and only if its graph is contained in the other's graph.

By the definition of a singularity, the origin can only be a singularity of an affine surface f if the tangent cone of f has degree at least 2. Thus, in order to smooth an isolated singularity at the origin, we can simply add a term of degree 1 or 0.

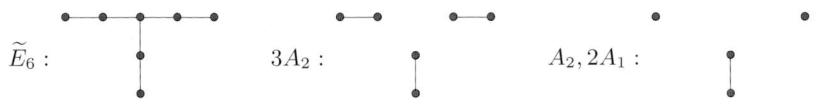

Fig. 6. There exists a cubic surface with three singularities of type A_2, because the disjoint union of three graphs of the A_2 singularity is a subgraph of the graph $\widetilde{E}_6$. A cubic surface with an A_2 singularity and two A_1 singularities can be specialized into one with three A_2 singularities as can be seen from the graphs. See table 1 on page 122 for the Coxeter Diagrams of the singularities of the rational double points on cubic surfaces.

But which terms can we add to the equation of f without changing the type of a singularity at the origin? For A_1 singularities, this is very easy: These singularities are characterized by the fact that their tangent cone also defines an A_1 singularity.[4] So, we can add any term of degree greater than two and any term of degree two whose coefficient is small enough. E.g. $x^2 + y^2 - z^2 + \frac{1}{10}z^2 + \frac{1}{13}xy + x^3$ has a singularity of type A_1^- at the origin.

Using the preceding facts we can deform a cubic surface with four singularities of type A_1^- into one with only three such singularities:

Example 15 (Smoothing one of four A_1 Singularities). Let KM$_{27}$ be the cubic surface with four A_1^--singularities (see table 2 on page 125). Three of its singularities lie in the plane $\{z = 0\}$. Using SURFEX, it is easy to find an ε, s.t. the surface KM$_{27} + \varepsilon z^2$ has the desired topology (see fig. 7):

Go to the SURFEX web-page [10], start the SURFEX program, and enter the equation of KM$_{27}$. Then add a term **+0.1*z^2** and check the **permanently** checkbox – this will premanently recompute raytraced images of your surface. Drag the computer mouse over the green ball to rotate the surface until you see all singularities. You can scale the image by pressing **s** on your keyboard while dragging. Now your SURFEX screen should look similar to fig. 7 on the following page. The singularity in the middle has been smoothed in such a way that the neighborhood of the singularity looks like a hyperboloid of one sheet. Adding **-0.1*z^2** leads to a neighborhood which looks like a hyperboloid of two sheets. □

It is a little more subtle to keep singularities of type A_j^- or $A_j^+, j > 1$, while deforming others. Forgetting about the sign for a moment, these singularities have the equation $x^{j+1} + y^2 + z^2$ in a suitable coordinate system. $A_j, j > 1$, singularities are characterized by the property that their tangent cone is of degree two and consists of the union of two different planes.[5]

---

[4]This is also the reason why the geometers of the 19th century called the A_1 singularities conical singularities or singularities of type C_2. Other names are *proper node, ordinary double point*.

[5]This is the reason why the classical geometers called a singularity of type A_j a biplanar node B_{j+1}. A singularity whose tangent cone consists of a single multiple plane was called a uniplanar node.

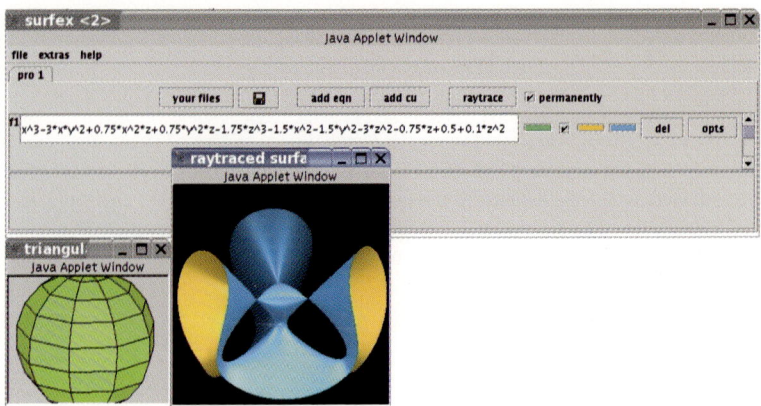

Fig. 7. Smoothing one of the four singularities of the cubic surface KM$_{27}$.

Let f be a polynomial in three variables x, y, z defining a singularity of type $A_j, j \geq 2$, at the origin. By the finite determinacy theorem (see, e.g., [5]), we can add an element of the ideal $I := \mathfrak{m}^2 \cdot J_f$ to f without changing the type of the singularity. Here, $\mathfrak{m}$ denotes the maximal ideal (x, y, z) of the origin and thus $\mathfrak{m}^2 = (x^2, xy, xz, y^2, yz, z^2)$. $J_f := (\frac{\partial f}{\partial x}, \frac{\partial f}{\partial y}, \frac{\partial f}{\partial z})$ is the so-called jacobian ideal generated by the partial derivatives of f.

Example 16. We take the singularity of type A_3^- at the origin, defined by $f := x^4 + y^2 - z^2 = 0$. Its jacobian ideal is $J_f = (x^3, y, z)$. If we choose $g_1 := xy \in \mathfrak{m}^2$ and $g_2 := y \in J_f$ we get $g := g_1 g_2 = xy^2$. Then $f + g$ still defines a singularity of type A_3 at the origin. Furthermore, $f + \varepsilon g$ is an A_3^- singularity for ε small enough.

We now come to the global situation of a cubic surface f with only isolated singularities of type $A_j, j \geq 1$. The following example describes how to use the techniques above to deform some of its singularities while keeping others:

Example 17 (Deforming two of three A_2^- Singularities to A_1^- Singularities). We start with the surface KM$_{43}$ which has exactly three singularities of type A_2^- (fig. 8(a)). The surface $tl + z^3 + z^2$ (fig. 8(b)) has three singularities of type A_1^- at the same coordinates, because the tangent cone is a cone of the form $x^2 - y^2 + z^2$ locally at each of these points. One of these singularities has the coordinates $Q := (-2, 0, 0)$. To get a surface with a singularity of type A_2^- at Q and two singularities of type A_1^-, we need to adjust the construction slightly.

Our general remarks from the beginning of this subsection tell us that we have to look at the jacobian ideal $J_{\text{KM}_{43}}$ at Q. Over the rational numbers, SINGULAR gives the following primary decomposition: $J_{\text{KM}_{43}} = (x, y, z^2) \cap (x - 1, y^2 - 3, z^2) \cap (x + 2, y, z^2)$. Locally at Q, the relevant primary component is

$(x+2, y, z^2)$. We choose $E := x+2 \in (x+2, y, z^2)$. As $z^2 \in \mathfrak{m}^2$, we then know that $KM_{43} + z^2 \cdot E$ has a singularity of type A_2 at Q.

Locally at the other two singularities (which both have x-coordinate 1), E takes the value $1 + 2 = 3$. Thus, at these singularities, $KM_{43} + z^2 \cdot E$ behaves like $KM_{43} + z^2 \cdot 3$, which has A_1^- singularities at these points as already seen above.

To check that our choices of planes and constants were reasonable and to understand the construction a little better, we can again use SURFEX. We type the equation of KM_{43} into SURFEX as f1. Then we add another two equations using the add eqn button and choose f2 to be $x + 2$ and f3 to be z. If the

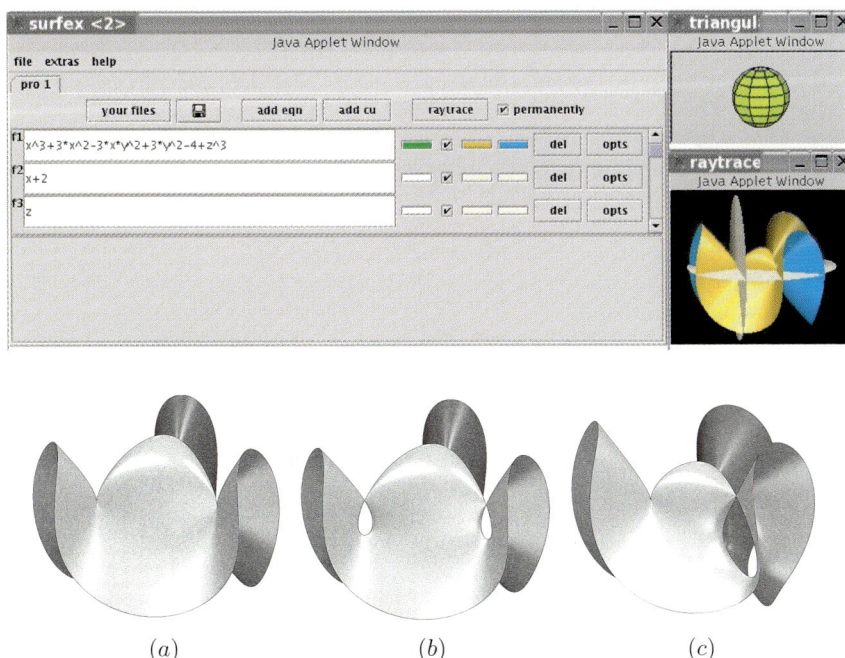

Fig. 8. Deforming the surface KM_{43} (image (a)) with three singularities of type A_2^- into KM_{28} (image (c)) with one such singularity and two A_1^- singularities.

permanently checkbox is activated we already see the three surfaces in one picture. When adjusting the colors by clicking at the right of the equations, we get a result similar to fig. 8. We can hide some of the surfaces by deselecting the checkbox at the right of the equations. When typing into f1 the changes described above, we obtain successively the three lower images shown in the figure. We can produce the black/white images used for the present publication in the following way: We press the button showing the small disk, select the dithered checkbox, choose an appropriate resolution, and then click on save. A small dialog shows up, where we can give some filename. The high-resolution image is then computed on the webserver. From there, it can then be downloaded using the `your files` button in the SURFEX window. □

References

1. V.I. Arnold, S.M. Gusein-Zade, and A.N. Varchenko. *Singularities of Differentiable Maps*. Birkhäuser, 1985.
2. J.W. Bruce and C.T.C. Wall. On the Classification of Cubic Surfaces. *J. Lond. Math. Soc., II. ser.*, (19):245–259, 1979.
3. A. Cayley. A Memoir on Cubic Surfaces. *Philos. Trans. Royal Soc.*, CLIX:231–326, 1869.
4. M. Demazure. Surfaces de Del Pezzo, I, II, III, IV et V. In M. Demazure, H. Pinkham, and B. Teissier, editors, *Séminaire sur les singularités des surfaces*, Lecture Notes in Math. 777, pages 21–69. Springer-Verlag, 1980.
5. A. Dimca. *Topics on Real and Complex Singularities*. Vieweg, 1987.
6. S. Endraß. *Symmetrische Flächen mit vielen gewöhnlichen Doppelpunkten*. PhD thesis, Fr.-A.-Universität Erlangen-Nürnberg, 1996.
7. S. Endraß. SURF 1.0.3. `http://surf.sourceforge.net`, 2001. A Computer Software for Visualising Real Algebraic Geometry.
8. G. Fischer. *Mathematische Modelle/Mathematical Models*. Vieweg, 1986. Bildband, Kommentarband.
9. G.-M. Greuel, G. Pfister, and H. Schönemann. SINGULAR 2.0. A Computer Algebra System for Polynomial Computations, Univ. Kaiserslautern, 2001. `http://www.singular.uni-kl.de`.
10. S. Holzer, O. Labs, and R. Morris. SURFEX – Visualization of Real Algebraic Surfaces. `www.surfex.AlgebraicSurface.net`, 2005.
11. F. Klein. Über Flächen dritter Ordnung. *Math. Annalen*, VI:551–581, Tafeln I–VI, 1873. also in: [12], p. 11–62.
12. F. Klein. *Gesammelte Mathematische Abhandlungen*, volume II. Verlag von Julius Springer, Berlin, 1922.
13. H. Knörrer and T. Miller. Topologische Typen reeller kubischer Flächen. *Mathematische Zeitschrift*, 195, 1987.
14. O. Labs. Algebraic Surface Homepage. Information, Images and Tools on Algebraic Surfaces. `www.AlgebraicSurface.net`, 2003.
15. O. Labs and D. van Straten. The Cubic Surface Homepage. `www.CubicSurface.net`, 2000.
16. O. Labs and D. van Straten. A Visual Introduction to Cubic Surfaces Using the Computer Software Spicy. In M. Joswig and N. Takayama, editors, *Algebra, Geometry, and Software Systems*, pages 225–238. Springer, 2003.
17. R. Morris. LSMP: Liverpool surface modeling package. `http://www.singsurf.org`, 2003.
18. K. Polthier. JAVAVIEW. `www.JavaView.de`, 2001.
19. L. Schläfli. On the Distribution of Surfaces of the Third Order into Species, in Reference to the Presence or Absence of Singular Points and the Reality of their Lines. *Philos. Trans. Royal Soc.*, CLIII:193–241, 1863.
20. B. Segre. *The non-singular Cubic Surfaces*. Clarendon, Oxford, 1942.

Bézier patches on almost toric surfaces

Rimvydas Krasauskas

Faculty of Mathematics and Informatics, Vilnius University,
Naugarduko 24, 03225 Vilnius, Lithuania
rimvydas.krasauskas@maf.vu.lt

Summary. The paper is devoted to the parametrization extension problem: given a loop composed of 3 (resp. 4) rational Bézier curves on a rational surface X, find a triangular (resp. tensor product) Bézier patch on X of optimal degree bounded by these curves. The constructive solution to the formulated problem is presented in details for cases where X is a sphere or a hyperbolic paraboloid. Then X is an almost toric surface the general solution is outlined using the universal rational parametrization theory. Also an open problem of a linear precision property for toric Bézier patches is discussed.

1 Introduction

Rational surfaces are widely used in geometric modeling in the form of rational Bézier patches, which have been tested in a variety of practical applications. On the other hand, potential of arbitrary rational surfaces are not fully realized, since in general it is difficult to control real rational parametrizations: they can have basepoints, they are not necessary one-to-one or even onto.

We focus on the *parametrization extension problem*: how to find Bézier patches of optimal degree on the given surface with given boundary curves. Our idea is to study all rational parametrizations of a given rational surface in order to find suitable Bézier patches on it. Here we restrict our considerations to the simple cases of quadric surfaces and then sketch the extension of this theory to a large important subclass of rational surfaces in $\mathbb{P}^3(\mathbb{R})$, so called *real almost toric surfaces*, that include many real rational surfaces that are popular in geometric modeling. The reason is that a constructive description of all rational parametrizations exactly for this class of surfaces was developed in terms of the Universal Rational Parametrization (URP) [2]. This is a natural extension of the results obtained by Dietz, Hoschek and Jüttler [6, 7] and announced by the author in [10].

In Section 2 the parametrization extension problem is discussed for the simplest case of Bézier patches in $\mathbb{P}^n(\mathbb{R})$. Section 3 is devoted to surfaces

$\mathbb{P}^1(\mathbb{C})$ and $\mathbb{P}^1(\mathbb{R}) \times \mathbb{P}^1(\mathbb{R})$ as abstract versions of an oval and a double ruled quadrics. We survey the URP theory for real almost toric surfaces and sketch the solution to the parametrization extension problem for this class of surfaces in Section 4. The last Section 5 is devoted to an open problem of a linear precision property for toric Bézier patches.

2 Bézier curves and surface patches in $\mathbb{P}^n(\mathbb{R})$

2.1 Definitions

A real n-dimensional projective space $\mathbb{P}^n(\mathbb{R})$ is defined as a space of all lines in $\mathbb{R}^{n+1}$ going through the origin $0 = (0, \ldots, 0)$. Alternatively $\mathbb{P}^n(\mathbb{R})$ can be regarded as a quotient of $\mathbb{R}^{n+1} \setminus 0$ with respect to the multiplicative group $\mathbb{R}^* = \mathbb{R} \setminus 0$ action $\lambda \cdot (x_0, \ldots, x_n) = (\lambda x_0, \ldots, \lambda x_n)$. This means that $\mathbb{P}^n(\mathbb{R})$ is a factor set of $\mathbb{R}^{n+1} \setminus 0$ with respect to the following equivalence relation: $x \sim x'$ if there exists such $\lambda \in \mathbb{R}^*$ that $x' = \lambda \cdot x$ (denote also $x'/x = \lambda$). There is a natural projection

$$\Pi : \mathbb{R}^{n+1} \setminus 0 \longrightarrow \mathbb{P}^n(\mathbb{R}), \tag{2.1}$$

which maps $(x_0, \ldots, x_n)$ to its equivalence class $[x_0, \ldots, x_n]$. These x_i are homogeneous coordinates of a corresponding point in $\mathbb{P}^n(\mathbb{R})$.

Complex projective spaces $\mathbb{P}^d(\mathbb{C})$ are similarly defined: just change all appearances of $\mathbb{R}$ to $\mathbb{C}$ (complex numbers) in the previous definition. In particular, $\mathbb{C}^* = \mathbb{C} \setminus 0$ denotes the multiplicative group of non-zero complex numbers.

Let us denote a closed interval $[0,1] \subset \mathbb{R}$ by I, and define a standard affine triangle $\triangle$ and a unit square $\square$ as follows

$$\triangle = \{(s,t) \in \mathbb{R}^2 \mid s \geq 0,\ t \geq 0,\ s+t \leq 1\}, \tag{2.2}$$

$$\square = \{(s,t) \in \mathbb{R}^2 \mid 0 \leq s \leq 1,\ 0 \leq t \leq 1\}. \tag{2.3}$$

We consider a rational Bézier curve $c : \mathrm{I} \to \mathbb{P}^n(\mathbb{R})$, a Bézier rectangular (=tensor-product) patch $c : \square \to \mathbb{P}^n(\mathbb{R})$ or a Bézier triangular patch $c : \triangle \to \mathbb{P}^n(\mathbb{R})$ in a real projective space $\mathbb{P}^n(\mathbb{R})$ as a composition of the polynomial map γ and the projection Π:

$$D \xrightarrow{\gamma} \mathbb{R}^{n+1} \setminus 0 \xrightarrow{\Pi} \mathbb{P}^n(\mathbb{R}), \quad D = \mathrm{I}, \square, \triangle. \tag{2.4}$$

Here γ is a polynomial Bézier curve $\gamma(t)$, a Bézier triangle or a Bézier rectangle $\gamma(s,t)$ defined as linear combinations of homogeneous control points $p_i, p_{ij}, p_{ijk} \in \mathbb{R}^{n+1}$

$$\sum_{i=0}^{d} p_i B_i^d(t), \quad \sum_{i=0}^{d_1}\sum_{j=0}^{d_2} p_{ij} B_i^{d_1}(s) B_j^{d_2}(t), \quad \sum_{i+j+k=d} p_{ijk} B_{ijk}^d(s,t) \tag{2.5}$$

with coefficients expressed in terms of Bernstein polynomials

$$B_i^d(t) = \binom{d}{i}(1-t)^{d-i}t^i, \quad B_{ijk}^d(s,t) = \frac{d!}{i!j!k!}(1-s-t)^i s^j t^k. \quad (2.6)$$

In all these cases polynomial maps $\gamma : D \to \mathbb{R}^{n+1}$ are *liftings* of rational maps $c : D \to \mathbb{P}^n(\mathbb{R})$ in the sense that $\gamma \circ \Pi = c$.

Definition 1. Let γ and γ' be two polynomial maps.
(i) γ and γ' are called *equivalent* $\gamma \sim \gamma'$ if $\Pi \circ \gamma = \Pi \circ \gamma'$.
(ii) γ is called an *irreducible* if $\gcd(\gamma_0, \ldots, \gamma_n) = 1$ for all coordinates $(\gamma_0, \ldots, \gamma_n)$ of γ.

Lemma 2. *Let γ and γ' be two polynomial maps. If $\gamma \sim \gamma'$ and γ is irreducible then $\gamma' = \lambda \cdot \gamma$ for some polynomial $\lambda = \lambda(t)$ (i.e. polynomial map $\lambda : D \to \mathbb{R}^*$ in the group $\mathbb{R}^*$).*

Proof. For every t vectors $\gamma(t)$ and $\gamma'(t)$ are proportional, and a function $\lambda(t) = \gamma_i'(t)/\gamma_i(t)$ is defined for some their non-zero coordinates. So λ is a rational function, and it can be represented in an irreducible form $\lambda = p/q$, where p and q are some mutually prime polynomials. Hence $p\gamma_i = q\gamma_i'$, and q divides γ_i for all i. Since γ is irreducible, q is constant.

Corollary 3. *If $\gamma \sim \gamma'$ and both polynomial maps γ and γ' are irreducible then $\gamma' = \lambda \cdot \gamma$, for some constant $\lambda \in \mathbb{R}^*$. Therefore, for any rational map $c : D \to \mathbb{P}^n(\mathbb{R})$ there exists the unique irreducible lifting $\tilde{c} : D \to \mathbb{R}^{n+1}$ up to constant multiplier in $\mathbb{R}^*$.*

2.2 Parametrization extension problem

Let X be one of the previous spaces where some kind of curves in polynomial representation make sense, i.e. $X = \mathbb{R}^n$, $\mathbb{R}^{n+1} \setminus 0$, $\mathbb{P}^n(\mathbb{R})$, $\mathbb{R}^*$, $\mathbb{P}^n(\mathbb{C})$, any rational surface in $\mathbb{P}^n(\mathbb{R})$, etc.

Definition 4. A *chain* of curves $\langle c_i \rangle_0^k = \langle c_0, \ldots, c_k \rangle$ in X is a sequence $c_i : I \to X$ such that $c_i(1) = c_{i+1}(0)$, $i = 0, \ldots, k-1$. The chain $\langle c_i \rangle_0^k$ is *closed* if $c_0(0) = c_k(1)$.

Example 5. Let $\sigma : \triangle \to X$ be a Bézier triangle of degree d. Then there is a natural closed chain of Bézier curves of degree d called its boundary $\partial\sigma = \langle \partial_0\sigma, \partial_1\sigma, \partial_2\sigma \rangle$: $\partial_0\sigma(t) = \sigma(t, 1-t)$, $\partial_1\sigma(t) = \sigma(1-t, 0)$, $\partial_2\sigma(t) = \sigma(0, t)$.

Example 6. Let $\sigma : \square \to X$ be a Bézier quadrangle of bi-degree (d_1, d_2). Then there is a natural closed chain of Bézier curves of degrees d_1, d_2, d_1, d_2 called its boundary $\partial\sigma = \langle \partial_0\sigma, \partial_1\sigma, \partial_2\sigma, \partial_3\sigma \rangle$: $\partial_0\sigma(t) = \sigma(t, 0)$, $\partial_1\sigma(t) = \sigma(1, t)$, $\partial_2\sigma(t) = \sigma(1-t, 1)$, $\partial_3\sigma(t) = \sigma(0, 1-t)$.

Now we are ready to formulate the *parametrization extension problem*.

Problem 7. For a given closed chain $\langle c_i \rangle_0^k$ of curves in X ($k = 2, 3$) find a Bézier patch $c : D \to \mathbb{P}^n(\mathbb{R})$ ($D = \triangle, \square$) such that $\partial c = \langle c_i \rangle_0^k$. The minimal possible parametrization degree would be desirable.

When X is an affine space $\mathbb{R}^n$ or $\mathbb{R}^n \setminus 0$ a solution of Problem 7 for polynomial patches is almost obvious. Let us start from the first non-trivial case $X = \mathbb{P}^n(\mathbb{R})$ in this section.

Any solution to Problem 7 $c : D \to \mathbb{P}^n(\mathbb{R})$ will have an irreducible lifting $\tilde{c} : D \to \mathbb{R}^{n+1}$. The boundary $\partial \tilde{c}$ should be closed (not necessary irreducible) lifting of the given chain $\langle c_i \rangle_0^k$. This is the necessary condition for existence of the solution.

Therefore we consider the following smaller problem first: how to find closed liftings of closed chains of curves. From Corollary 3 it follows that any chain has a unique lifting of irreducible curves in $\mathbb{R}^{n+1} \setminus 0$ with a given beginning point. Unfortunately, a lifting of a closed chain is not necessary closed.

Let $\langle c_i \rangle_0^k$ be a closed chain of Bézier curves in $\mathbb{P}^n(\mathbb{R})$, and let $\langle \gamma_i \rangle_0^k$ be a chain of arbitrary liftings of c_i, i.e. $\Pi \circ \gamma_i = c_i$, for all $i = 0, \ldots, k$. Define a nonzero number

$$\Omega(\langle \gamma_i \rangle_0^k) = \frac{\gamma_0(0)}{\gamma_k(1)} \in \mathbb{R}^*. \tag{2.7}$$

It is clear that any proportional lifting $\langle \gamma_i' \rangle_0^k$, $\gamma_i' = \lambda \cdot \gamma_i$, gives the same number $\Omega \langle \gamma_i \rangle_0^k = \Omega \langle \gamma_i' \rangle_0^k$.

Suppose all $\tilde{c}_i$ are irreducible liftings of c_i for $i = 0, \ldots, k$. Then the definition $\Omega \langle c_i \rangle_0^k := \Omega \langle \tilde{c}_i \rangle_0^k$ is justified by Corollary 3. In fact $\Omega \langle c_i \rangle_0^k$ is an obstruction to the existence of a closed irreducible lifting in the following sense.

Lemma 8. (i) *A closed chain $\langle c_i \rangle$ of Bézier curves in $\mathbb{P}^n(\mathbb{R})$ has closed irreducible lifting to $\mathbb{R}^{n+1} \setminus 0$ if and only if $\Omega \langle c_i \rangle = 1$.*
(ii) *All closed liftings of $\langle c_i \rangle$ are in 1–1 correspondence with chains in the group $\mathbb{R}^*$ connecting 1 with $\Omega \langle c_i \rangle$.*

Proof. The part (i) is clear, so we will check only (ii). Fix an irreducible lifting $\langle \gamma_i \rangle$ of a closed chain $\langle c_i \rangle$ in $\mathbb{P}^n(\mathbb{R})$. For any closed lifting $\langle \gamma_i' \rangle$ of the same chain with the same initial point $\gamma_0'(0) = \gamma_0(0)$ define a chain $\langle \lambda_i \rangle = \langle \gamma_i'/\gamma_i \rangle$ in the group $\mathbb{R}^*$. Since $\lambda_0(0) = \gamma_0'(0)/\gamma_0(0) = 1$, $\lambda_k(1) = \gamma_k'(1)/\gamma_k(1) = \Omega \langle c_i \rangle$, the chain $\langle \lambda_i \rangle$ connects 1 and $\Omega \langle c_i \rangle$. It is clear that inverse correspondence is defined by the formula $\langle \gamma_i' \rangle = \langle \lambda_i \cdot \gamma_i \rangle$.

Now one can solve the parametrization extension problem for a Bézier triangle in $\mathbb{P}^n(\mathbb{R})$ as stated the following theorem (the Bézier rectangle case is similar).

Theorem 9. *Let $\langle c_0, c_1, c_2 \rangle$ be a closed chain of curves of degrees $\deg c_i = d_i$, $d = d_0 \geq d_1 \geq d_2$, in $\mathbb{P}^n(\mathbb{R})$. Then a Bézier triangle $c : \triangle \to \mathbb{P}^n(\mathbb{R})$ of the minimal degree such that $\partial c = \langle c_i \rangle$ can be found or its non-existence can be proved in two steps:*

(1) Find a closed lifting $\langle \gamma_i \rangle$ of the minimal degree d' of the chain $\langle c_i \rangle$ in $\mathbb{R}^{n+1} \setminus 0$. There are three cases depending on $\Omega = \Omega \langle c_i \rangle \in \mathbb{R}^*$
  (i) if $\Omega < 0$ then such a lifting does not exist;
  (ii) if $\Omega = 1$ then $d' = d$;
  (iii) if $\Omega > 0$ and $\Omega \neq 1$ then $d' = d$, if $d_2 \leq d-1$, and $d' = d+1$, otherwise;
(2) Find a Bézier triangle $\sigma : \triangle \to \mathbb{R}^{n+1} \setminus 0$ with a boundary $\partial \sigma = \langle \gamma_i \rangle$. Its existence and degree d'' depends on the dimension n and the position of the origin $0 \in \mathbb{R}^{n+1}$:
  – ($n = 1$): the *sufficient* condition for the existence: 0 does not belong to a convex hull of all control points of all curves γ_i; then $d'' = d'$;
  – ($n > 1$): the triangle σ exists, $d'' = d'$, except the case $0 \in \sigma(\triangle)$ and $d' = 1, 2$ when $d'' = 3$.

Finally, if both steps are successful, $c = \Pi \circ \sigma$ is the solution.

2.3 Reparametrizations of Bézier curves

Definition 10. *Call two curves $c, c' : \mathrm{I} \to \mathbb{P}^n(\mathbb{R})$ similar if they have the same endpoints and $c(\mathrm{I}) = c'(\mathrm{I})$, i.e. they have the same images and orientations.*

There is a useful method how to obtain a new polynomial Bézier curve from an old one with the same initial point and a similar projection:

$$\Phi_\rho^d : \sum_{i=0}^d p_i B_i^d(t) \mapsto \sum_{i=0}^d (\rho^i p_i) B_i^d(t), \quad \rho \in \mathbb{R}^*. \tag{2.8}$$

In fact $\Pi \circ \Phi_\rho^d \gamma$ is a composition of $\Pi \circ \gamma$ and some projective transformation of the domain $\mathrm{I} = [0, 1]$ fixed on the endpoints (see details in [15]). Note that one endpoint is fixed $\Phi_\rho^d \gamma(0) = \gamma(0)$ and the other $\Phi_\rho^d \gamma(1) = \rho^d \cdot \gamma(1)$ depends on degree d and the parameter ρ.

In practice it is useful to relax requirements on the boundary curves and to formulate the following variant of Problem 7.

Problem 11. *For a given closed chain $\langle c_i \rangle_0^k$ of curves in X ($k = 2, 3$) find a Bézier patch $c : D \to \mathbb{P}^n(\mathbb{R})$ ($D = \triangle, \square$) such that the chain ∂c is similar to $\langle c_i \rangle_0^k$. The minimal possible parametrization degree would be desirable.*

This reparametrization trick can help to find better solutions. For example, we can skip the step (1)(iii) in Theorem 9, since the obstruction $\Omega > 0$ can be reduced to the unit 1 after the transformation Φ_ρ^d, $\rho = \Omega^{1/d}$.

3 Bézier patches in $\mathbb{P}^1(\mathbb{C})$ and $\mathbb{P}^1(\mathbb{R}) \times \mathbb{P}^1(\mathbb{R})$

One can study Bézier patches also in more general spaces $\mathbb{P}^n(\mathbb{C})$ or $\mathbb{P}^n(\mathbb{R}) \times \mathbb{P}^k(\mathbb{R})$ using the similar approach. The cases $\mathbb{P}^1(\mathbb{C})$ and $\mathbb{P}^1(\mathbb{R}) \times \mathbb{P}^1(\mathbb{R})$ are most interesting, since they are directly related to oval and double ruled quadrics in $\mathbb{P}^3(\mathbb{R})$.

Define Bézier curves and surface patches on $\mathbb{P}^1(\mathbb{C})$ as compositions

$$D \xrightarrow{\gamma} \mathbb{C}^2 \setminus 0 \xrightarrow{\Pi} \mathbb{P}^1(\mathbb{C}), \quad D = \mathrm{I}, \square, \triangle, \qquad (3.1)$$

where the map $\gamma : D \to \mathbb{C}^2 \setminus 0$ is polynomial with complex coefficients. We call γ a lifting of a rational map $c : D \to \mathbb{P}^1(\mathbb{C})$ if $\gamma \circ \Pi = c$.

Now we can repeat the similar definitions and lemmas as in Section 2.

Definition 12. Let γ and γ' be two complex polynomial maps.
(i) γ and γ' are called *S-equivalent* $\gamma \sim_S \gamma'$ if $\Pi \circ \gamma = \Pi \circ \gamma'$.
(ii) γ is called *S-irreducible* if $\gcd(\gamma_0, \gamma_1) = 1$ for complex coordinates (γ_0, γ_1) of γ.

Lemma 13. *Let γ and γ' be two complex polynomial maps. If $\gamma \sim_S \gamma'$ and γ is S-irreducible then $\gamma' = \lambda \cdot \gamma$ for some complex polynomial $\lambda = \lambda(t)$ (i.e. polynomial map $\lambda : D \to \mathbb{C}^*$ in the group $\mathbb{C}^*$).*

Corollary 14. *If $\gamma \sim \gamma'$ and both complex polynomial maps γ and γ' are S-irreducible then $\gamma' = \lambda \cdot \gamma$, for some constant $\lambda \in \mathbb{C}^*$. Therefore, for any rational map $c : D \to \mathbb{P}^1(\mathbb{C})$ there exists the unique S-irreducible lifting $\tilde{c} : D \to \mathbb{C}^2$ up to constant multiplier in $\mathbb{C}^*$.*

Let $\langle c_i \rangle_0^k$ be a closed chain of Bézier curves in $\mathbb{P}^1(\mathbb{C})$, and let $\langle \gamma_i \rangle_0^k$ be a chain of arbitrary liftings of c_i. Then we define a nonzero complex number

$$\Omega(\langle \gamma_i \rangle_0^k) = \frac{\gamma_0(0)}{\gamma_k(1)} \in \mathbb{C}^*. \qquad (3.2)$$

Similarly using the S-irreducible lifting we define (cf. Corollary 14) the obstruction $\Omega \langle c_i \rangle_0^k := \Omega \langle \tilde{c}_i \rangle_0^k$. The following lemma can be proved similarly to Lemma 8.

Lemma 15. *(i) A closed chain $\langle c_i \rangle$ of Bézier curves in $\mathbb{P}^1(\mathbb{C})$ has closed S-irreducible lifting to $\mathbb{C}^2 \setminus 0$ if and only if $\Omega \langle c_i \rangle = 1$.
(ii) All closed liftings of $\langle c_i \rangle$ are in 1–1 correspondence with chains in the group $\mathbb{C}^*$ connecting 1 with $\Omega \langle c_i \rangle$.*

A unit sphere S with the equation $x_0^2 = x_1^2 + x_2^2 + x_3^2$ in $\mathbb{P}^3(\mathbb{R})$ can be parametrized by $\mathbb{P}^1(\mathbb{C})$ using the map

$$P_S(z_0, z_1) = (z_0 \bar{z}_0 + z_1 \bar{z}_1, \mathrm{Re}(2\bar{z}_0 z_1), \mathrm{Im}(2\bar{z}_0 z_1), z_0 \bar{z}_0 - z_1 \bar{z}_1), \qquad (3.3)$$

which is homogeneous in the following sense $P_S(\lambda z_0, \lambda z_1) = |\lambda|^2 P_S(z_0, z_1)$.

Finally we use the key observation that follows from Section 4: there is 1–1 correspondence $\gamma \mapsto P_S \circ \gamma$ between S-irreducible maps $\gamma : D \to \mathbb{C}^2 \setminus 0$ and such irreducible maps $\gamma' : D \to \mathbb{R}^4 \setminus 0$, that the image $\Pi(\gamma'(D))$ is contained on the sphere $S \subset \mathbb{P}^3(\mathbb{R})$. Hence we can apply obtained results on $\mathbb{P}^1(\mathbb{C})$ to the sphere case, using the simple formula for degrees $\deg(P_S \circ \gamma) = 2 \deg \gamma$.

Theorem 16. Let $\langle c_0, c_1, c_2 \rangle$ be a closed chain of curves of degrees $\deg c_i = 2d_i$, $d = d_0 \geq d_1 \geq d_2$, on the sphere S^2. Then there exists a Bézier triangle c with ∂c equal (or similar) to this chain such that at least $\deg c = 2d'$, where d' depends on $\Omega = \Omega \langle c_0, c_1, c_2 \rangle \in \mathbb{C}^*$ and d_i according to the following cases (except a few degenerated cases when $d = 1, 2$):

(i) $\Omega = 1$ then $d' = d$;
(ii) $\Omega \in \mathbb{R}_{>0}$, $\Omega \neq 1$ then $d' = d$ (with reparametrization), otherwise see (iii);
(iii) $\Omega \in \mathbb{C} \setminus \mathbb{R}$ then $d' = d$ if $d_2 \leq d - 1$, otherwise $d' = d + 1$;
(iv) $\Omega \in \mathbb{R}_{<0}$ then $d' = d$ if $d_2 \leq d - 2$ or $d_1 \leq d - 1$, otherwise $d' = d + 1$;

Proof. By Lemma 15 we need to find a chain in $\mathbb{C}^*$ connecting 1 and Ω in $\mathbb{C}^*$. The case (i) is obvious. In cases (ii) (without reparametrization) and (iii) we can connect 1 and Ω just by the line $(1-t) \cdot 1 + t \cdot \Omega$. In the case (iv) we have to make a detour round the complex zero via an arc of some parabola or two line segments.

Consider reparametrizations of spherical curves. For any polynomial curve $\gamma \colon I \to \mathbb{C}^2 \setminus 0$ the following formula can be easy verified

$$P_S \circ \Phi_\rho^d(\gamma) = \Phi_\rho^{2d}(P_S \circ \gamma). \tag{3.4}$$

Therefore reparametrization paths coincide with lines in $\mathbb{C}^*$ that go through the origin. So, if $\Omega \in \mathbb{R}_{>0}$ then we can move it to 1 using reparametrization of any curve γ_i.

Finally we have built a lifting $\langle \gamma_i' \rangle$ of degree d' in $\mathbb{C}^2 \setminus 0$ of the given chain $\langle c_i \rangle$ on the sphere S. Now we can almost always fill it by a Bézier triangle σ (i.e. $\partial \sigma = \langle \gamma_i' \rangle$) of the same degree d'. Then put $c = \Pi \circ P_S \circ \sigma$, $\deg c = 2d'$. There are just a few degenerated cases when the triangle σ is going through the origin $0 \in \mathbb{C}^2$ (cf. Theorem 9, Step (2) then $n = 1$).

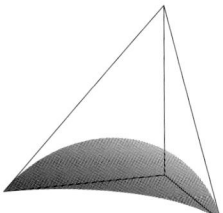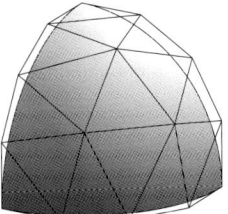

Fig. 1. A biangle and a symmetric Bézier triangle on the sphere.

Example 17. The obstruction Ω has a clear geometrical meaning. Consider a biangle patch on the sphere bounded by two circular arcs c_1 and c_2 as in Fig. 1(left). Add one constant curve $* \colon t \mapsto c_1(0)$ and obtain a closed chain $\langle *, c_1, c_2 \rangle$. Calculate $\Omega = \Omega \langle *, c_1, c_2 \rangle$, and suppose that $|\Omega| = 1$ (if not then reparametrize c_2 according to (3.4) with $d = 1$ and $\rho = \sqrt{|\Omega|}$, i.e. $\Omega = e^{i\varphi}$. An angle φ is nothing else than an angle of the biangle. See [9] for details.

Example 18. Suppose we would like to find a symmetric Bézier representation of an octant of the sphere as in Fig. 1(right). It is easy to calculate S-irreducible representations for all three circular sides (as quadratic Bézier curves with weights: $1, 1/\sqrt{2}, 1$). The obstruction Ω is equal to $e^{i\varphi}$, $\varphi = \pi/4$. Indeed, for a spherical triangle with circular edges φ means a half of a difference between sum of all three angles and π (cf. [6]). Taking into account the symmetry requirement we build the chain of three lines connecting points 1, $e^{i\pi/12}$, $e^{i\pi/6}$ and $e^{i\pi/4} = \Omega$ in $\mathbb{C}^*$. In Fig. 1(right) we see the resulting quartic Bézier triangle with its control net.

Let us consider the case of $\mathbb{P}^1(\mathbb{R}) \times \mathbb{P}^1(\mathbb{R})$ now. Define Bézier curves and surface patches on $\mathbb{P}^1(\mathbb{R}) \times \mathbb{P}^1(\mathbb{R})$ as compositions

$$D \xrightarrow{\gamma} (\mathbb{R}^2 \setminus 0) \times (\mathbb{R}^2 \setminus 0) \xrightarrow{\Pi^2} \mathbb{P}^1(\mathbb{R}) \times \mathbb{P}^1(\mathbb{R}), \quad D = \mathrm{I}, \square, \triangle, \tag{3.5}$$

where the map $\gamma : D \to (\mathbb{R}^2 \setminus 0) \times (\mathbb{R}^2 \setminus 0)$ is in fact a pair $(\gamma^{(0)}, \gamma^{(1)})$ of polynomial maps $\gamma^{(i)} : D \to \mathbb{R}^2 \setminus 0$, $i = 0, 1$. We call γ a lifting of a rational map $c : D \to \mathbb{P}^1(\mathbb{R}) \times \mathbb{P}^1(\mathbb{R})$ if $\Pi^2 \circ \gamma = c$.

Now we can repeat similar definitions and lemmas as in case of $\mathbb{P}^1(\mathbb{C})$.

Definition 19. Let γ and γ' be two polynomial maps $D \to \mathbb{R}^2 \times \mathbb{R}^2$.
(i) γ and γ' are called *H-equivalent* $\gamma \sim_H \gamma'$ if $\Pi^2 \circ \gamma = \Pi^2 \circ \gamma'$.
(ii) $\gamma = (\gamma_0, \gamma_1)$ is called *H-irreducible* if γ_0 and γ_1 are irreducible as polynomial maps $D \to \mathbb{R}^2$.

The group $\mathbb{R}^* \times \mathbb{R}^*$ acts on $(\mathbb{R}^2 \setminus 0) \times (\mathbb{R}^2 \setminus 0)$ by the formula $(\lambda, \mu) \cdot (s_0, s_1, t_0, t_1) = (\lambda s_0, \lambda s_1, \mu t_2, \mu t_3)$.

Lemma 20. *Let γ and γ' be two polynomial maps $D \to \mathbb{R}^2 \times \mathbb{R}^2$. If $\gamma \sim_H \gamma'$ and γ is H-irreducible then $\gamma' = (\lambda, \mu) \cdot \gamma$ for some polynomial pair $\lambda(t), \mu(t)$ (i.e. a polynomial map $\lambda : D \to \mathbb{R}^* \times \mathbb{R}^*$ in the group $\mathbb{R}^* \times \mathbb{R}^*$).*

Corollary 21. *If $\gamma \sim_H \gamma'$ and both polynomial maps γ and γ' are H-irreducible then $\gamma' = \lambda \cdot \gamma$, for some constant $(\lambda, \mu) \in \mathbb{R}^* \times \mathbb{R}^*$. Therefore, for any rational map $c : D \to \mathbb{P}^1(\mathbb{R}) \times \mathbb{P}^1(\mathbb{R})$ there exists the unique H-irreducible lifting $\tilde{c} : D \to (\mathbb{R}^2 \setminus 0) \times (\mathbb{R}^2 \setminus 0)$ up to constant multiplier in $\mathbb{R}^* \times \mathbb{R}^*$.*

Let $\langle c_i \rangle_0^k$ be a closed chain of Bézier curves in $\mathbb{P}^1(\mathbb{R}) \times \mathbb{P}^1(\mathbb{R})$, and let $\langle \gamma_i \rangle_0^k$ be a chain of arbitrary liftings of c_i. Then we define a pair of real nonzero numbers

$$\Omega(\langle \gamma_i \rangle_0^k) = \left(\frac{\gamma_0^{(0)}(0)}{\gamma_k^{(0)}(1)}, \frac{\gamma_0^{(1)}(0)}{\gamma_k^{(1)}(1)} \right) \in \mathbb{R}^* \times \mathbb{R}^*. \tag{3.6}$$

Similarly using the H-irreducible lifting we define the obstruction $\Omega \langle c_i \rangle_0^k := \Omega \langle \tilde{c}_i \rangle_0^k$ (cf. Corollary 21). The following lemma is a consequence of Lemma 8.

Lemma 22. (i) *A closed chain $\langle c_i \rangle$ of Bézier curves in $\mathbb{P}^1(\mathbb{R}) \times \mathbb{P}^1(\mathbb{R})$ has closed H-irreducible lifting to $(\mathbb{R}^2 \setminus 0) \times (\mathbb{R}^2 \setminus 0)$ if and only if $\Omega \langle c_i \rangle = 1$.*
(ii) *All closed liftings of $\langle c_i \rangle$ are in 1–1 correspondence with chains in the group $\mathbb{R}^* \times \mathbb{R}^*$ connecting 1 with $\Omega \langle c_i \rangle$.*

A hyperbolic paraboloid $H \colon x_0 x_3 = x_1 x_2$ in $\mathbb{P}^3(\mathbb{R})$ can be parametrized by $\mathbb{P}^1(\mathbb{R}) \times \mathbb{P}^1(\mathbb{R})$ using the map

$$P_H(s_0, s_1, t_0, t_1) = (s_0 t_0, s_1 t_0, s_0 t_1, s_1 t_1), \tag{3.7}$$

which is bi-homogeneous $P_H(\lambda s_0, \lambda s_1, \mu t_0, \mu t_1) = \lambda \mu P_H(s_0, s_1, t_0, t_1)$.

Finally we use the key observation that follows from Section 4: there is 1–1 correspondence $\gamma \mapsto P_H \circ \gamma$ between H-irreducible maps $\gamma : D \to (\mathbb{R}^2 \setminus 0) \times (\mathbb{R}^2 \setminus 0)$ and such irreducible maps $\gamma' : D \to \mathbb{R}^4 \setminus 0$, that the image $\Pi(\gamma'(D))$ is contained on the hyperbolic paraboloid $H \subset \mathbb{P}^3(\mathbb{R})$. Hence we can apply the results obtained in case of $\mathbb{P}^1(\mathbb{R}) \times \mathbb{P}^1(\mathbb{R})$ to this case as well, using the formula for degrees $\deg(P_S \circ \gamma) = \deg \gamma^{(0)} + \deg \gamma^{(1)}$.

Theorem 23. *Let $\langle c_0, c_1, c_2 \rangle$ be a closed chain of curves on H of bi-degrees*

$$(\deg_0 c_i, \deg_1 c_i) = (d_{i0}, d_{i1}), \quad d_* = (d_{*0}, d_{*1}) = (\max_i d_{i0}, \max_i d_{i1}).$$

Then a triangle $c \colon \triangle \to H \subset \mathbb{P}^3(\mathbb{R})$ of the minimal degree d' such that ∂c is equal (or similar) to $\langle c_i \rangle$ can be found or its non-existence can be proved in two steps:

(1) *Find a closed lifting $\langle \gamma_i \rangle$ of the minimal bi-degree $d' = (d'_0, d'_1)$ of the chain $\langle c_i \rangle$ in $(\mathbb{R}^2 \setminus 0)^2$ (after possible reparametrization). There are three cases depending on the obstruction $(\Omega_0, \Omega_1) = \Omega = \Omega \langle c_0, c_1, c_2 \rangle \in (\mathbb{R}^*)^2$*
  (i) *if $\Omega_0 < 0$ or $\Omega_1 < 0$ then such a lifting does not exist;*
  (ii) *if $\Omega = (1,1)$ then $d' = d_*$;*
  (iii) *if not all d_i are equal to d_* or $d_{*1} \log \Omega_0 = d_{*0} \log \Omega_1$ then $d' = d_*$, otherwise $d' = (d_{*0} + 1, d_{*1})$ or $d' = (d_{*0}, d_{*1} + 1)$.*
(2) *Find a polynomial Bézier triangle σ in $(\mathbb{R}^2 \setminus 0)^2$ with a boundary $\partial \sigma = \langle \gamma_i \rangle$, i.e. find a pair of such triangles $(\sigma^{(0)}, \sigma^{(1)})$ with a respective pair of boundaries in $\mathbb{R}^2 \setminus 0$. This is exactly the step (2) in Theorem 9 when $n = 1$.*

Finally, if both steps are successful then put $c = \Pi_H \circ \sigma$.

Proof. Consider only the most different from Theorem 9 case (iii) of the step (1). The key idea is to reparametrize curves c_i and to simplify the obstruction Ω. For a curve $\gamma = (\gamma^{(0)}, \gamma^{(1)})$ in $(\mathbb{R}^2 \setminus 0)^2$ with a bi-degree (d_0, d_1) reparametrization of $P_H \circ \gamma$ can be calculated by a formula

$$P_H \circ \left(\Phi_\rho^{d_0}(\gamma^{(0)}), \Phi_\rho^{d_1}(\gamma^{(1)}) \right) = \Phi_\rho^{d_0 + d_1}(P_H \circ \gamma).$$

Hence, the reparametrization of an H-irreducible lifting $\tilde{c}_i = (\tilde{c}_{i0}, \tilde{c}_{i1})$ with different values of the parameter ρ traces a curve

$$\rho \mapsto \left(\rho^{-d_{i0}}\Omega_0, \rho^{-d_{i1}}\Omega_1\right) \in \mathbb{R}^* \times \mathbb{R}^*.$$

In logarithmic coordinates $(\log, \log)$: $(\mathbb{R}^*)^2 \to \mathbb{R}^2$ such curve converts to a line going through a point $(\log \mathcal{O}_0, \log \mathcal{O}_1)$ with a direction vector $d_i = (d_{i0}, d_{i1})$. Now our goal is to reach the origin $(0,0) = (\log 1, \log 1)$ when we are allowed to travel only in given fixed directions d_i, $i = 0, 1, 2$. Distinguish two cases: all d_i are equal to d_* or not. In the latter case we can find at least two different directions $d_* = (d_{*0}, d_{*1})$ and, for example, $(d_{*0} - 1, d_{*1})$ or $(d_{*0}, d_{*1} - 1)$ after possible degree raising (we skip trivial case when d_{*0} or d_{*1} is zero). It is sufficient for our success. In the former case there is only one direction $d_* = (d_{*0}, d_{*1})$. So we can reach the origin if and only if $d_{*1} \log \Omega_0 = d_{*0} \log \Omega_1$. Otherwise, we move until we intersect any of the coordinate axes and then apply Theorem 9, step (1).

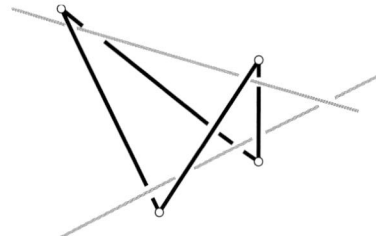

Fig. 2. A closed chain of line segments in $\mathbb{P}^3(\mathbb{R}) \setminus (L_1 \cup L_2)$.

Remark 24. The step (2) in Theorem 23 we have only sufficient conditions for existence. In practice this works in most cases. Homotopical triviality of the closed chain is an obvious necessary condition but it is not sufficient in general for given degrees. Consider for example a closed linear chain in $X = (\mathbb{R}^2 \setminus 0) \times (\mathbb{R}^2 \setminus 0)$ that is visualized in Fig. 2. In order to show the picture in 3D we reduced dimension 4 to 3 by including X in $\mathbb{R}^4 \setminus 0$ and then projecting to $\mathbb{P}^3(\mathbb{R})$. The complement $\mathbb{P}^3(\mathbb{R}) \setminus \Pi(X)$ is a union of two lines L_1 and L_2 shown in grey. Our goal is to fill the contour with Bézier rectangular patch of bi-degree $(1,1)$ at the same time avoiding intersections with lines. Any such patch should be contained in the convex hull of control points, i.e., in the tetrahedron. Unfortunately, this is even topologically impossible: the contour is contractible in $\mathbb{P}^3(\mathbb{R}) \setminus (L_1 \cup L_2)$, but it is non-contractible in the affine (i.e. visible) part $\mathbb{R}^3 \setminus (L_1 \cup L_2)$.

4 General case: real almost toric surfaces

Here we just survey the theory of universal rational parametrizations in case of almost toric surfaces with canonical real structures for simplicity. We adapt for our needs a part of more general results obtained for more general toric varieties [2] (also [13] for non-canonical real structures).

Let $\mathbb{Z}^2$ be the lattice of points with integer coordinates in the plane $\mathbb{R}^2$ and $\Delta \subset \mathbb{R}^2$ be a *lattice polygon*, i.e. a convex polygon with vertices in the lattice. Let the edges of the polygon Δ be on the lines $L_i(x) = 0$, $i = 1, \ldots, r$, defined by the affine forms $L_i(x) = \langle n_i, x \rangle + a_i$, where $\langle , \rangle$ is a scalar product on $\mathbb{Z}^2$, the normals n_i of edges are inward oriented shortest vectors with integer coordinates. Define the rational map (where $q_m \in \mathbb{C}^{d+1}$ are called control points):

$$P_\Delta(u_1, \ldots, u_r) = \sum_{m \in \Delta \cap \mathbb{Z}^2} q_m u_1^{L_1(m)} \cdots u_r^{L_r(m)}, \qquad (4.1)$$

Suppose that the control points q_m are linearly independent. Then P_Δ defines a rational map to $\mathbb{P}^N(\mathbb{R})$, $N = \#(\Delta \cap \mathbb{Z}^2) - 1$.

Definition 25. *A toric surface X_Δ is a subset in $\mathbb{P}^N(\mathbb{C})$ which is parametrized by a rational map P_Δ with linearly independent control points.*

In this case it is well known that an implicit degree of X_Δ in $\mathbb{P}^N(\mathbb{C})$ is equal to the normalized area $\mathrm{Area}(\Delta)$ of the polygon Δ, i.e. twice its usual euclidean area (cf. [3]). Also P_Δ is a rational map $\mathbb{C}^r \dashrightarrow \mathbb{P}^N(\mathbb{C})$ which is well defined outside of the *exceptional set* $Z = P_\Delta^{-1}(0) \subset \mathbb{C}^r$, $0 \in \mathbb{C}^{N+1}$.

In the general case P_Δ can be decomposed $\mathbb{C}^r \to \mathbb{P}^N(\mathbb{C}) \to \mathbb{P}^d(\mathbb{C})$, where the last map is some central projection $\pi : \mathbb{P}^N(\mathbb{C}) \to \mathbb{P}^d(\mathbb{C})$.

Definition 26. *The image $X = \pi X_\Delta \subset \mathbb{P}^d(\mathbb{C})$ of the toric surface X_Δ is called* almost toric *if their implicit degrees coincide* $\deg X = \deg X_\Delta$, *i.e.* $\deg X$ *is equal to the normalized area* $\mathrm{Area}(\Delta)$ *of the lattice polygon Δ.*

We define a *real almost toric surface* as a real part $\mathbb{R}_c X$ of some complex almost toric surface X with respect to certain real structure c. The canonical real structure means that all coefficients must be real. References for non-canonical real structures (NCRS) are [13, 11].

Fix a lattice polygon $\Delta \subset \mathbb{R}^2$ bounded by lines L_i with normals n_i, $i = 1, \ldots, r$, and the rational map P_Δ well defined on $\mathbb{C}^r \setminus Z$.

Then a subgroup $G \subset (\mathbb{C}^*)^r$, defined by two equations $\prod_{i=0}^r \lambda_i^{\langle n_i, e_j \rangle} = 1$ ($\{e_j\}$, $j = 1, 2$, is a basis of $\mathbb{Z}^2$) naturally acts on the space $\mathbb{C}^r \setminus Z$. The exceptional set Z, the group G and the group action depends only on the collection of normals $\{n_1, \ldots, n_r\}$ which encode all the information about the *normal fan* Σ of the polygon Δ (see details in [1]). A fan Σ is *regular* if all pairs of adjacent normals $\{n_i, n_{i+1}\}$, $i = 1, \ldots, r-1$, and $\{n_r, n_0\}$ define a basis in $\mathbb{Z}^2$. An *abstract* toric surface X_Σ is defined as a factor $(\mathbb{C}^r \setminus Z)/G$ by the action of the group G. X_Σ is non-singular if and only if Σ is regular.

The map P_Δ is homogeneous in the following sense: if $(\lambda_1, \ldots, \lambda_r) \in G$ then $P_\Delta(\lambda_1 u_1, \ldots, \lambda_r u_r) = \theta_\Delta(\lambda) P_\Delta(u_1, \ldots, u_r)$, $\theta_\Delta(\lambda) = \prod_{i=1}^r \lambda^{a_i}$.

We say that a polynomial map $f = (f_1, \ldots, f_r) : \mathbb{C}^k \to \mathbb{C}^r$ is Σ-*irreducible* if $\gcd(f_{i_1}, \ldots, f_{i_s}) = 1$ whenever no vertex of Δ is incident with its edges defined by the lines $L_{i_1}, \ldots, L_{i_s}$ for all subsets $\{i_1, \ldots, i_s\} \subset \{1, \ldots, r\}$.

The following theorem is valid also for the singular case if the fan Σ is substituted with its regular refinement $\widetilde{\Sigma}$ [2].

Theorem 27. *Let $X \subset \mathbb{P}^d(\mathbb{C})$ be an almost toric surface, defined as an image of P_Δ associated with a lattice polygon Δ with a regular fan Σ as above. Denote by $C(X) = \Pi^{-1} X \subset \mathbb{C}^{d+1}$ the cone over X, and by $\operatorname{Sing} X$ the singular locus of X. Then P_Δ is a **universal rational parametrization** of X in the following sense:*

(1) for any Σ-irreducible map $f : \mathbb{C}^k \to \mathbb{C}^r$ the composition $P_\Delta \circ f$ is an irreducible map $\mathbb{C}^k \to C(X) \subset \mathbb{C}^{d+1}$;

(2) conversely, given any irreducible map $g : \mathbb{C}^k \to CX \subset \mathbb{C}^{d+1}$ such that its image $\Pi(g(\mathbb{C}^k))$ in X is not contained in $\operatorname{Sing} X$, there is a Σ-irreducible $f : \mathbb{C}^k \to \mathbb{C}^r$ such that $g = P_\Delta \circ f$.

(3) If f and f' are Σ-irreducible, then $P_\Delta \circ f = P_\Delta \circ f'$ if and only if $f' = \lambda \cdot f$ for some $\lambda \in G$, $\theta_\Delta(\lambda) = 1$.

The following corollary (cf. [2, Theorem 5.1]) will be most important for our applications.

Corollary 28. *Any polynomial map $h : \mathbb{C}^k \to \mathbb{C}^r$ can be represented in the form $h = g \cdot f$, where $f : \mathbb{C}^k \to \mathbb{C}^r$ is Σ-irreducible polynomial map and $g : \mathbb{C}^k \to G$ is in general a rational map to the group G.*

Indeed, call maps g in the group G *admissible* w.r.t a polynomial map f if $h = g \cdot f$ is also a polynomial map (cf. [2, Example 5.2]).

Now we switch to real coefficients and polygonal domains $D = \mathrm{I}, \triangle, \square \subset \mathbb{R}^2$ as in Sections 2 and 3. Consider a real almost toric surface X and its parametrization $\Pi_\Delta = \Pi \circ P_\Delta : \mathbb{R}^r \setminus Z \to X \subset \mathbb{P}^d(\mathbb{R})$. Bézier curves and surface patches on X are rational maps $c : D \to X$ that can be lifted to some polynomial maps $\gamma : D \to \mathbb{R}^r \setminus Z$, i.e. $c = \Pi_\Delta \circ \gamma$.

Let $\langle c_i \rangle_0^k$ be a closed chain of Bézier curves on X, and let $\langle \tilde{c}_i \rangle_0^k$ be a chain of Σ-irreducible liftings of c_i. Since it is unique up to the action of G, we can define the obstruction $\Omega(\langle c_i \rangle_0^k) = \tilde{c}_0(0)/\tilde{c}_k(1) \in G$. The following lemma is similar to Lemma 8.

Lemma 29. *(i) A closed chain $\langle c_i \rangle$ of Bézier curves in X has closed Σ-irreducible lifting to $\mathbb{R}^r \setminus Z$ if and only if $\Omega\langle c_i \rangle = 1$.*
(ii) All closed liftings of $\langle c_i \rangle$ are in 1–1 correspondence with admissible chains in the group G connecting 1 with $\Omega\langle c_i \rangle$.

Now we are ready to sketch a parametrization extension algorithm which should solve Problems 7 and 11 for the most general case of a real almost toric surface $X \subset \mathbb{P}^d(\mathbb{R})$ parametrized by $\Pi_\Delta = P_\Delta \circ \Pi : \mathbb{R}^r \setminus Z \to X$.

4.1 Parametrization extension algorithm

Input: a closed chain of curves $\langle c_i \rangle$ on X.
Output: a Bézier patch $c : D \to X$ of minimal degree such that $\partial c = \langle c_i \rangle$.
(1) lift the boundary chain $\langle c_i \rangle$ to Σ-irreducible chain $\langle \gamma_i \rangle$ in the space $\mathbb{R}^s \setminus Z$, and calculate $\Omega = \Omega \langle \gamma_i \rangle$; IF $\Omega = 1$ THEN go to (4);
(2) simplify $\Omega := \Phi \cdot \Omega$ by using reparametrization paths of curves γ_i;
   IF $\Omega = 1$ THEN go to (4);
(3) find a closed chain of *admissible* curves $\langle g_i \rangle$ in the group G connecting 1 and Ω, and calculate the closed lifting $\langle \gamma_i \rangle$, $\gamma_i := g_i \cdot \gamma_i$;
(4) fill this closed contour by some Bézier patch $\sigma : D \to \mathbb{R}^s \setminus Z$, i.e. $\partial \sigma = \langle \gamma_i \rangle$;
(5) calculate a Bézier patch $c : D \to X$ as the projection of σ: $c = \Pi_\Delta \circ \sigma$.

4.2 Examples of real almost toric surfaces

A class of real almost toric surfaces includes many rational surfaces that are popular in geometric modeling:

- quadric surfaces in $\mathbb{P}^3(\mathbb{R})$:
  - double ruled quadric, in particular a hyperbolic paraboloid H (Section 3);
  - oval quadric (with NCRS), in particular a sphere S (Section 3) [13]), a quadratic biangle patch [9].
  - quadratic cone (singular case) (see [13])
- toric Bézier patches (with general control points) [12], in particular:
  - Bézier triangular and rectangular patches (images of $\mathbb{P}^2(\mathbb{R})$ and $\mathbb{P}^1(\mathbb{R}) \times \mathbb{P}^1(\mathbb{R})$);
  - cones over arbitrary rational curves (singular case) [13];
  - *all* ruled surfaces $X \subset \mathbb{P}^3(\mathbb{R})$: according to [16, Theorem 3], there are two directrix curves on X of degrees d_1 and d_2, $d_1 + d_2 = \deg X$, and the corresponding lattice trapezoid has the same normalized area;
- other cases with NCRS:
  - quartic ring, 1-horn and 2-horn Dupin cyclides [11], in particular a torus [13];
  - surfaces of revolution with conical directrices [11];
  - a pear shaped surface of degree 6 [11];
  - higher degree biangle patches with general control points [9].

5 Linear precision for toric Bézier patches

In Section 2 a rational Bézier patch is defined as a map $c : D \to \mathbb{P}^n(\mathbb{R})$ that is defined on a polygon $D = \triangle, \square$ in an affine plane $\mathbb{R}^2$ and takes its values in $\mathbb{P}^n(\mathbb{R})$. The map $c : D \to \mathbb{P}^n(\mathbb{R})$ can be considered as defined on the whole affine plane $\mathbb{R}^2$ as well. From the algebraic geometric point of view it is more natural to consider projective extensions of Bézier patches. Perhaps for the first time this idea was used in geometric modeling in [5].

Let us look at the details of this construction. We substitute the Bernstein polynomials $B_i^d(t)$ and $B_{ijk}^d(s,t)$ (2.6) by their homogeneous versions

$$\widetilde{B}_i^d(t_0, t_1) = \binom{d}{i} t_0^{d-i} t_1^i, \quad \widetilde{B}_{ijk}^d(t_0, t_1, t_2) = \frac{d!}{i!j!k!} t_0^i t_1^j t_2^k, \tag{5.1}$$

into formulas (2.5) in the definition for polynomial Bézier patches. Then the collections of variables (t_0, t_1) and (t_0, t_1, t_2) are treated as homogeneous coordinates on $\mathbb{P}^1(\mathbb{R})$ and $\mathbb{P}^2(\mathbb{R})$, respectively. Finally we obtain yet other decompositions of Bézier patches:

- $D = \triangle$: a Bézier triangular patch is the composition $\triangle \subset \mathbb{P}^2(\mathbb{R}) \to \mathbb{P}^n(\mathbb{R})$, where coordinates (s, t) of the affine triangle $\triangle$ goes to homogeneous coordinates $[1 - s - t, s, t] \in \mathbb{P}^2(\mathbb{R})$.
- $D = \square$: a Bézier rectangular patch is the composition $\square \subset \mathbb{P}^1(\mathbb{R}) \times \mathbb{P}^1(\mathbb{R}) \to P^n(\mathbb{R})$, where $(s, t) \in \square$ goes to $([1 - s, s], [1 - t, t]) \in \mathbb{P}^1(\mathbb{R}) \times \mathbb{P}^1(\mathbb{R})$.

Notice, that in Section 4 $\mathbb{P}^2(\mathbb{R})$ and $\mathbb{P}^1(\mathbb{R}) \times \mathbb{P}^1(\mathbb{R})$ were identified as particular cases of toric surfaces. Now we are ready to generalize these classical Bézier constructions to more general toric surfaces.

Consider a toric map P_Δ (see (30)) associated with an arbitrary lattice polygon $\Delta \subset \mathbb{R}^2$, bounded by lines $L_i(t) = 0$, $i = 1, \ldots, r$, and define the inclusion

$$L : \Delta \to \mathbb{R}^r, \quad t \mapsto (L_1(t), \ldots, L_r(t)). \tag{5.2}$$

The composition $T_\Delta = \Pi \circ P_\Delta \circ L$ defines a toric generalization of Bézier patches. Recall that Π is a natural projection $\mathbb{R}^{n+1} \to \mathbb{P}^n(\mathbb{R})$.

Definition 30 (cf. [12]). *A toric Bézier patch $T_\Delta : \Delta \to \mathbb{P}^n(\mathbb{R})$ with homogeneous control points $q_m \in \mathbb{R}^{n+1}$, $m \in D \cap \mathbb{Z}^2$, is defined by the formula*

$$T_\Delta(t) = \Pi \left(\sum_{m \in D \cap \mathbb{Z}^2} q_m L_1(t)^{L_1(m)} \cdots L_r(t)^{L_r(m)} \right).$$

There is just one difference from [12, Definition 1]: the coefficients c_m are skipped. Here we include them in weights w_m of the corresponding control points $p_m \in \mathbb{R}^n$, i.e. $q_m = (w_m, w_m p_m) \in \mathbb{R}^{n+1}$.

Definition 31. *A rational patch $c : D \to \mathbb{P}^n(\mathbb{R})$, $c(t) = \Pi(\sum_i (w_i, w_i p_i) F_i(t))$, $D \subset \mathbb{R}^2$, has the* linear precision *property if for some collection of control points $p_i \in \mathbb{R}^2$ and their weights $w_i \in \mathbb{R}$, it is an identity on the domain D.*

Many important examples of rational surface patches have this linear precision property which is important for practice: Bézier triangular and rectangular surfaces, Wachspress patches [4], S-patches [14], M-patches [8], etc. Hence it is natural to raise the following problem.

Problem 32. Find out what toric Bézier patches or their modifications have the linear precision property.

Here by modification we mean different parametrizations of the same patch.

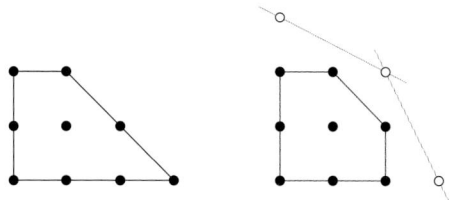

Fig. 3. Lattice polygons for Hirzebruch and pentagonal patches.

The following example is easily derived from Bézier rectangular patch using the degree elevation procedure.

Example 33. Consider the Hirzebruch patch associated with the lattice trapezoid Δ shown in Fig. 3 (left) and slightly modify the affine map L (see (5.2)) by taking $L'(s,t) = (t, 1-s, 1-t, s)$. Then $P_\Delta(L'(s,t))$ has linear precision with the following points p_m and weights w_m, $m \in \Delta \cap \mathbb{Z}^2$,

$$p_m = \begin{pmatrix} (0,1) & (1,1) & & \\ (0,1/2) & (1/2,1/2) & (1,1/2) & \\ (0,0) & (1/3,0) & (2/3,0) & (1,0) \end{pmatrix}, \quad w_m = \begin{pmatrix} 1 & 1 & & \\ 2 & 4 & 2 & \\ 1 & 3 & 3 & 1 \end{pmatrix}.$$

The natural domain D will be defined by inequalities $L'_i(s,t) \geq 0$, $i = 1, \ldots, 4$, i.e. it will be a unit square $\square$.

Example 34. Consider a toric Bézier patch associated with a lattice pentagon Δ shown in Fig 3(right). Denote the bounding lines $L_1 = t$, $L_2 = 2-s$, $L_3 = 3-s-t$, $L_4 = 2-t$, $L_5 = s$, and consider two additional lines: $H_1 = 3-s-t/2$, $H_2 = 3 - s/2 - t$ (shown in grey). Define $L' = (L_1 H_2, L_2, L_3, L_4, L_5 H_1)$ and modify the patch T_Δ as follows $T'_\Delta = \Pi \circ P_\Delta \circ L'$ (with the same domain Δ). Then one can check linear precision by straightforward calculations with the following control points and weights:

$$p_m = \begin{pmatrix} (0,2) & (1,2) & & \\ (0,6/5) & (8/7,8/7) & (2,1) & \\ (0,0) & (6/5,0) & (2,0) & \end{pmatrix}, \quad w_m = \begin{pmatrix} 2 & 2 & \\ 5 & 7 & 2 \\ 3 & 5 & 2 \end{pmatrix}.$$

It is interesting that the total parametrization degree of T'_Δ and T_Δ are the same.

This example was derived from the pentagonal M-patch [8]. It is a surprise that weights and coordinates of control points are such simple rational numbers.

Problem 32 is open even for a toric patch associated with the hexagon that has vertices $(0,0)$, $(1,0)$, $(2,1)$, $(2,2)$, $(1,2)$, $(0,1)$.

References

1. D. Cox: What is a toric variety? In: Goldman, R., Krasauskas, R. (eds.) Topics in algebraic geometry and geometric modeling. Contemp. Math. **334** (2003), 203–223.
2. D. Cox, R. Krasauskas, M. Mustaţă: Universal rational parametrizations and toric varieties. In: Goldman, R., Krasauskas, R. (eds.) Topics in algebraic geometry and geometric modeling. Contemp. Math. **334** (2003), 241–265.
3. D. Cox, J. Little, D. O'Shea: Using algebraic geometry. GTM 185, Springer, Berlin (1998).
4. W. Dahmeh, H. P. Dikshit, A. Ojha: On Wachspress quadrilateral patches. Comput. Aided Geom. Design **17** (2000), 879–890.
5. T. DeRose: Rational Bezier curves and surfaces on projective domains. In: G. Farin, (ed.) NURBS for Curve and Surface Design, SIAM, Philadelphia, PA, 35–45 (1991).
6. R. Dietz, R., Hoschek, J., and Jüttler, B.: An algebraic approach to curves and surfaces on the sphere and other quadrics. Comput. Aided Geom. Design **10** (1993), 211–229.
7. Dietz, J. Hoschek, B. Jüttler: Rational patches on quadric surfaces. Computer-aided Design, **27** (1995), 27–40.
8. K. Karciauskas: Rational M-patches and tensor-border patches. In: R. Goldman, R. Krasauskas (eds.), Topics in Algebraic Geometry and Geometric Modeling. Contemp. Math. **334** (2003), 101–128.
9. K. Karciauskas, R. Krasauskas: Rational biangle surface patches. In: Proceedings of VI Intern. Conf. Central Europe on Computer Graphics and Visualization, Plzen (1998), 165–170.
10. R. Krasauskas: Universal parameterizations of some rational surfaces. In: A. Le Méhauté, C. Rabut, L. L. Schumaker (eds.), Curves and Surfaces with Applications in CAGD. Vanderbilt Univ. Press, Nashville (1997), 231–238.
11. R. Krasauskas: Shape of toric surfaces. In: R. Durikovic, S. Czanner (eds.), Proceedings of the Spring Conference on Computer Graphics SCCG 2001, IEEE (2002), 55–62.
12. R. Krasauskas: Toric surface patches. Advances in Computational Mathematics, **17** (2002), 89–113.
13. R. Krasauskas, M. Kazakevičiūtė: Universal rational parametrizations and spline curves on toric surfaces. In: Computational Methods for Algebraic Spline Surfaces, ESF Exploratory Workshop, Springer (2005), 213–231.
14. C. T. Loop, T. DeRose: A multisided generalization of Bézier surfaces. ACM Transactions on Graphics, **8** (1989), 204–234.
15. R. R. Patterson: Projective transformations of the parameter of a Bernstein–Bézier curve. ACM Transactions on Graphics, **4** (1985), 276–290.
16. M. Peternell, H. Pottmann, B. Ravani: On the computational geometry of ruled surfaces. Computer-aided Design, **13** (1999), 17–32.

On parametric surfaces of low degree in $\mathbb{P}^3(\mathbb{C})$

Mohamed Elkadi, André Galligo, Thi Ha Lê

Laboratoire J-A. Dieudonné
Université de Nice Sophia-Antipolis
Parc Valrose, 06108 Nice Cedex 2, France
{elkadi, galligo, lethiha}@math.unice.fr

Summary. Parametrized surfaces of low degrees are very useful in applications, especially in Computer Aided Geometric Design and Geometric Modeling. The precise description of their geometry is not easy in general. Here we study some of the corresponding projective complex surfaces of low implicit degree (i.e., smaller than 12). We show that, generically up to linear changes of coordinates, they are classified by a few number of continuous parameters (called moduli). We present normal forms and provide compact implicit equations for these surfaces and for their singular locus together with a geometric interpretation.

1 Introduction

The common representation of surfaces in Solid Modeling and Computer Aided Geometric Design is via parametrized patches, i.e. images of maps

$$\phi : [0,1] \times [0,1] \to \mathbb{R}^3$$

$$(t,u) \mapsto \phi(t,u) = \left(\frac{\phi_1(t,u)}{\phi_0(t,u)}, \frac{\phi_2(t,u)}{\phi_0(t,u)}, \frac{\phi_3(t,u)}{\phi_0(t,u)}\right),$$

where $\phi_0, \phi_1, \phi_2, \phi_3$ are polynomials in two variables t and u, with real coefficients. Such a ϕ is classified by the maximal degree (or bidegree) of $\phi_0, \phi_1, \phi_2, \phi_3$; for instance, a bicubic is a surface of bidegree $(3,3)$. Surface patches are encountered in many applications [16, 14, 9]. However self-intersection points of these patches are often difficult to detect [2, 15, 22]. More generally the intersection curve of two such surfaces can be very complicated and plays an important role in the understanding of the geometry of the surface and also in the applications. Therefore it is worthwhile to study systematically parametrized surfaces of low degrees in order to have at our disposal mastered geometric models together with their singular loci. Our study of generic cases in the complex setting is a necessary first step towards a finer and more complete analysis in the real setting. We build some tools which hopefully will be useful also for this last task.

The surfaces of total degree 1 are planes, while surfaces with parametrization of bidegree $(1,1)$ are planes or quadrics. The surfaces with a parametrization of total degree 2 (including Steiner surfaces) and have been extensively studied [8, 3, 1]. The next classes to understand are those with "low" bidegrees (m,n) (i.e. with $m \leq 3$ and $n \leq 3$), or total degree at most 3, and this is the subject of our study.

During the XIX and the beginning of the XX century, many algebraic surfaces have been studied and several interesting classification have been achieved. However theses surfaces were, in general, classified by their properties or by the total degree of their implicit equation. In this paper we concentrate on the production of normal forms, in the generic case (here generic means in particular that these surfaces are base-point free), of a complex projective surface which is the image $\mathcal{S}$ of a polynomial map

$$(t:s;u:v) \in \mathbb{P}^1(\mathbb{C}) \times \mathbb{P}^1(\mathbb{C}) \mapsto (\phi_0:\phi_1:\phi_2:\phi_3) \in \mathbb{P}^3(\mathbb{C})$$

(resp. $(t:u:v) \in \mathbb{P}^2(\mathbb{C}) \to (\phi_0:\phi_1:\phi_2:\phi_3) \in \mathbb{P}^3(\mathbb{C}))$ of bidegree (m,n) with $m \leq 3$ and $n \leq 3$ (resp. degree n with $n \leq 3$). So we can apply computer algebra techniques to the study of these surfaces; our main target being the description of their singular loci. In [11], we obtained these results for the case when $m = 1$ and $n = 2$, that we briefly recall in subsection 3.3. We also quickly recall the case of degree 2 which is well known, and treat the case of degree 3. Let us emphasize that the complexity of these classes of surfaces highly increases with the integers m and n: For $m = 1$ and $n = 2$, we found that the family of surfaces depend on 2 moduli (observe that 2 is the difference between the number of coefficients 24 describing ϕ and the fix number 22). While for $m = 3$ and $n = 3$, the number of these coefficients is 64 and we find $64 - 22 = 42$ moduli. The implicit total degree of a $(1,2)$-surface is generically 4 while it is 18 for a $(3,3)$-surface. Also the degree of the singular locus of a generic $(1,2)$-surface is 3 while it is 132 for a $(3,3)$-surface. These observations point out that the algebraic study of these surfaces and their geometry is a challenging question not only for the applications in Computer Aided Geometric Design and Solid Modeling but also for the Computer Algebra community. Indeed, such a big number of parameters (moduli) and high degrees require expertise to represent, compute, store and re-use the involved polynomials, formulas and drawing devices. In this paper we describe a method for computing the number of moduli and the corresponding normal forms with the optimal number of coefficients, and then we use these normal forms to get geometric invariants and features of the considered surfaces. We treat in details the cases of bidegree $(1,2),(1,3),(2,2),(2,3)$, and degrees 2 and 3. All are of implicit degree smaller than 12, but the geometry is yet rather complicated. Next interesting questions will be the study of non-generic complex cases and the trace in the projective real space $\mathbb{P}^3(\mathbb{R})$ of these surfaces and their singular loci which is truly important for applications.

The considered surfaces $\mathcal{S}$ can be seen either as the total space of a family of curves $\mathcal{C}_t$ or of curves $\mathcal{D}_u$: $\mathcal{S} = \cup_t \mathcal{C}_t$ or $\mathcal{S} = \cup_u \mathcal{D}_u$. Both interpretations will be useful to describe the geometry of $\mathcal{S}$.

When we restrict our setting to the affine case, we let

$$x = \frac{\phi_1}{\phi_0}, \quad y = \frac{\phi_2}{\phi_0}, \quad z = \frac{\phi_3}{\phi_0},$$

and we consider them as functions of $\frac{t}{s}$ and $\frac{u}{v}$, that we rename t and u.

The paper is organized as follows. In section 2, we describe the group action on the parameterizations of surfaces that we will study, and we introduce the notion of a normal form. We recall also some known results. In section 3, we give normal forms of tensor product surfaces of bidegrees (m, n) with m and n smaller than 3, and triangular surfaces of degrees at most 3. In section 4, we describe some methods for converting the studied parameterizations into their implicit equations in order to study their singular loci. In the last section, we discuss the local singularities with respect to parameterizations and double point loci of the studied surfaces.

2 Algebro-geometric preliminaries

In this section, we will introduce the basic notions and notations together with some known results.

2.1 Parameter space and genericity

We consider two kinds of parametrized surfaces in the complex projective 3-space $\mathbb{P}^3 = \mathbb{P}^3(\mathbb{C})$. The first one, often called tensor product of curves, is the image of a rational map

$$\phi : \mathbb{P}^1 \times \mathbb{P}^1 \to \mathbb{P}^3,$$

defined by 4 bihomogeneous polynomials $\phi_0, \phi_1, \phi_2, \phi_3$, of bidegree (m, n) in the couple of homogeneous variables $(t : s; u : v)$ (this means that these polynomials are homogeneous in $(t : s)$ of degree m and homogeneous in $(u : v)$ of degree n. Each such ϕ_k is the sum of $(m+1)(n+1)$ terms $a_{k,i,j} t^i s^{m-i} u^j v^{n-j}$, $i = 0, \ldots, m$, $j = 0, \ldots, n$.

The complex coefficients $a_{k,i,j}$, $k = 0, \ldots, 3$, $i = 0, \ldots, m$, $j = 0, \ldots, n$, are the entries of a $4 \times (m+1)(n+1)$ matrix $\mathcal{A}$. This nonzero matrix is defined up to a multiplicative scalar factor in $\mathbb{C}$. So the corresponding parameter space $\mathcal{P}$ is $\mathbb{P}(\mathbb{C}^{4(m+1)(n+1)})$.

We say that a surface represented by a matrix $\mathcal{A}$ is generic, with respect to a property (or an algorithm) if the set of matrices which satisfy this property (or algorithm) form a dense set in $\mathcal{P}$; in our context it will be a Zariski open subset of $\mathcal{P}$. The image $\mathcal{S}$ of the map ϕ is a surface in $\mathbb{P}^3$ of degree $2mn$. Therefore $\mathcal{S}$ admits, in the generic case (more precisely where it is base-point free), an implicit homogeneous equation $F(X, Y, Z, T) = 0$ of total degree $2mn$. An important task, called the implicitization problem, consists in calculating this polynomial F from the parametric representation ϕ (see section 4).

The second kind of surfaces, often called triangular, is the image of a rational map
$$\phi : \mathbb{P}^2 \to \mathbb{P}^3,$$
defined by 4 homogeneous polynomials $\phi_0, \phi_1, \phi_2, \phi_3$, of degree n in three homogeneous variables $(t : u : v)$. Each such ϕ_k is the sum of $\frac{(n+1)(n+2)}{2}$ terms $a_{k,i,j} t^i u^j v^{n-i-j}$. The coefficients $a_{k,i,j}$ are the entries of a $4 \times \frac{(n+1)(n+2)}{2}$ matrix $\mathcal{A}$. The corresponding parameter space $\mathcal{P}$ is $\mathbb{P}(\mathbb{C}^{2(n+1)(n+2)})$. Generically (more precisely where it is base-point free), the image $\mathcal{S}$ of the map ϕ is a surface in $\mathbb{P}^3$ of degree n^2.

2.2 Group action and normal forms

We consider an equivalence relation between parametrized surfaces with prescribed bidegree (resp. degree), induced by the group of projective changes of coordinates in the source and in the target of such a ϕ. This action respects the degrees. More precisely, two surfaces $\mathcal{S}_1$ and $\mathcal{S}_2$ defined by
$$\phi_1 : \mathbb{P}^1 \times \mathbb{P}^1 \to \mathbb{P}^3 \quad \text{and} \quad \phi_2 : \mathbb{P}^1 \times \mathbb{P}^1 \to \mathbb{P}^3$$
are equivalent if and only if there exist $g_1 \in \mathrm{PGL}(2, \mathbb{C}) \times \mathrm{PGL}(2, \mathbb{C})$ and $g_2 \in \mathrm{PGL}(4, \mathbb{C})$ such that $\phi_2 = g_2 \circ \phi_1 \circ g_1$. The set
$$\mathcal{G} = \mathrm{PGL}(2, \mathbb{C}) \times \mathrm{PGL}(2, \mathbb{C}) \times \mathrm{PGL}(4, \mathbb{C})$$
is a Lie group of dimension $\dim \mathcal{G} = 3 + 3 + 15 = 21$. Note that a projective change of coordinates on $\mathbb{P}^1$ (also called a homography) is determined by the image of 3 points, while it is determined by 4 points for $\mathbb{P}^2$, and 5 points for $\mathbb{P}^3$.

For surfaces with triangular parametrization $\phi : \mathbb{P}^2 \to \mathbb{P}^3$, the corresponding group of projective changes of coordinates is the Lie group $\mathcal{G} = \mathrm{PGL}(3, \mathbb{C}) \times \mathrm{PGL}(4, \mathbb{C})$ which has dimension $\dim \mathcal{G} = 8 + 15 = 23$.

For both kinds of parametric surfaces, $\mathcal{G}$ acts (as described in subsection 2.1) on the parameter space $\mathcal{P}$; so $\mathcal{P}$ is an union of orbits. Each such orbit is an algebraic subset of $\mathcal{P}$ and its dimension is smaller or equal to $\dim(\mathcal{G})$; indeed when we fix ϕ_1 in the equivalence relation above, all the ϕ_2 form a parametrized variety in $\mathcal{P}$, image of $\mathcal{G}$. Moreover, it can be proved that the union of orbits of maximal dimension (denoted by N_1) is dense in $\mathcal{P}$. This union corresponds to what we called the generic case. The quotient space $\mathcal{P}/\mathcal{G}$ might be complicated, but if we let $N = \dim(\mathcal{P}) - N_1$, so $N \geq \dim(\mathcal{P}) - \dim(\mathcal{G})$, the family of orbits of maximal dimension (the generic case) is described by N parameters, called moduli. We expect that N is indeed equal to $\dim(\mathcal{P}) - \dim(\mathcal{G})$ and we will prove this equality for each case that we will consider and study.

We call a normal form for a parametrization ϕ, a "simple" representative ϕ_0 of the equivalence class of ϕ modulo the action of $\mathcal{G}$ (here simple means that ϕ_0 depends on a few number of coefficients). The parametrization ϕ_0 bears a special format that we specify for each bidegree or degree surface considered below. The notions of normal form and moduli go back to Riemann. It was also popularized by Arnold (in singular theory) who provided long lists of normal forms of singularities of germs of functions (see [4]).

2.3 Singular and double point loci

We suppose that ϕ is one-to-one on an open subset of its source set. Then a double point of ϕ is a couple of different values (t_1, u_1) and (t_2, u_2) in $\mathbb{P}^1 \times \mathbb{P}^1$ such that $\phi(t_1, u_1) = \phi(t_2, u_2)$. So we define the double point locus $\mathcal{D}$ of ϕ to be the closure in $\mathbb{P}^1 \times \mathbb{P}^1$ of the set of double points:

$$\mathcal{D} = \overline{\{(t, u) \in \mathbb{P}^1 \times \mathbb{P}^1 : \exists (t_1, u_1) \neq (t, u), \phi(t_1, u_1) = \phi(t, u)\}},$$

and similarly when we replace $\mathbb{P}^1 \times \mathbb{P}^1$ by $\mathbb{P}^2$. The double point locus $\mathcal{D}$ is generically a curve.

The image by ϕ of $\mathcal{D}$ is denoted by $\mathcal{F}$, it is a curve in $\mathbb{P}^3$ drawn on $\mathcal{S}$. The points of $\mathcal{F}$ are singular points of $\mathcal{S}$, but they are not necessarily the only singular points of $\mathcal{S}$. The surface $\mathcal{S}$ may also contain isolated singularities which can be detected locally with respect to the parametrization. Indeed one has just to express that the differential $D_{(t,u)}\phi : \mathbb{C}^2 \to \mathbb{C}^3$ of ϕ at the point (t, u) is not injective, i.e. the corresponding matrix has rank at most 1. Moreover, when we cut $\mathcal{S}$ by a generic plane Π of $\mathbb{P}^3$ we get a curve whose only singularities are nodes, and these nodes are the intersection point of $\mathcal{F}$ with Π (see section 5).

2.4 Enumerative properties

Intersection theory in Algebraic Geometry allows to derive formulas to compute the degrees of various algebraic subsets of interest. Let us explain how one can, in the generic case, obtain the degrees of the subvarieties $\mathcal{S}, \mathcal{D}$ and $\mathcal{F}$ introduced in subsection 2.3.

First we recall how to get the implicit degree of $\mathcal{S}$: This is the number of intersection points between $\mathcal{S}$ and a generic line in $\mathbb{P}^3$, such a line is the intersection of two generic planes Π_1 and Π_2. Recall that ϕ is generically one-to-one on $\mathcal{S}$, $\phi^{-1}(\Pi_1)$ and $\phi^{-1}(\Pi_2)$ are two curves in $\mathbb{P}^1 \times \mathbb{P}^1$ of bidegree (m, n) (resp. in $\mathbb{P}^2$ of degree n). Bézout's theorem provides the number of intersection points between these curves, hence the aimed implicit degree.

We recall that Bézout's theorem in $\mathbb{P}^2$ says that two curves of degrees n_1 and n_2 without common component have $n_1 n_2$ intersection points (counted with multiplicities). In $\mathbb{P}^1 \times \mathbb{P}^1$, two curves of bidegrees (m_1, n_1) and (m_2, n_2) without common component have $m_1 n_2 + m_2 n_1$ intersection points (see [20] for details).

Now the reasoning for calculating the degree of $\mathcal{F}$ goes as follows: The intersection of $\mathcal{S}$ with a generic plane Π is a curve $\mathcal{C}$ of known degree d with δ nodes and genus g. These numbers are linked by the classical following formula

$$g = \frac{(d-1)(d-2)}{2} - \delta. \tag{2.1}$$

But g is also the genus of the smooth curve $\phi^{-1}(\mathcal{C})$ which is of bidegree (m,n) in $\mathbb{P}^1 \times \mathbb{P}^1$ (resp. of degree n in $\mathbb{P}^2$) and can be computed in a different way, so we complete the computation in each case (see section 5).

3 Number of moduli and normal forms

We saw in subsection 2.2 that the dimension of the Lie group

$$\mathcal{G} = \text{PGL}(2,\mathbb{C}) \times \text{PGL}(2,\mathbb{C}) \times \text{PGL}(4,\mathbb{C})$$

acting on the parameter space $\mathcal{P}$ is $\dim \mathcal{G} = 21$. Each orbit has at most dimension 21, therefore the number of moduli required to describe the class of generic surfaces is at least $\dim(\mathcal{P}) - 21$. The parameter space $\mathcal{P}$ of surfaces of bidegree (m, n) has dimension $4(m+1)(n+1)-1$, so this number of moduli is at least $4(m+1)(n+1) - 22$.

In order to prove that it is exactly this number for $m \leq 3$ and $n \leq 3$, we will exhibit in each (generic) case a projective change of coordinates in $\mathcal{G}$ such that it reduces the expression of the parametrization ϕ to a normal form depending only on this bound on the number of moduli. Our proof, in each case, will rely on the specific geometry attached to the considered (m, n)-surface.

We believe that this property is also true for any $(m,n) \in \mathbb{N} \times \mathbb{N}$, but this would require a more abstract proof; it is beyond the scope of this paper and will be addressed elsewhere.

Our approach also applies to recover the generic cases of parameterizations $\phi : \mathbb{P}^2 \to \mathbb{P}^3$ of degrees 2 and 3 that we briefly present (see subsections 3.1 and 3.2).

3.1 Quadrics and Roman surface

- A parametrization of bidegree $(1, 1)$, with 16 coefficients, is given by a square matrix $\mathcal{A}$ of type 4×4. Generically, $\mathcal{A}$ is invertible so defines a change of coordinates in $\mathbb{P}^3$, after performing this change we obtain the normal form:

$$X = tu \, , \ Y = t \, , \ Z = u \, , \ T = 1.$$

Here the number of moduli is 0. We deduce the implicit equation $XT - YZ = 0$ of a quadric.

- A parametrization $\phi : \mathbb{P}^2 \to \mathbb{P}^3$ of degree 2 is defined by $4 \times 6 = 24$ coefficients in the monomials basis of $\mathbb{C}[t, u, v]$. Its image S is generically a surface of degree 4 in $\mathbb{P}^3$, and the image of a generic line l in $\mathbb{P}^2$ is a conic $\mathcal{C}_l$ in $\mathbb{P}^3$ contained in a plane Π_l. This plane cuts the surface S along a curve of degree 4, so $\mathcal{C}_l$ and another conic. The line l varies in $\mathbb{P}^{2*} \sim \mathbb{P}^2$. We can easily write a system of algebraic equations to express the geometric constraint that $\Pi_l \cap S$ consists of twice the conic $\mathcal{C}_l$. It turns out that this system of equations has generically 4 distinct solutions l_1, l_2, l_3, l_4 corresponding to 4 distinct planes in general position. So they define a change of coordinates in $\mathbb{P}^3$. After performing this change of coordinates in $\mathbb{P}^3$, we get the parametrization:

$$X = l_1^2, \ Y = l_2^2, \ Z = l_3^2, \ T = l_4^2.$$

Performing a change of coordinates in $\mathbb{P}^2$ we can assume that

$$l_1 = u_1 + u_2 + u_3, l_2 = u_1 - u_2 + u_3, l_3 = u_1 - u_2 - u_3, l_4 = u_1 + u_2 - u_3.$$

If we set

$$\begin{cases} X_1 = \frac{1}{4}(X + T) - \frac{1}{8}(X + Y + Z + T) \\ Y_1 = \frac{1}{4}(X + Y) - \frac{1}{8}(X + Y + Z + T) \\ Z_1 = \frac{1}{4}(X + Z) - \frac{1}{8}(X + Y + Z + T) \\ T_1 = \frac{1}{4}(X + Y + Z + T) \end{cases}$$

we obtain the normal form:

$$X_1 = u_1 u_2, \ Y_1 = u_1 u_3, \ Z_1 = u_2 u_3, \ T_1 = u_1^2 + u_2^2 + u_3^2.$$

Here, the number of moduli is $N = 0$.

This is the classical form of the Roman surface of Steiner, which admits the simple implicit equation $X_1^2 Y_1^2 + X_1^2 Z_1^2 + Y_1^2 Z_1^2 - X_1 Y_1 Z_1 T_1 = 0$.

3.2 Case of a parametrization of degree 3

A parametrization $\phi : \mathbb{P}^2 \to \mathbb{P}^3$ of degree 3 is defined by a matrix $\mathcal{A}$ of $4 \times 10 = 40$ coefficients. The image S of ϕ in $\mathbb{P}^3$ is a surface of degree 9. The group $\mathcal{G} = \mathrm{PGL}(3, \mathbb{C}) \times \mathrm{PGL}(4, \mathbb{C})$ acting on the parameter space $\mathcal{P}$ has dimension $\dim \mathcal{G} = 8 + 15 = 23$. So we expect at least $39 - 23 = 16$ moduli.

Proposition 1. *The number of moduli is exactly the expected one $N = 16$.*

Proof. We first observe that the image by ϕ of a generic line $l \in \mathbb{P}^{2*}$ is a cubic curve $\mathcal{C}_l$ in $\mathbb{P}^3$. The vanishing of a determinant of size 4 provides an algebraic condition on l which expresses that $\mathcal{C}_l$ is a plane cubic curve; if so let Π_l be the plane containing this curve. This algebraic condition defines a curve $\mathcal{E}$

in $\mathbb{P}^{2*}$ which is generically of degree 6. For any $l \in \mathcal{E}$, the plane Π_l cuts $\mathcal{S}$ along a curve of degree 9 consisting of $\mathcal{C}_l$ and a sextic. This sextic is the image via ϕ of a quadric Q_l in the parameter space. This quadric decomposes into a product of two lines if and only if its determinant vanishes. This happens generically for a finite number (greater than 4) of cases. We choose 4 such lines and denote them by l_1, l_2, l_3, l_4. The corresponding quadratic forms can be written $Q_1 = l_5 l_6, Q_2 = l_7 l_8, Q_3 = l_9 l_{10}, Q_4 = l_{11} l_{12}$, for some linear forms $l_i, i = 5, \ldots, 12$.

Performing a projective change of coordinates in $\mathbb{P}^2$, we can fix 4 among these 12 linear forms $l_1, \ldots, l_{12}$. Let us choose $l_4 = t, l_{11} = u, l_{12} = v$, and $l_{10} = t + u + v$. We also choose a change of coordinates in $\mathbb{P}^3$ such that the plane containing the cubic curve attached to l_1 (resp. l_2, l_3, l_4) is $X = 0$ (resp. $Y = 0, Z = 0, T = 0$). Then we obtain the following normal form for the parameterization ϕ of degree 3:

$$X = l_1 l_5 l_6 \ , \ Y = l_2 l_7 l_8 \ , \ Z = l_3 l_9 (t + u + v) \ , \ T = tuv.$$

Generically, we can write the others $l_i = t + a_i u + b_i v, i = 1, 2, 3, 5, 6, 7, 8, 9$. So the moduli are a_i, b_i and their number $N = 16$.

3.3 Case of a surface of bidegree $(1, 2)$

A generic surface $\mathcal{S}$ of bidegree $(1, 2)$ in $\mathbb{P}^3$ has implicit degree 4. So a plane Π_t containing a conic $\mathcal{C}_t$ of the family describing $\mathcal{S}$ will cut $\mathcal{S}$ along a (plane) curve of degree 4 and bidegree $(1, 2)$ in (t, u). Because of the bidegree $(1, 2)$ of the parameterization, this curve is formed by $\mathcal{C}_t$ and a pair of lines $\mathcal{L}_{u_1}$ and $\mathcal{L}_{u_2}$ belonging to the family of lines $\mathcal{L}_u$ which also describe $\mathcal{S}$.

Lemma 2. *For a generic surface $\mathcal{S}$ of bidegree $(1, 2)$ in $\mathbb{P}^3$ there are 4 distinct values of t in $\mathbb{P}^1$, say t_1, t_2, t_3, t_4, such that the corresponding distinct planes Π_{t_i} cut $\mathcal{S}$ along a conic $\mathcal{C}_{t_i}$ and a double line $\mathcal{L}_{u_i}$, for $i = 1, 2, 3, 4$.*

Proof. We give the sketch of the proof, for more details see [11]. The family of planes Π_t describing $\mathcal{S}$ is defined by 3×3 minors with degree 1 entries in t. Substituting X, Y, Z, T in this equation by the parametrization ϕ, we get a polynomial condition on t such that there is a double root in u. A closer inspection shows that, generically there are exactly 4 such values for t.

Lemma 3. *There exists a unique element in $\mathcal{G}$ which sends the couples $(t_i, u_i), i = 1, 2, 3, 4$, on $(0, 0), (1, 1), (\infty, \infty), (a, b)$, and the planes Π_{t_i} on the coordinates planes $X = 0, Y = 0, Z = 0, T = 0$. Moreover, a (resp. b) is the cross ratio of (t_1, t_2, t_3, t_4) (resp. (u_1, u_2, u_3, u_4)).*

The proof is immediate. We deduce the following proposition:

Proposition 4. *After a suitable change of coordinates in $\mathcal{G}$, a generic surface of bidegree $(1, 2)$ is the image of a parametrization of the following type:*

$$X = tu^2 \ , \ Y = (t-1)(u-1)^2 \ , \ Z = (t-a)(u-b)^2 \ , \ T = 1.$$

The parameterization given by proposition 4 is the normal form of a generic parametrization of a surface of bidegree $(1,2)$. It depends on two moduli a and b which are different from $0, 1, \infty$. So here the number of moduli is $N = 2$, the expected one.

3.4 Case of a surface of bidegree $(1,3)$

A generic surface of bidegree $(1,3)$ is described by a family of cubic curves $\mathcal{C}_t$. We will show that this family contains 4 planar cubic curves $\mathcal{C}_{t_i}, i = 1, 2, 3, 4$, contained in planes $\Pi_{t_i}, i = 1, 2, 3, 4$, which define a projective change of coordinates in $\mathbb{P}^3$ suitable to our purpose.

Lemma 5. *Let us fix t. The curve $\mathcal{C}_t$ is defined in $\mathbb{P}^3$ by*

$$\begin{cases} X = (a_1 t + b_1)u^3 + (c_1 t + d_1)u^2 + (e_1 t + f_1)u + (g_1 t + h_1) \\ Y = (a_2 t + b_2)u^3 + (c_2 t + d_2)u^2 + (e_2 t + f_2)u + (g_2 t + h_2) \\ Z = (a_3 t + b_3)u^3 + (c_3 t + d_3)u^2 + (e_3 t + f_3)u + (g_3 t + h_3) \\ T = (a_4 t + b_4)u^3 + (c_4 t + d_4)u^2 + (e_4 t + f_4)u + (g_4 t + h_4) \end{cases}$$

where $a_i, b_i, c_i, d_i, e_i, f_i, g_i, h_i$ are complex coefficients; it is planar when the determinant of the corresponding matrix

$$\begin{pmatrix} a_1 t + b_1 & c_1 t + d_1 & e_1 t + f_1 & g_1 t + h_1 \\ a_2 t + b_2 & c_2 t + d_2 & e_2 t + f_2 & g_2 t + h_2 \\ a_3 t + b_3 & c_3 t + d_3 & e_3 t + f_3 & g_3 t + h_3 \\ a_4 t + b_4 & c_4 t + d_4 & e_4 t + f_4 & g_4 t + h_4 \end{pmatrix}$$

vanishes. Generically, this happens for 4 distinct values t_1, t_2, t_3, t_4 of t.

As in subsection 3.3, we perform changes of coordinates in $\mathbb{P}^1 \times \mathbb{P}^1$ and $\mathbb{P}^3$ in order to send the values t_1, t_2, t_3, t_4 on $0, 1, \infty$ and a, and the corresponding planes $\Pi_{t_1}, \Pi_{t_2}, \Pi_{t_3}, \Pi_{t_4}$ on $X = 0, Y = 0, Z = 0, T = 0$. Now each of these planes cuts the surface $\mathcal{S}$ along a curve of degree 6 (which is the implicit degree of $\mathcal{S}$) consisting of a planar cubic curve and three lines. Hence $\mathcal{S}$ is determined by these 4 cubic curves and these 12 lines. From that description we deduce the following normal form of the parametrization ϕ of $\mathcal{S}$ in lemma 5:

Proposition 6. *After a change of coordinates in $\mathcal{G}$, a generic surface of bidegree $(1,3)$ is the image of a parametrization of the following type which depends on 10 moduli $a, b, c, d, e, f, g, h, i, j \in \mathbb{C}$:*

$$\begin{cases} X = t(u^3 + bu^2 + cu + d) \\ Y = (t-1)(u^3 + eu^2 + fu + g) \\ Z = (t-a)(u^3 + hu^2 + iu + j) \\ T = u^2 - 1 \end{cases}$$

So the number of moduli is $N = 10$, the expected one.

Proof. We follow the same reasoning as above; we choose a change of coordinates in u such that the roots in $\mathbb{P}^1$ of the last polynomial are $\infty, 1, -1$.

The number of moduli in proposition 6 is $N = 10 = 31 - 21$.

As $(n+1)(3+1) = (n-1+1)(3+1) + 4$, the previous argument generalizes to all classes of surfaces of bidegrees $(n, 3)$. To illustrate this claim, we now treat the cases $(2, 3)$ and $(3, 3)$.

3.5 Cases of surfaces of bidegrees $(n, 3)$

Lemma 5 immediately generalizes to the case of a surface of bidegree $(n, 3)$, as it is described by a family of cubic curves $\mathcal{C}_t$. Such a curve $\mathcal{C}_t$ is defined by a parametrization where each component X, Y, Z, T is a sum of 4 monomials $u^3, u^2, u, 1$ with polynomial coefficients of degree n. So $\mathcal{C}_t$ is a planar curve if the determinant of the corresponding 4×4 matrix vanishes. Generically, this happens for $4n$ distinct values of t; and we choose 4 of them that we denote by t_1, t_2, t_3, t_4. We follow the same treatment as in subsection 3.4 and we have:

Proposition 7. *After a change of coordinates in $\mathcal{G}$, a generic surface of bidegree $(2, 3)$ is the image of a parametrization of the following type which depends on $N = 26$ moduli, the expected number:*

$$\begin{cases} X = t(tu^3 + a_1tu^2 + b_1tu + c_1t + d_1u^3 + e_1u^2 + f_1u + g_1) \\ X = (t-1)(tu^3 + a_2tu^2 + b_2tu + c_2t + d_2u^3 + e_2u^2 + f_2u + g_2) \\ Z = (t-\alpha)(tu^3 + a_3tu^2 + b_3tu + c_3t + d_3u^3 + e_3u^2 + f_3u + g_3) \\ T = t(u^2 - 1) + d_4u^3 + e_4u^2 + f_4u + g_4. \end{cases}$$

Remark 8. Similarly a generic surface of bidegree $(3, 3)$ is the image of a parametrization which depends on exactly the expected number of moduli $N = 42$. One easily sees the pattern for the case $(n, 3)$.

3.6 Case of a surface of bidegree $(2, 2)$

A generic surface $\mathcal{S}$ of bidegree $(2, 2)$ has implicit degree 8 and is formed by two families of conics that we denote by $\mathcal{C}_t$ and $\mathcal{D}_u$. The curve $\mathcal{C}_t$ is contained in a plane Π_t and $\mathcal{D}_u$ in a plane P_u. The two planes contain the point $M(t, u)$ whose coordinates are given by the parametrization. Moreover, if we suppose that t and u are distinct from 0 and ∞, then Π_t is defined by 3 points $M(t, u), M(t, 0), M(t, \infty)$ and P_u is defined by $M(t, u), M(0, u), M(\infty, u)$. Then the planes Π_t and P_u coincide if and only if the two determinants

$$\begin{cases} \det\bigl(M(t, u), M(t, 0), M(t, \infty), M(0, u)\bigr) \\ \det\bigl(M(t, u), M(t, 0), M(t, \infty), M(\infty, u)\bigr) \end{cases}$$

vanish for $t \neq 0, t \neq \infty, u \neq 0, u \neq \infty$. In order to exclude these values, we consider the following determinants

$$\begin{cases} F_1(t,u) = \det\left(M(t,u), \frac{M(t,0)-M(t,u)}{u}, M(t,\infty), \frac{M(0,u)-M(t,u)}{t}\right) \\ F_2(t,u) = \det\left(M(t,u), \frac{M(t,0)-M(t,u)}{u}, M(t,\infty), M(\infty,u)\right) \end{cases}$$

and the affine common solutions of these two equations which are generically of bidegrees $(7,5)$ and $(6,5)$ in (t,u). These numbers are not symmetric because we assumed that $M(t,u), M(t,0), M(t,\infty)$ are linearly independent. Generically, these two equations have a finite number of couples (t_i, u_i) of solutions. An easy check shows that this number is at least 4, so we choose 4 values $(t_1, u_1), (t_2, u_2), (t_3, u_3), (t_4, u_4)$ such that for $i = 1, 2, 3, 4, \mathcal{C}_{t_i}$ and $\mathcal{D}_{t_i}$ are contained in the same plane P_{t_i}. Performing a change of coordinates in $\mathcal{G}$, we can assume that these 4 planes are $X = 0, Y = 0, Z = 0, T = 0$, and moreover that

$$(t_1, u_1) = (0,0), \quad (t_2, u_2) = (1,1), \quad (t_3, u_3) = (\alpha, \beta), \quad (t_4, u_4) = (\infty, \infty).$$

So we have the following normal form:

Proposition 9. *After a change of coordinates in $\mathcal{G}$, a generic surface of bidegree $(2,2)$ is the image of a parametrization of the following type which depends on $N = 14$ moduli:*

$$\begin{cases} X = tu(tu + a_1 t + b_1 u + c_1) \\ Y = (t-1)(u-1)(tu + a_2 t + b_2 u + c_2) \\ Z = (t-\alpha)(u-\beta)(tu + a_3 t + b_3 u + c_3) \\ T = tu + a_4 t + b_4 u + c_4. \end{cases}$$

The number $N = 14$ in proposition 9 is exactly the number $14 = 35 - 21$ of expected moduli for a $(2,2)$-surface.

4 Implicit equation and singular locus

Important problems in Computer Aided Geometric Design, such as the determination of the singular locus of a surface, the intersection problem of surfaces are better treated via an implicit equation. So in this section we discuss the implicitization problem of the different surfaces studied above (i.e. find methods and algorithms to convert parametric representations of these rational surfaces into implicit ones).

Let ϕ be a parametrization of bidegree (m,n) of the surface $\mathcal{S}$. To determine its implicit equation, we eliminate the variables (t,u) in the following polynomial system:

$$\begin{cases} F_1(t,u) = \phi_1(t,u) - x\,\phi_0(t,u) \\ F_2(t,u) = \phi_2(t,u) - y\,\phi_0(t,u) \\ F_3(t,u) = \phi_3(t,u) - z\,\phi_0(t,u). \end{cases} \quad (4.1)$$

As the degree of the implicit equation $f(x,y,z) = 0$ of $\mathcal{S}$ is generically $2mn$, the polynomial f contains $\frac{(2mn+1)(2mn+2)(2mn+3)}{6}$ monomials. Its expression in the monomial basis is huge and without apparent structure, so we want to give $f(x,y,z)$ in a more compact way using a matrix formulation. There are several methods to obtain such a matrix.

Determinantal Sylvester type matrices can be given using the multigraded resultant studied in [17], [21], [10]. If d and δ are two positive integers, we denote by $S(d,\delta)$ the vector space over $\mathbb{C}$ of polynomials in (t,u) of degree at most d (resp. δ) in t (resp. u), and let $\mathcal{M}$ be the linear map

$$\mathcal{M} : S(m-1, 2n-1)^3 \to S(2m-1, 3n-1)$$
$$(Q_1, Q_2, Q_3) \mapsto Q_1 F_1 + Q_2 F_2 + Q_3 F_3. \tag{4.2}$$

The matrix M of this map in the monomial bases is sparse and its nonzero entries are the coefficients of F_1, F_2, F_3. The size of M is $6mn \times 6mn$ and its determinant is exactly the implicit equation of $\mathcal{S}$. Similar matrices which give $f(x,y,z)$ can be obtained with slightly different source set and target set in $\mathcal{M}$ (see [21], [10]).

It is possible to use another matrix of smaller size to eliminate (t,u) in the system (4.1), namely the bezoutian matrix of F_1, F_2, F_3 viewed as polynomials in the variables (t,u): Let (t_1, u_1) be other parameters, we recall that the bezoutian of F_1, F_2, F_3 is the polynomial

$$\Theta_{(F_1,F_2,F_3)}(t,u,t_1,u_1) = \frac{1}{(t_1-t)(u_1-u)} \begin{vmatrix} F_1(t,u) & F_1(t_1,u) & F_1(t_1,u_1) \\ F_2(t,u) & F_2(t_1,u) & F_2(t_1,u_1) \\ F_3(t,u) & F_3(t_1,u) & F_3(t_1,u_1) \end{vmatrix}.$$

It is clear that $\Theta_{(F_1,F_2,F_3)}(t,u,t_1,u_1) \in \mathbb{C}[t,u,t_1,u_1]$. If we develop it as a polynomial in (t,u,t_1,u_1), we obtain

$$\Theta_{(F_1,F_2,F_3)}(t,u,t_1,u_1) = \sum_{\alpha=(\alpha_1,\alpha_2), \beta=(\beta_1,\beta_2)} c_{\alpha,\beta}(x,y,z)\, t^{\alpha_1} u^{\alpha_2} t_1^{\beta_1} u_1^{\beta_2},$$

with $c_{\alpha,\beta}(x,y,z) \in \mathbb{C}[x,y,z]$. The bezoutian matrix of F_1, F_2, F_3 is the matrix $\mathrm{B}_{(F_1,F_2,F_3)} = \left(c_{\alpha,\beta}(x,y,z)\right)_{\alpha,\beta}$ (with respect to any order on the monomials $t^{\alpha_1} u^{\alpha_2}$ and $t_1^{\beta_1} u_1^{\beta_2}$).

The coefficients $c_{\alpha,\beta}(x,y,z)$ of $\mathrm{B}_{(F_1,F_2,F_3)}$ are more complicated than those of the matrix M of the linear map (4.2), but the size of $\mathrm{B}_{(F_1,F_2,F_3)}$ is $2mn \times 2mn$ and its determinant is the implicit equation of $\mathcal{S}$ ([12, 7]). Other matrices which combine Sylvester type and Bézout type coefficients can be used (for more details, see [10]).

Other techniques to implicitize rational surfaces exist in the literature such as the moving quadrics method which gives a matrix of smaller size namely mn, but requires the computation of mn independent syzygies between the elements of I^2 (where I is the ideal generated by $\phi_0, \phi_1, \phi_2, \phi_3$), in bidegree $(m-1, n-1)$ (see [9]). In the case of ruled surfaces (i.e. surfaces of bidegree

$(1, n))$, we can obtain the implicit equation of $\mathcal{S}$ as the resultant in one variable of two syzygies in $\mathbb{C}[u]^4$ between the generators of I (called a μ-basis of the parametrization ϕ, for more details see [6] and [5]).

Example 10. To illustrate the implicitization problem of a rational surface, we compute the implicit equation of a surface $\mathcal{S}$ given by a parametrization of bidegree $(m, n) = (1, 3)$. This method can be used for every $(m, n) \in \mathbb{N}^* \times \mathbb{N}^*$. Using the normal form in proposition 6 and the determinantal Sylvester type matrix given by (4.2), the implicit equation is the determinant of the following matrix:

$$\begin{pmatrix}
-x & 0 & x & 0 & 0 & 0 & 0 & 0 & 0 & -d & -c & -b & -1 & 0 & 0 & 0 & 0 & 0 \\
0 & -x & 0 & x & 0 & 0 & 0 & 0 & 0 & 0 & -d & -c & -b & -1 & 0 & 0 & 0 & 0 \\
0 & 0 & -x & 0 & x & 0 & 0 & 0 & 0 & 0 & 0 & -d & -c & -b & -1 & 0 & 0 & 0 \\
0 & 0 & 0 & -x & 0 & x & 0 & 0 & 0 & 0 & 0 & 0 & -d & -c & -b & -1 & 0 & 0 \\
0 & 0 & 0 & 0 & -x & 0 & x & 0 & 0 & 0 & 0 & 0 & 0 & -d & -c & -b & -1 & 0 \\
0 & 0 & 0 & 0 & 0 & -x & 0 & x & 0 & 0 & 0 & 0 & 0 & 0 & -d & -c & -b & -1 \\
y_2 & f & y_1 & 1 & 0 & 0 & 0 & 0 & -g & -f & -e & -1 & 0 & 0 & 0 & 0 & 0 & 0 \\
0 & y_2 & f & y_1 & 1 & 0 & 0 & 0 & 0 & -g & -f & -e & -1 & 0 & 0 & 0 & 0 & 0 \\
0 & 0 & y_2 & f & y_1 & 1 & 0 & 0 & 0 & 0 & -g & -f & -e & -1 & 0 & 0 & 0 & 0 \\
0 & 0 & 0 & y_2 & f & y_1 & 1 & 0 & 0 & 0 & 0 & -g & -f & -e & -1 & 0 & 0 & 0 \\
0 & 0 & 0 & 0 & y_2 & f & y_1 & 1 & 0 & 0 & 0 & 0 & -g & -f & -e & -1 & 0 & 0 \\
0 & 0 & 0 & 0 & 0 & y_2 & f & y_1 & 1 & 0 & 0 & 0 & 0 & -g & -f & -e & -1 \\
z_2 & ai & z_1 & a & 0 & 0 & 0 & 0 & -j & -i & -h & -1 & 0 & 0 & 0 & 0 & 0 & 0 \\
0 & z_2 & ai & z_1 & a & 0 & 0 & 0 & 0 & -j & -i & -h & -1 & 0 & 0 & 0 & 0 & 0 \\
0 & 0 & z_2 & ai & z_1 & a & 0 & 0 & 0 & 0 & -j & -i & -h & -1 & 0 & 0 & 0 & 0 \\
0 & 0 & 0 & z_2 & ai & z_1 & a & 0 & 0 & 0 & 0 & -j & -i & -h & -1 & 0 & 0 & 0 \\
0 & 0 & 0 & 0 & z_2 & ai & z_1 & a & 0 & 0 & 0 & 0 & -j & -i & -h & -1 & 0 \\
0 & 0 & 0 & 0 & 0 & z_2 & ai & z_1 & a & 0 & 0 & 0 & 0 & -j & -i & -h & -1
\end{pmatrix}$$

where $y_1 = e + y, y_2 = -y + g, z_1 = ah + z, z_2 = -z + aj$. We notice that this matrix is very sparse and has a nice structure.

Once we have the homogeneous implicit equation $F(X, Y, Z, T)$ of $\mathcal{S}$, its singular locus is defined by the set of equations which expresses that all the first partial derivatives of F vanish on $\mathcal{S}$. Since the surface $\mathcal{S}$ is also given by a parametrization ϕ, we can substitute the variables X, Y, Z, T, in these partial derivatives by $\phi_0, \phi_1, \phi_2, \phi_3$, given by the normal form of $\mathcal{S}$. Then we get a system of 4 polynomial equations on $\mathbb{P}^1 \times \mathbb{P}^1$.

We will perform this computation below and solve this system for surfaces of bidegrees $(1,2), (1,3), (2,2)$, and the case of a surface of degree 3.

5 Pinch points and double point loci

Another (direct) approach would be to note that a singular point $M(t,u)$ of $\mathcal{S}$ can be detected either locally with respect to the parametrization, i.e. the cross-product $\frac{\partial M}{\partial t}(t,u) \wedge \frac{\partial M}{\partial u}(t,u)$ must vanish; or it corresponds to a double point of the parametrization, i.e., $\phi(t_1, u_1) = \phi(t_2, u_2)$ with $(t_1, u_1) \neq (t_2, u_2)$.

In our case, the computation will show that there is a curve of singular points in $\mathcal{S}$ but only a finite number of points which correspond to local singularities of the parametrization, those are called pinch points. Moreover, the double-point locus will be dense in the singular locus as this finite number of points will belong to the closure of the double point locus.

The surfaces $(1,2)$ and $(1,3)$ are ruled as they are described by a family of lines $\mathcal{L}_u$. For a $(1,2)$-surface the singular locus is a twisted cubic and each line $\mathcal{L}_u$ cuts twice this curve and crosses another line of the family at each intersection point. A $(1,3)$-surface also admits a similar, but more complicated, presentation. The geometry of $(2,2)$ and degree 3 surfaces is more intricate. However a classical result of Whitney states that the only locally stable singularities of the image of a map germ between $\mathbb{C}^2$ and $\mathbb{C}^3$ are double points and pinch points, the so called Whitney Umbrella. So the main task is to describe how many such singularities appear.

We recover in this section special cases of classical enumerative results. For (modern) rigorous proofs of these (old) formulas, we refer to [18] and more recently [19].

5.1 Surfaces of bidegree $(1,2)$

We recall here the main features of the case of $(1,2)$-surface, for more details we refer to [11]. Generically, the double point locus is dense in the singular locus which consists of a twisted cubic curve drawn on the surface. The local singularities are 4 distinct points. At each of these points the surface admits a singularity locally isomorphic to a Whitney Umbrella. This twisted cubic (in $\mathbb{P}^3$) is the image of a curve of bidegree $(2,2)$ in the parametric space. It admits also an involution (permutation of two couples of parameters which correspond to the same point in $\mathbb{P}^3$).

5.2 Surfaces of bidegree $(1,3)$

To study the case of a $(1,3)$-surface, we proceed as we have done in [11] for the case of a $(1,2)$-surface. We start with the affine parametrization in normal form given in proposition 6:

$$x = \frac{t(u^3+bu^2+cu+d)}{u^2-1}, \quad y = \frac{(t-1)(u^3+eu^2+fu+g)}{u^2-1}, \quad z = \frac{(t-a)(u^3+hu^2+iu+j)}{u^2-1}.$$

The local singularities are given by $\operatorname{rank}(\operatorname{Jac}(x,y,z)(t,u)) < 2$. This amounts to solve a polynomial system of 3 equations in (t, u) of bidegree $(1, 7)$. We can simplify further by a GCD computation. It turns out that, for generic values of the coefficients, the system has 8 solutions.

The degree of the double point locus of a generic $(1, 3)$-surface $\mathcal{S}$ can be determined as follows: As the implicit degree of $\mathcal{S}$ is 6, a generic plane section $\mathcal{C}$ of $\mathcal{S}$ is a plane curve of degree 6 with δ nodal singularities. By formula (2.1), its genus $g = 10 - \delta$. This curve $\mathcal{C}$ is the image by ϕ of a curve in $\mathbb{P}^1 \times \mathbb{P}^1$ of bidegree $(1, 3)$ and therefore is the graph of a map from $\mathbb{P}^1$ to $\mathbb{P}^1 \times \mathbb{P}^1$, so its genus is 0, and we deduce that $\delta = 10$. Then the double point locus $\mathcal{F}$ in $\mathcal{S} \subset \mathbb{P}^3$ cuts a generic plane in 10 points, so $\mathcal{F}$ is a curve of degree 10.

Now let us consider the number of double points in $\{X = 0\} \cap \mathcal{S}$. This special plane section of $\mathcal{S}$ consists of a cubic $\mathcal{C}_0$ corresponding to $t = 0$, and 3 lines corresponding to $u = b, u = c, u = d$. We get $1 + 3 + 9 = 13$ points. Among them, 3 intersections of $\mathcal{C}_0$ with a line $\mathcal{L}$ (where $\mathcal{L} = \mathcal{L}_b, \mathcal{L}_c$ or $\mathcal{L}_d$) are not double points of the parametrization ϕ, so it remains 10 points in the double locus $\mathcal{F}$: The 3 mutual intersections of the 3 lines, the node of the cubic $\mathcal{C}_0$, and 2 of the intersections of each (of the 3) line with the cubic $\mathcal{C}_0$. Hence on $\mathcal{L}$, we find 4 points of $\mathcal{F}$ and 7 points of $\mathcal{F}$ on $\mathcal{C}_0$. As $\mathcal{F}$ is the set of double points of the parametrization, $\mathcal{F}$ is the image by ϕ of a curve $\mathcal{D}$ of bidegree (a, b) such that by Bézout theorem $3a + b = 2 \times 10 = 20$. From the previous discussion, we guess that generically $a = 4$ and $b = 8$. This is easily checked using the computational approach.

When we substitute, in the 4 partial derivatives of the implicit equation of $\mathcal{S}$, (X, Y, Z, T) by ϕ given in proposition 6, we obtain 4 equations of bidegree $(5, 15)$ in (t, u) whose gcd defines the closure of the double point locus $\mathcal{D}$ in $\mathbb{P}^1 \times \mathbb{P}^1$.

Another approach to obtain the double point locus is to consider coplanar couples of lines $\mathcal{L}_{u_1}$ and $\mathcal{L}_{u_2}$ in the family describing $\mathcal{S}$. For that purpose we choose 2 points $M_1 = M(0, u_1)$ and $M_3 = M(1, u_1)$ on $\mathcal{L}_{u_1}$ and 2 points $M_2 = M(0, u_2)$ and $M_4 = M(1, u_2)$ on $\mathcal{L}_{u_2}$ and require that these 4 points belong to the same plane. This amounts, as $u_1 \neq u_2$, to consider the vanishing of the determinant

$$R(u_1, u_2) = \det\left(OM_1 + OM_2, \frac{M_1 M_2}{u_2 - u_1}, OM_3 + OM_4, \frac{M_3 M_4}{u_2 - u_1}\right).$$

Note that the polynomial $R(u_1, u_2)$ is symmetric in (u_1, u_2). We denote by $\sigma = u_1 + u_2$ and $\tau = u_1 u_2$ the first elementary symmetric functions and get a new polynomial Q equals to R when we perform the substitution. Using a computer algebra system (e.g. Maple), we see that generically Q has degree 4 in σ and τ, and degree 6 in (σ, τ). The condition $Q(\sigma, \tau) = 0$ defines a curve Γ and the double point locus $\mathcal{D}$ is a double covering of Γ. Indeed, from each (σ, τ), we compute two values u_1 and u_2 such that $\mathcal{L}_{u_1}$ intersect

$\mathcal{L}_{u_2}$. Moreover, the parameters t_1 and t_2 of the intersection point $M(t_1, u_1) = M(t_2, u_2)$ are obtained by considering 4 minors of size 3 of the matrix defined by (OM_1, OM_2, OM_3, OM_4) which provide a linear relation between these 4 vectors; these minors are polynomials in (σ, τ) and $u_2 - u_1$. Therefore to each generic point of $\varGamma$ is attached 2 points (t_1, u_1) and (t_2, u_2) of $\mathcal{D}$ which map to the same point of $\mathcal{F}$ via ϕ; so there is a birational map between $\varGamma$ and $\mathcal{F}$.

Finally, the double lines $\mathcal{L}_u$ correspond to $u_1 = u_2$ or $\tau = \frac{\sigma^2}{4}$. Substituting in Q we obtain a polynomial in σ of degree 8 which has generically 8 solutions. Those give rise to 8 local Whitney Umbrellas on $\mathcal{S}$.

5.3 Surfaces of bidegree $(2, 2)$

We proceed as in the previous case. We start with the affine parametrization in normal form given in proposition 9 and we compute its partial derivatives. The local singularities with respect to the parametrization are given by a system of 3 polynomial equations of bidegree $(4, 4)$ in (t, u). It turns out that for generic values of the parameters they have 21 solutions.

Since the degree of $\mathcal{S}$ is 8, by formula (2.1) the genus of a generic section of $\mathcal{S}$ is $g = 21 - \delta$. Using the adjunction formula [13], we obtain $g = 1$, hence $\delta = 20$.

The double point locus $\mathcal{D}$ is a curve in $\mathbb{P}^1 \times \mathbb{P}^1$ of bidegree (a, b) which must satisfy $2a + 2b = 2 \times 20$. By symmetry, we get $a = b = 10$.

From the implicit equation $F(X, Y, Z, T)$ of $\mathcal{S}$ which has total degree 8, we derive the 4 partial derivatives of F (of degree 7). Substituting the parametrization ϕ given in proposition 9 in these equations, we obtain a system of 4 polynomials of bidegree $(14, 14)$ in (t, u), which defines generically the inverse image $\mathcal{D}$ in $\mathbb{P}^1 \times \mathbb{P}^1$, by ϕ, of the singular locus of $\mathcal{S}$. We check that the gcd of these 4 polynomials has bidegree $(10, 10)$ as expected.

5.4 Surfaces of bidegree $(2, 3)$

The local singularities of a surface $\mathcal{S}$ of bidegree $(2, 3)$ are given by a system of 3 polynomial equations of bidegree $(4, 7)$. This system has generically 36 solutions. Since the implicit degree of $\mathcal{S}$ is 12, the genus of a generic plane section of $\mathcal{S}$ is $g = 55 - \delta = (2-1) \times (3-1) = 2$, so $\delta = 53$. We should have in the parameter space a curve $\mathcal{D}$ of bidegree (a, b) such that $3a + 2b = 2 \times 53$. The computation shows that generically $a = 18$ and $b = 26$.

5.5 Surfaces of degree 3

From the parametrization given in subsection 3.2 in the affine space, we deduce that the local singularities are solutions of a system of 3 polynomial equations of bidegree $(6, 6)$. Generically, this system has 27 solutions.

As $\mathcal{S}$ is a surface of implicit degree 9, its generic plane section $\mathcal{H}$ has also degree 9. The genus of this plane curve is $g = 28 - \delta$. The section $\mathcal{H}$ is the image of a (generic) curve $\mathcal{C}$ of degree 3 in $\mathbb{P}^2$, its genus $g = 1$. Hence $\delta = 27$, and the degree of $\mathcal{F}$ is 27. The intersection number between $\mathcal{C}$ and $\mathcal{D}$ is $27 \times 2 = 54 = 3\deg(\mathcal{D})$, then we deduce that degree of $\mathcal{D}$ is 18. It is exactly what we find by the computational approach.

References

1. D. Anderson, T. Sederberg, *Steiner surface patches*, IEEE Computer Graphics and Applications **5** (1985), 23–36.
2. L. Andersson, J. Peters, N. Stewart, *Self-intersection of composite curves and surfaces*, Comp. Aided Geom. Design **15** (1998), 507–527.
3. F. Aries, B. Mourrain, J.-P. Técourt, *Algorithmic of quadratically parametrizable surfaces*, to appear, (2005).
4. V. I. Arnold, S. M. Guseĭn-Zade, A. N. Varchenko, *Singularities of differentiable maps. Vol. I*, vol. 82 of Monographs in Mathematics, Birkhäuser Boston Inc., Boston, MA, 1985.
5. F. Chen, W. Wang, *Revisiting the μ-basis of a rational ruled surface*, J. Symb. Comput. **36** (2003), 699–716.
6. F. Chen, J. Zheng, and T. Sederberg, *The μ-basis of a rational ruled surface*, Comp. Aided Geom. Design **18** (2001), 61–72.
7. A. Chtcherba, D. Kapur, *Exact resultants for corner-cut unmixed multivariate polynomial systems using the dixon formulation*, J. Symb. Comput. **36** (2003), 289–315.
8. A. Coffman, A. J. Schwartz, C. Stanton, *The algebra and geometry of Steiner and other quadratically parametrizable surfaces*, Comp. Aided Geom. Design **13** (1996), 257–286.
9. D. Cox, *Equations of parametric curves and surfaces via syzygies*, in Symbolic computation: Solving equations in algebra, geometry, and engineering, Contemp. Math. **286** (2001), 1–20.
10. A. Dickenstein, I. Emiris, *Multihomogeneous resultant formulae by means of complexes*, J. Symb. Comp. **36** (2003), 317–342.
11. M. Elkadi, T. Lê, and A. Galligo, *Parametrized surfaces in $\mathbb{P}^3$ of bidegree (1,2)*, in International Symposium on Symbolic and Algebraic Computation, 2004, 141–148.
12. M. Elkadi, B. Mourrain, Algorithms for residues and Lojasiewicz exponents, J. Pure and Appl. Algebra, **153** (2000), 27–44.
13. W. Fulton, *Intersection theory*, Springer-Verlag, 1984.
14. G. Farin, *Curves and surfaces for computer aided geometric design. A practical guide*, Academic Press, Inc., Boston, MA, 1993.
15. A. Galligo, J.-P. Pavone, *A sampling algorithm computing self-intersection of parametric surfaces*, this book, (2005).
16. A. Galligo, M. Stillman, *The geometry of bicubic surfaces and splines*, preprint (2005).
17. M. Kapranov, B. Sturmfels, A. Zelevinsky, *Chow polytopes and general resultants*, Duke Math. J. **67** (1992), 189–218.
18. R. Piene, *Some formulas for a surface in* $\mathbf{P}^3$, in Algebraic geometry (Proc. Sympos., Univ. Tromsø, Tromsø, 1977), vol. 687 of Lecture Notes in Math., Springer, Berlin, 1978, pp. 196–235.

19. R. Piene, *Singularities of some projective rational surfaces*, in Computational methods for algebraic spline surfaces, Springer, Berlin, 2005, pp. 171–182.
20. I. Shafarevitch, *Basic Algebraic Geometry*, New-York, Springer-Verlag, 1974.
21. B. Sturmfels, A. Zelevinsky, *Multigraded resultants of Sylvester type*, J. Algebra **163** (1994), 115–127.
22. J. B. Thomassen, *Self-intersection problems and approximate implicitization*, in Computational methods for algebraic spline surfaces, Springer, Berlin, 2005, pp. 155–170.

On the intersection with revolution and canal surfaces

Mario Fioravanti [*], Laureano Gonzalez–Vega[*,**] and Ioana Necula [†]

Departamento de Matemáticas, Estadística y Computación
Universidad de Cantabria, Spain
{fioravam,gonzalezl,ioana.necula}@unican.es

Summary. This paper presents a new algorithm for computing the intersection of a rational revolution surface or a canal surface, given in parametric or implicit form, and another surface given in parametric form. The problem is reduced to finding the zero set of a bivariate equation which represents the parameter values of the intersection curve, as a subset of one of the surfaces. The algorithm involves both symbolic and numerical computations, and follows three steps: implicitization of the first surface, determination of the topology of the intersection curve, and computation of the curve. The algorithm applies equally well to any other type of surfaces whose parametric equations can be reduced to a set of two equations with only one parameter.

Introduction

Computing the intersection curve of two surfaces is a fundamental process in many areas, such as visualization and CAD/CAM treatment of complicated shapes, design of 3D objects, computer animation, NC machining (milling) and creation of Boundary Representation in solid modelling (see [8], [11], [12], [14], [15]). The main goal concerning the surface–to–surface intersection problem is to develop a robust, accurate and fast algorithm for computing the intersection curve, with the least user intervention needed. The algorithm proposed in this paper is an improvement in these aspects with respect to general surface–to–surface intersection algorithms, due to the properties of the class of surfaces considered.

Given two surfaces S_1 and S_2, if S_1 is given in implicit form, and S_2 is given in parametric form, one may substitute each variable of S_1 implicit equation for the parametric expression of the corresponding component of S_2, reducing the problem to the resolution of an equation with two variables. The

---

[*]Partially supported by the spanish grant BFM 2002–04402–C02–02.
[†]Partially supported by the European Union funded projects GAIA II (IST–2002–35512) and RAAG (HPRN–CT–2001–00271)

real solutions of this equation are the parameter values corresponding to the intersection curve, as part of $\mathbf{S}_2$. If both surfaces are given parametrically, the implicitization of one of them is not an easy task in general. However, we will show that, when dealing with rational revolution surfaces or with rational canal surfaces, the implicitization can be done in a straightforward manner.

In this paper, we consider the intersection of a revolution or canal surface $\mathbf{S}_1(u,s)$ and an arbitrary surface $\mathbf{S}_2(v,t)$, given by their rational parametric equations. The parameters u, s, v, and t take values in $\mathbb{R}$. In some applications the image of $\mathbf{S}_2(v,t)$ will be part of the surface one is interested in, so other patches may be necessary to complete the surface. First, we will show how to obtain the implicit equation $H(x,y,z) = 0$ of $\mathbf{S}_1$, by just computing one single resultant, so the intersection problem is reduced to the solution of an equation $G(v,t) = 0$. Then the efficient algorithm described in [5] is used to determine the topology of the curve defined by $G(v,t)$.

Finally, solution points of $G(v,t) = 0$ are computed numerically, which allows to generate the points of $\mathbf{S}_2$ that form the intersection curve. The algorithm can be used as well for other types of surfaces whose parametric equations can be reduced to a set of two equations with only one parameter.

The computations in the implicitization process may produce instability errors when the revolution or canal surface is defined using double precision floating point arithmetic. In this case the algorithm should be adapted following the ideas in [1], [9] and [10].

In some cases, the implicitization process produces some extraneous components which do not correspond to real points of the intersection curve (neither to the corresponding surface). Points in such a component verify the equation $H(x,y,z) = 0$ but they do not belong to the surface $\mathbf{S}_1$, and they can not be eliminated algebraically. The algorithm checks suitable points on each component in order to eliminate these extraneous components from the solution intersection curve.

There are two main features of the algorithm. First, it provides a simple way to obtain the implicit equation of a revolution or canal surface. Second, the ability to determine the topology of the intersection curve is a very useful guidance for the numerical computation of the curve.

Compared to the algorithms in [6] and [7], which reduce the problem to the solution of a set of bivariate equations, the present algorithm has the following advantages. In [6] and [7] a plane curve is obtained but there is not one to one correspondence between the plane curve and the desired intersection curve; there may be a point in the plane curve which corresponds to a whole line in the intersection curve. Moreover, degenerate cases may occur and they must be treated separately. With the present algorithm a one to one correspondence between the plane curve and the intersection curve is obtained, with the only possible exception of singular points, in which case a line in the plane curve may correspond to a single point in the intersection curve (but not the other way around). In addition, the present algorithm can be used to obtain the intersection of a revolution or canal surface with any other surface given parametrically. The condition of being a revolution or canal surface is used in the implicitization step only. On the other hand, the algorithms in [6] and [7] compute the intersection of a revolution, ringed or ruled surface with another surface of one of these classes.

For initial testing the algorithm has been implemented in the Computer Algebra System MAPLE. The graphic interface of the algorithm has been created with the MAPLE package Maplets. The implicitization process is done symbolically, the topology determination involves both symbolic and numerical methods, and the final computations of the intersection curve points are performed numerically.

The paper is divided in four sections. In the first one it is shown how to compute the implicit equation of a rational revolution surface or a canal surface whose central curve and radius are rational functions. The reasons why the implicit equation of a parametric surface may have geometric extraneous components is explained. The second section shows how to obtain useful information about the intersection of two surfaces when the implicit equation of one of them is available. The third section presents a collection of examples illustrating the practical use of the algorithm. The last section is devoted to conclusions.

1 Computing the implicit equation

If one starts with the implicit equation of $\mathbf{S}_1$, this step is, of course, unnecessary. First we study the case of computing the implicit equation of revolution surfaces, which is of special interest in many applications, and then we study the case of canal surfaces.

1.1 On the implicit equation of a revolution surface

Consider a revolution surface $\mathbf{S}$. Choosing a system of coordinates such that the z–axis is the axis of revolution of $\mathbf{S}$, its parametrization will be of the form

$$\mathbf{S}(u,s) = [\varphi(s)\frac{1-u^2}{1+u^2}, \varphi(s)\frac{2u}{1+u^2}, \psi(s)], \tag{1.1}$$

where $\varphi = \frac{n_\varphi}{d_\varphi}$, and $\psi = \frac{n_\psi}{d_\psi}$, and $n_\varphi, d_\varphi, n_\psi, d_\psi$ are polynomials in s. We assume that neither n_φ and d_φ, nor n_ψ and d_ψ, have common factors. The curve $[\varphi(s), \psi(s)]$, in the xz–plane is the curve that generates the surface. The parameter u takes all real values. The image of this parametrization is the surface minus a meridian. In order to complete the surface one may either take the limit as $u \to \infty$, or define a similar patch that covers the missing meridian. However, the implicit equation that will be obtained next holds for the whole surface.

The components x, y, z of $\mathbf{S}$ obey the equations

$$x^2 + y^2 = \varphi^2(s) = \left(\frac{n_\varphi(s)}{d_\varphi(s)}\right)^2, \qquad (1.2)$$
$$z = \psi(s) = \frac{n_\psi(s)}{d_\psi(s)}.$$

Assuming $d_\varphi(s) \neq 0$ and $d_\psi(s) \neq 0$, then:

$$\begin{aligned} f(x,y,s) &\equiv d_\varphi^2(s)(x^2+y^2) - n_\varphi^2(s) = 0, \\ g(z,s) &\equiv d_\psi(s)z - n_\psi(s) = 0. \end{aligned} \qquad (1.3)$$

In order to obtain the implicit equation of **S**, Sylvester resultant of f and g with respect to s, $H(x,y,z)$, is computed. Since n_φ and d_φ have no common factors, and n_ψ and d_ψ have no common factors, f and g are irreducible in $\mathbb{R}[x,y,z,s]$. Hence, H is not identically zero.

Moreover, the degrees of H are described in the following lemma.

Lemma 1.

$$\deg_x(H) = \deg_y(H) = 2\deg_s(g) = 2\max\{\deg(d_\psi), \deg(n_\psi)\}.$$

$$\deg_z(H) = \deg_s(f) = 2\max\{\deg(d_\varphi), \deg(n_\varphi)\}.$$

Thus, $H(x,y,z) = 0$ contains the implicit equation of **S**. In case that $H(x,y,z)$ has factors of multiplicity bigger than one, they are reduced to multiplicity one. This cases may happen even if the parametrization is one to one; for example, if

$$\varphi(s) = \frac{s}{1+s^2}, \quad \psi(s) = \frac{1}{1+s^2}, \quad \text{then } \operatorname{Res}_s(f,g) = (z^2 - z + x^2 + y^2)^2.$$

The factorization of a polynomial in three variables required to reduce multiplicities can be done in polynomial time, and any computer algebra package is able to do it in a few seconds.

There may also be some algebraic factors in the resultant which do not correspond to points in S. The appearance of such factors depends on how the parametric equations are algebraically operated before the computation of the resultant, especially the denominators, as the following example shows.

Example 2. Consider hyperboloid $\mathbf{S}(u,s) = (x(u,s), y(u,s), z(u,s))$, where

$$x = \frac{1-s^2-2su}{1+s^2}, \quad y = \frac{2s+u(1-s^2)}{1+s^2}, \quad z = u. \qquad (1.4)$$

Replacing u by z in x and y one gets

$$p \equiv x(1+s^2) - (1-s^2-2zs) = 0, \quad q \equiv y(1+s^2) - (2s+z(1-s^2)) = 0,$$

and

$$\operatorname{Res}_s(p,q) = (1+z^2)(x^2+y^2-z^2-1),$$

contains the extraneous factor $1 + z^2$.

Going back to equations (1.4) and solving u in terms of y and s, one gets

$$u = \frac{y(1+s^2) - 2s}{1 - s^2}.$$

This provides two equations in x, y, z and s, and the corresponding resultant is now

$$H_2(x, y, z) = (y-1)(y+1)(x^2 + y^2 - z^2 - 1),$$

which contains the extraneous factors $(y-1)$ and $(y+1)$. This extraneous factors come from the values of s ($s = 1$ and $s = -1$) annihilating the denominator of the obtained expression when solving u in terms of y and s.

The real values of s such that $d_\varphi(s) = 0$ or $d_\psi(s) = 0$ must be checked separately in order to determine if they provide points in the surface.

1.2 On the implicit equation of a canal surface

A *canal surface* $\mathbf{S}(u, s)$ with central curve $\mathbf{c}(s)$ and radius function $r(s)$ is the envelope of a 1-parameter family of spheres with center in $\mathbf{c}$ and radius r. Therefore, $\mathbf{S}$ obeys the following equations

$$\begin{aligned} |\mathbf{S}(u, s) - \mathbf{c}(s)|^2 - r^2(s) &= 0, \\ (\mathbf{S}(u, s) - \mathbf{c}(s)) \cdot \mathbf{c}'(s) + r(s)r'(s) &= 0. \end{aligned} \quad (1.5)$$

Pipe surfaces (or tubes) are particular cases of canal surfaces, with r constant.

Assume that $\mathbf{c}(s)$ and $r(s)$ are rational. It has been proved in [13] that such a canal surface has rational parametrizations. If x, y and z are the components of $\mathbf{S}$, and $c_x(s)$, $c_y(s)$ and $c_z(s)$ are the components of $\mathbf{c}(s)$, then

$$\begin{aligned} f(x,y,z,s) &\equiv (x - c_x(s))^2 + (y - c_y(s))^2 + (z - c_z(s))^2 - r(s)^2 = 0 \\ g(x,y,z,s) &\equiv (x - c_x(s))c_x'(s) + (y - c_y(s))c_y'(s) + \\ &\quad + (z - c_z(s))c_z'(s) + r(s)r(s)' = 0. \end{aligned} \quad (1.6)$$

Let n_f and d_f be the numerator and denominator of f, n_g and d_g the numerator and denominator of g. Notice that d_f and d_g depend only on s. If n_f and n_g have common factors depending on s only, let $J(s) = \gcd(n_f, n_g)$. Let $H(x, y, z)$ be the resultant of (n_f/J) and (n_g/J) with respect to s. The division by J is necessary for the resultant to be different from zero. Then $H(x, y, z) = 0$ contains the implicit equation of $\mathbf{S}$.

The following estimate for the degree of H holds.

Lemma 3.

$$\deg(H) \leq \deg_s(n_f) + 2\deg_s(n_g).$$

Proof. The resultant of n_f and n_g with respect to s is the determinant of a square matrix with $\deg_s(n_f) + \deg_s(n_g)$ rows and columns. In the first $\deg_s(n_g)$ rows, each of x, y and z appear with degree at most 2, and there are not crossed terms. In the following $\deg_s(n_f)$ rows, x, y and z appear with degree at most 1, and there are not crossed terms. □

As before, the factors in $H(x, y, z)$ of multiplicity bigger than one are reduced to multiplicity one. The real values of s such that $d_f(s) = 0$ or $d_g(s) = 0$ together with $J(s) = 0$, need a separate analysis to determine if they produce points in the surface **S**.

1.3 Computational issues

In case of high degrees of the equations in (1.3) or (1.6), to avoid difficulties when computing the implicit equation, other formulations such as the Bezout expression for the resultant should be used (see, for example, [2]).

The algorithm is accurate when working with exact arithmetic. On the other hand, when the coefficients of the equations in (1.3) or (1.6) are floating point real numbers, rounding errors may occur in the computation of the topology of the curve explained in the next section. The appropriate way of obtaining the topology of the curve in these cases is explained in [5] and [1]. In order to avoid other numerical instabilities arising from the computation of the determinant of a polynomial matrix, ad–hoc techniques such as polynomial interpolation should be used instead of the classical ways of computing the determinant (such as Gauss Method). Note that the use of interpolation techniques, as the ones presented in [9, 10], requires the knowledge of bounds on the degree of $H(x, y, z)$ such as those provided by the lemmas 1 and 3.

1.4 Geometric extraneous components

There may be some real solution points of $H(x, y, z) = 0$ which are not points of the surface parameterized by $\mathbf{S}(u, s)$, for real values of u and s (see examples 6 and 7 below). A connected component formed by such points is called a *geometric extraneous component* (see [3] for a detailed study and characterization of these components).

The points in a geometric extraneous component correspond to values of the parameters in $\mathbb{C} \setminus \mathbb{R}$ providing real points. They are singular points as real points belonging to the surface defined by the equation $H(x, y, z) = 0$, i.e. they are zeros of $H(x, y, z)$, $\partial H/\partial x$, $\partial H/\partial y$ and $\partial H/\partial z$, simultaneously.

As part of the algorithm, we check singular points in order to verify if they belong to a geometric extraneous component, which must be removed.

2 Computing the intersection curve

If a point $\mathbf{S}_2(v, t)$ is also on the surface $\mathbf{S}_1$ with implicit equation $H(x, y, z) = 0$, its components must verify the equation

$$G(v,t) \equiv H(\mathbf{S}_2(v,t)) = 0. \tag{2.1}$$

The seminumerical algorithm presented in [5] (based on the Sturm-Habicht sequence and on the concept of generic position) is used to determine the topology of the plane curve $G(v,t) = 0$. The algorithm generates the critical points (i.e. those points either with a vertical tangent or singular) and the regular points over the critical points of the curve. It also determines the number of branches on each side (left and right) of a critical point and the branch connections between the critical and regular points (see the topology graph in figures 1, 2, 3, 4 and 5 to see concrete examples of the graphical representation of the output). Using this information, for each branch connection of the curve, certain number of intermediate points of the (v,t)-plane curve are computed by using standard numerical methods (for instance, Newton's method), and the corresponding points of $\mathbf{S}_2$ are obtained after the corresponding lifting.

If one is interested in some particular components of the intersection curve lying within certain bounds (as it happens in some applications), one may do the numerical computation of intermediate points only for the chosen components.

If a point (v_0, t_0) of the plane curve is non singular, the tangent vector to the curve at this point is proportional to

$$\left(\frac{\partial G}{\partial t}(v_0, t_0), -\frac{\partial G}{\partial v}(v_0, t_0)\right). \tag{2.2}$$

Hence, the corresponding tangent vector to the curve on the surface $\mathbf{S}_2$ at $\mathbf{S}_2(v_0, t_0)$ will be

$$w = \frac{\partial G}{\partial t}\frac{\partial \mathbf{S}_2}{\partial v} - \frac{\partial G}{\partial v}\frac{\partial \mathbf{S}_2}{\partial t}. \tag{2.3}$$

Thus, if (v_0, t_0) is a nonsingular point of the plane curve and $\mathbf{S}_2(v_0, t_0)$ is a regular point of $\mathbf{S}_2$, then the tangent vector defined in (2.3) is different from zero. Hence, there is a one to one correspondence between the plane curve and the surface-to-surface intersection curve, in a neighborhood of (v_0, t_0).

3 Experimental examples

The results of the application of the algorithm to some illustrative examples are shown in this section.

Example 4. Intersection of a sphere and a revolution surface.
Consider a sphere of radius 5, parameterized by

$$\mathbf{S}_1(u,s) = \left(\frac{5(1-u^2)(1-s^2)}{(1+u^2)(1+s^2)}, \frac{10u(1-s^2)}{(1+u^2)(1+s^2)}, \frac{10s}{1+s^2}\right),$$

and a revolution surface with rotation axis $[0, 1, t]$,

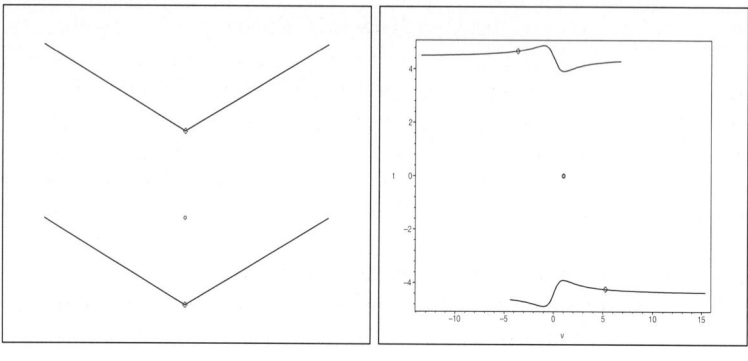

Example 4. Topology graph   Plane curve

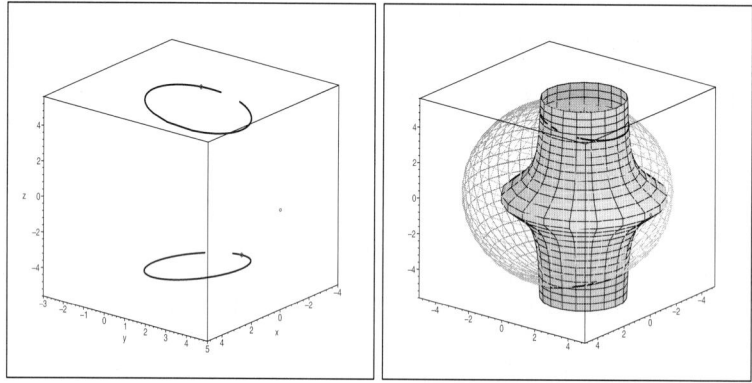

Fig. 1. Sphere of radius 5 and revolution surface

$$\mathbf{S}_2(v,t) = \left(\frac{5(1-v^2)(2t^2+4)}{(1+v^2)(t^2+1)}, 1 + \frac{2v(2t^2+4)}{(1+v^2)(t^2+1)}, t \right).$$

There is a meridian missing in each parametrization, corresponding to the limits $u \to \infty$ and $v \to \infty$, respectively (see Figure 1).

In this example equations (1.3) take the form

$$\begin{aligned} f &\equiv (1+s^2)^2(x^2+y^2) - (5(1-s^2))^2 = 0 \\ g &\equiv (1+s^2)z - 10s = 0 \end{aligned} \quad (3.1)$$

and $\mathrm{Res}_s(f,g) = (100x^2 + 100y^2 + 100z^2 - 2500)^2$, so we take

$$H = x^2 + y^2 + z^2 - 25.$$

After substitution we obtain

$$\begin{aligned} G(v,t) = (t^6v^2 - 18t^4v^2 - 8v^2 - 31v^2t^2 + 16v + 8t^4v + 24vt^2 + t^6 \\ - 18t^4 - 8 - 31t^2)(v^2+1) = 0. \end{aligned} \quad (3.2)$$

The results of the computation are shown in Figure 1. The graph of the first picture indicates the topology of the real algebraic plane curve $G(v,t) = 0$; there are two components that close when v tends to infinity and an isolated point at which the revolution surface touches the sphere tangentially. The second picture shows the curve in the (v,t)-plane, the third picture shows the intersection curve in $\mathbb{R}^3$, and the last one shows both surfaces together with the intersection curve.

Example 5. Intersection of a canal surface and a torus.
Let $\mathbf{S}_1$ be the canal surface with central curve $C(s) = [0, -s^2, 8s]$ and radius $r(s) = 1 + s/3$, so equations (1.6) take the form

$$f \equiv x^2 + (y+s^2)^2 + (z-8s)^2 - (1+s/3)^2 = 0$$
$$g \equiv \tfrac{1}{3} - 2(y+s)^2 s + 8z - \tfrac{575s}{9} = 0.$$

The torus is parameterized by

$$\mathbf{S}_2 = \left[\frac{2t}{1+t^2}, \left(4 + \frac{1-t^2}{1+t^2}\right)\left(\frac{1-v^2}{1+v^2}\right) + 1, 2\left(4 + \frac{1-t^2}{1+t^2}\right)\left(\frac{v}{1+v^2}\right) + 1\right].$$

Then $H \equiv \mathrm{Res}_s(f, g) =$

$= -13719194838y - 109719028863 - 732965219z^2 + 428490000y^4$
$+2239488zy^3 + 109099868401y^2 + 104976z^6 + 53654400zy^2 - 40450752z^3$
$-859499424zy - 9205941552z + 109743372253x^2 + 857887848z^2y^2$
$+3400761600yz^2 + 214035048x^2y^2 + 536659344x^2z^2 + 13714754688yx^2$
$+104976z^4y^2 + 13460256z^2y^3 + 16819488z^4y + 2519424z^3y + 25625160z^4$
$+13701335256y^3 + 104976x^4y^2 + 314928x^4z^2 + 314928x^2z^4$
$-13413600yx^4 - 13413600x^2y^3 + 80481600x^2z + 104976x^6$
$-214559928x^4 + 209952x^2z^2y^2 + 3405888x^2yz^2 + 2519424x^2zy.$

The numerator of $G(v,t) \equiv H(\mathbf{S}_2(v,t))$ is a polynomial of degree 20, and its denominator is $(1+t^2)^4(1+v^2)^4$.

The topology graph for $G(v,t) = 0$, the (v,t)–plane curve and the intersection curve are shown in Figure 2. When v tends to infinity, four components homeomorphic to a circle are obtained.

Example 6. Intersection of a pipe surface and a cylinder.
Let $\mathbf{S}_1$ be the pipe surface with central curve $C(s) = [2s, 0, s^2]$ and radius $r = 1$. Hence, equations (1.6) take the form

$$f \equiv (x-2s)^2 + y^2 + (z-s^2)^2 - 1 = 0$$
$$g \equiv 2x - 4s + 2(z-s^2)s = 0.$$

The parametrization of the cylinder is the following

$$\mathbf{S}_2 = \left[1 + \frac{2v}{1+v^2}, \frac{1-v^2}{1+v^2}, t\right].$$

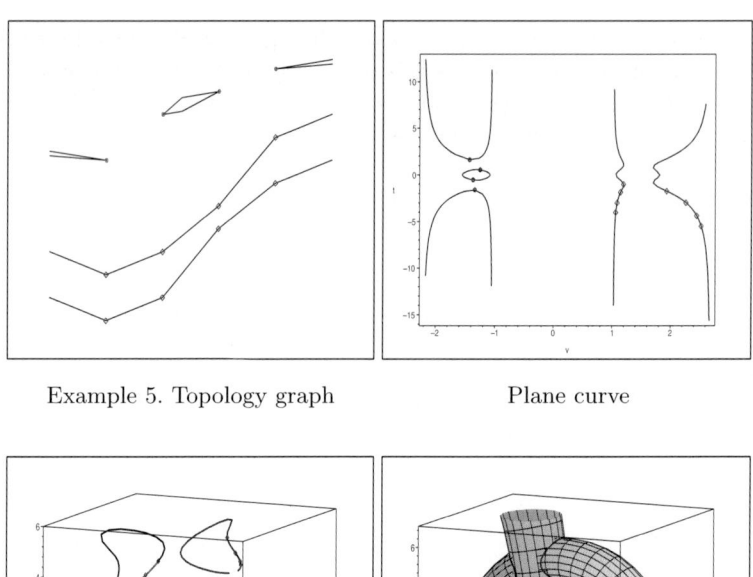

Example 5. Topology graph   Plane curve

Fig. 2. Canal surface and torus

Then

$$H \equiv \mathrm{Res}_s(f,g) = -25 + 40z + y^6 + x^6 + 16z^4 - 17x^2 - 40z^3 + 9z^2 \\ + 35y^2 + 8zy^4 + 8y^2z^3 - 11y^4 - 2x^4 + 2x^2y^2z^2 \\ -2x^2zy^2 + 30x^2z^2 - 6zx^2 + 14x^2y^2 - 10x^4z - 8x^2z^3 \\ + 6y^2z^2 - 48zy^2 + 3x^4y^2 + 3x^2y^4 + y^4z^2 + x^4z^2.$$

Substitution for $\mathbf{S}_2$ components gives

$$G(v,t) = 16(1+v^2)^2(2 - 64v^6 - 164v^5 + 408t^2v^5 + 20v - 18t + 2v^8 \\ + 20v^7 + 308t^2v^6 - 64v^2 + 16t^4 - 40t^3 + 49t^2 + 136t^2v + 49t^2v^8 \\ + 136t^2v^7 + 308t^2v^2 + 518t^2v^4 + 408t^2v^3 - 208tv^6 - 624tv^5 \\ - 112tv - 18tv^8 - 112tv^7 - 208tv^2 - 380tv^4 - 624tv^3 - 564v^4 \\ - 164v^3 - 224t^3v^6 - 96t^3v^5 - 40t^3v^8 - 32t^3v^7 - 32t^3v - 224t^3v^2 \\ - 368t^3v^4 - 96t^3v^3 + 96t^4v^4 + 64t^4v^6 + 16t^4v^8 + 64t^4v^2).$$

The topology graph for $G(v,t) = 0$, the (v,t)–plane curve and the intersection curve are shown in Figure 3. In this example we find a geometric extraneous component: the point of coordinates $(0, 0, 5/4)$ is a solution of $H(x, y, z) = 0$ and belongs to the cylinder $\mathbf{S}_2$ but it is not a point of $\mathbf{S}_1$. It corresponds to the complex parameter values $u = 1$ and $s = i\sqrt{3}/2$. This point will be removed from the intersection curve by the algorithm.

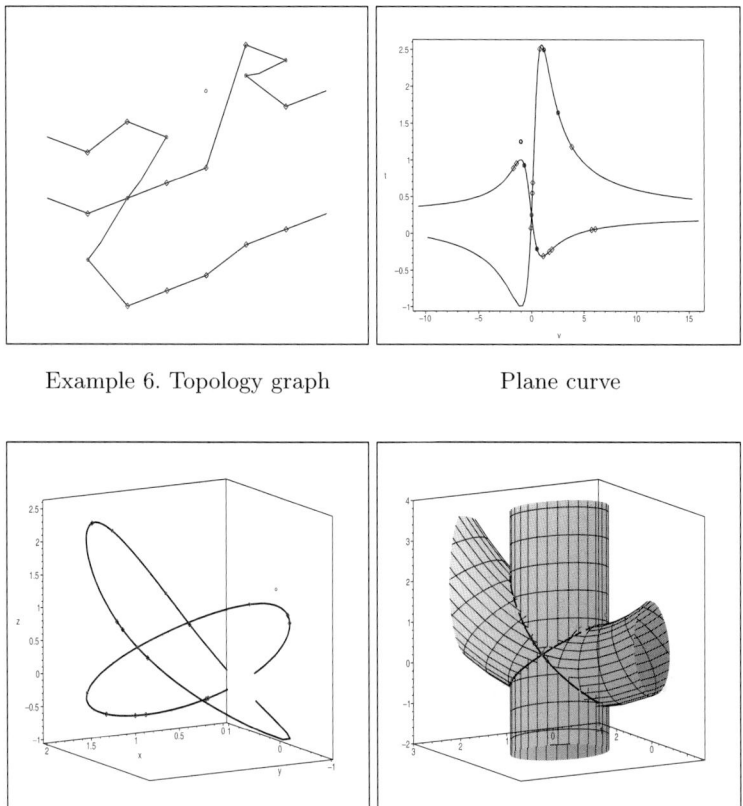

Fig. 3. Pipe surface and cylinder

Example 7. Intersection of a pipe surface and a ringed surface.
Let $\mathbf{S}_1$ be the pipe surface with central curve $C(s) = [2s, 1, s^2]$ and radius $r = 1$. Hence, equations (1.6) take the form

$$f \equiv (x - 2s)^2 + (y - 1)^2 + (z - s^2)^2 - 1 = 0$$
$$g \equiv 2x - 4s + 2(z - s^2)s = 0.$$

The parametrization of the cylinder is the following

$$\mathbf{S}_2 = \left[t^3 + \frac{1 - v^2}{1 + v^2}, t, t^3 + \frac{2v}{1 + v^2} \right].$$

The resultant of f and g is

$$H \equiv \text{Res}_s(f,g) = -512y - 512y^3z + 1024zy - 128x^2z^3 - 160x^4z \\
-256yz^2 - 128zx^2 + 16x^4z^2 - 96x^4y + 48x^4y^2 + 128y^4z + 128y^2z^3 \\
+48x^2y^4 - 192x^2y^3 + 192y^2z^2 + 512x^2z^2 - 640x^2y + 512x^2y^2 - 256yz^3 \\
+16y^4z^2 - 64y^3z^2 - 256y^2 + 256z^2 + 32x^2y^2z^2 - 64x^2yz^2 - 32x^2zy^2 \\
+64x^2zy + 16x^4 + 64y^4 + 384y^3 - 512z^3 + 16y^6 - 96y^5 + 256z^4 + 16x^6.$$

Introducing in H the components of $\mathbf{S}_2$ we obtain a rational expression whose denominator is $(1+v^2)^5$, and whose numerator is a polynomial $G(v,t)$ with 172 terms, total degree 28, $\deg_v(G) = 10$, and $\deg_t(G) = 18$. The results of the solution of $G(v,t)$ are represented in the two first pictures of Figure 4.

In this example we find four points belonging to geometric extraneous components, that is, they obey the equation $H(x,y,z) = 0$, and they belong to $\mathbf{S}_2$, but they do not belong to $\mathbf{S}_1$.

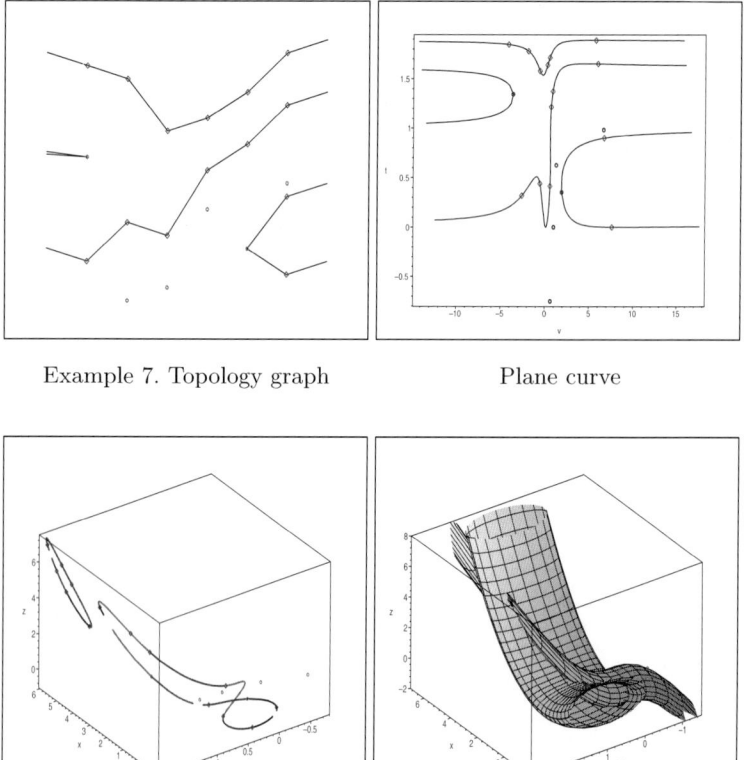

Example 7. Topology graph   Plane curve

Fig. 4. Pipe surface and ringed surface

Example 8. Intersection of a torus and another revolution surface.
Consider a torus, parameterized by

$$\mathbf{S}_1(u,s) = \left(\frac{2(1-u^2)(3+s^2)}{(1+u^2)(1+s^2)}, \frac{4u(3+s^2)}{(1+u^2)(1+s^2)}, \frac{4s}{1+s^2} \right),$$

and a revolution surface with rotation axis $[0, 3, t]$,

$$\mathbf{S}_2(v,t) = \left(\frac{(1-v^2)(2t^2+5)}{(1+v^2)(t^2+1)}, 3 + \frac{2v(2t^2+5)}{(1+v^2)(t^2+1)}, t \right).$$

In this example equations (1.3) for the torus take the form

$$\begin{aligned} f &\equiv (1+s^2)^2(x^2+y^2) - (2(3+s^2))^2 = 0 \\ g &\equiv (1+s^2)z - 4s = 0. \end{aligned} \quad (3.3)$$

The implicit equation for the torus is then

$$H = \mathrm{Res}_s(f,g) = -40x^2 + x^4 + y^4 - 40y^2 + 2x^2z^2 + 2y^2z^2 \\ +2x^2y^2 + z^4 + 24z^2 + 144 = 0. \quad (3.4)$$

The implicit equation for the algebraic plane curve is

$$G(v,t) = (-60 + 1680v + t^1 2v^4 - 602t^6v^4 - 602t^6 - 988t^2 - 1419t^4 \\ +3480v^2 + 2828t^6v^2 - 60v^4 + 1680v^3 + 23t^8 - 768t^6v - 768t^6v^3 \\ +2232t^2v + 23t^8v^4 + 622t^8v^2 + 2232t^2v^3 - 72t^8v - 72t^8v^3 + 2t^1 2v^2 \quad (3.5) \\ +t^{12} + 8104t^2v^2 + 54t^{10} + 48t^{10}v + 48t^{10}v^3 + 108t^{10}v^2 - 1419t^4v^4 \\ +7098t^4v^2 + 54t^{10}v^4 - 96t^4v - 96t^4v^3 - 988t^2v^4)(v^2+1) = 0.$$

The first picture in Figure 5 shows the topology of the plane algebraic curve $G(v,t) = 0$. The two apparent components in the graph, represent only one eight shaped component, because they match when v tends to infinity. The second picture shows the curve in the (v,t)–plane, the third picture shows the intersection curve in $\mathbb{R}^3$, and the last one shows both surfaces together with the intersection curve. Here we have a branch point at a tangential intersection of the surfaces.

4 Conclusions

We have presented a new algorithm for the computation of the intersection curve of a revolution or canal surface and another arbitrary surface. It is based on the use of the implicit equation of the first surface, which is easy to calculate for this kind of surfaces. The problem is then reduced to the study of a real algebraic plane curve defined implicitly. This algebraic curve is computed using the seminumerical algorithm proposed in [5]. Once the curve is computed, and geometric extraneous components that may appear

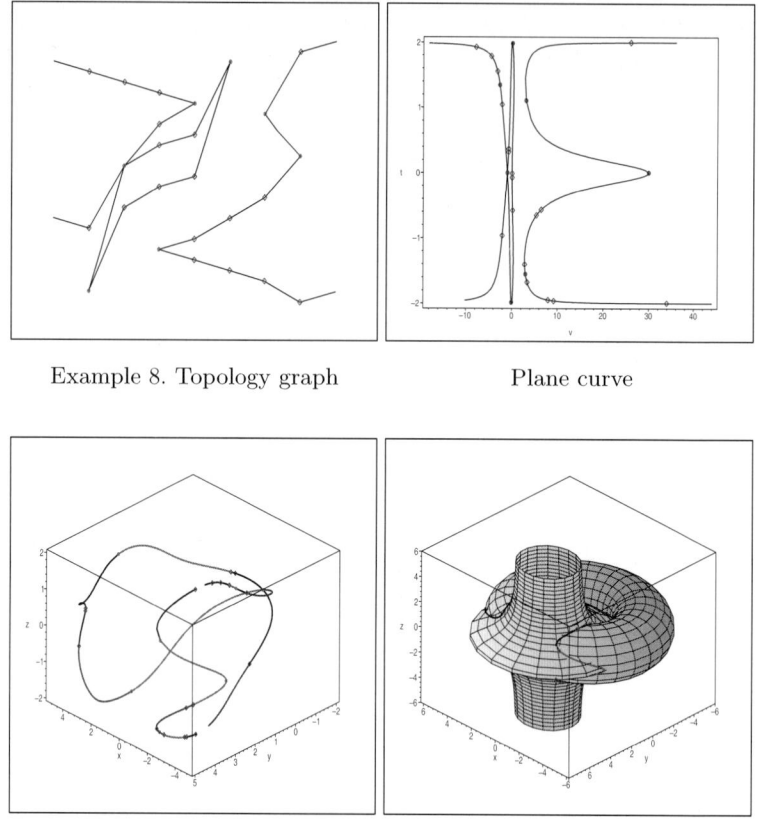

Example 8. Topology graph   Plane curve

Fig. 5. Torus and revolution surface

are removed, the curve is lifted to the intersection curve in three dimensional space. There is a one-to-one correspondence between the algebraic plane curve and the intersection curve, with the only possible exception of singular points, in which case, a component of the plane curve may correspond to a single point in the intersection curve.

Since the implicitization process may involve the computation of a greatest common divisor, it remains to analyze the behavior of the algorithm when the revolution or canal surface is defined using floating point real numbers.

According to the initial tests the algorithm is reasonably efficient. Using a Pentium 2800 Mhz processor, required 2 seconds to process example 4, approximately 25 minutes for example 5, 9 minutes for example 6, 30 minutes for example 7, and one minute for example 8. Note that the high computing time of example 7 is due to the analysis of the topology of a real algebraic plane curve defined by a polynomial $G(v,t)$ with 172 terms and total degree 28.

The implicitization technique we have used is based essentially in the fact that revolution and canal surfaces can be defined by two equations, depending on one single parameter. Therefore, the algorithm may be used to any other type of surfaces whose parametric equations can be reduced to a set of two equations with only one parameter. This is the case, for instance, of ruled surfaces (see [4]).

References

1. Corless, R. M.; Gonzalez–Vega, L.; Necula, I.; Shakoori, A., *Topology determination of implicitly defined real algebraic plane curves*, An. Univ. Timişoara Ser. Mat.-Inform. 41 (2003), Special issue, 83–96.
2. Diaz–Toca, G.; Gonzalez–Vega, L., *Various New Expressions for Subresultants and Their Applications*, Applicable Algebra in Engineering, Communication and Computing 15 (2004), 233–266.
3. Fioravanti, M.; Gonzalez-Vega, L., *On the geometric extraneous components appearing when using implicitization*, in: Dæhlen, M., Mørken, K., Schumaker, L. (eds.), Mathematical Methods for Curves and Surfaces, Nashboro Press, 2005, 157-168.
4. Fioravanti, M.; Gonzalez-Vega, L.; Necula, I., *Computing the intersection of two ruled surfaces by using a new algebraic approach*, Journal of Symbolic Computation (2006), to appear.
5. Gonzalez-Vega, L.; Necula, I., *Efficient topology determination of implicitly defined algebraic plane curves*, Computer Aided Geometric Design 19 (2002), 719–743.
6. Heo, H.–S.; Kim, M-S.; Elber, G., *The intersection of two revolution surfaces*, Computer–Aided Design 31 (1999), 33–50.
7. Heo, H.–S.; Hong, S. J., *The intersection of two ringed surfaces and some related problems*, Graphical Models 63 (2001), 228–244.
8. Hoschek, J.; Lasser, D., Fundamentals of Computer Aided Geometric Design, A.K. Peters, Wellesley, MA, 1993.
9. Marco, A.; Martínez, J.–J., *Implicitization of rational surfaces by means of polynomial interpolation*, Computer Aided Geometric Design 19 (2002), 327–344.
10. Marco, A.; Martínez, J.–J., *Parallel computation of determinants of matrices with polynomial entries*, Journal of Symbolic Computation 37 (2004), 749–760.
11. Miller, J.R.; Goldman, R.N., *Geometric algorithms for detecting and calculating all conic sections in the intersection of any two natural quadric surfaces*, Graphical Models and Image Processing 57(1) (1995), 55–66.
12. Mortenson, M. E., Geometric Modeling, John Wiley and Sons, New York, 1985.
13. Peternell, M.; Pottmann, H., *Computing Rational Parametrizations of Canal Surfaces*, Journal of Symbolic Computation 23 (1997), 255–266.
14. Patrikalakis, N.M., *Surface-to-surface intersections*, IEEE Computer Graphics and Applications 13(1) (1993), 89–95.
15. Rossignac, J.R.; Requicha, A.G., *Constructive non-regularized geometry*, Computer-Aided Design 23(1) (1991), 21–32.

A sampling algorithm computing self-intersections of parametric surfaces

A. Galligo[1] and J.P. Pavone[2]

[1] UMR CNRS 6621 Universite de Nice-Sophia Antipolis. Parc Valrose 06108 Nice Cedex 02, France. `galligo@math.unice.fr`
[2] INRIA. 2004 route des lucioles - BP 93 FR-06902 Sophia Antipolis. `jppavone@math.unice.fr`

Summary. In this paper we present a sampling algorithm which detects and describes the self-intersection locus of a parametric surface. We provide several criteria of injectivity, they serve to decompose the domain of the parametrization to get a family of smaller patches. The organization of our algorithm relies on a segmentation of the surface based on simple informations such as partial derivatives. Its implementation makes an important use of $2d$ and $3d$ bounding boxes trees. We provide some examples with timings and illustrations.

1 Introduction

We consider a surface S given by a parametrization (at least of class C^1) also denoted by S:

$$S : U \subset \mathbb{R}^2 \to \mathbb{R}^3,$$

defined over an open subset U of $\mathbb{R}^2$. A double point of S is defined as a pair of points $(p, q) \in U^2$, such that:

$$S(p) - S(q) = 0.$$

Any surface has a trivial double point set given by $D = \{(p,p) | p \in U\}$, so we consider that a surface is self-intersecting if there exists at least one $p \neq q$ such that $S(p) = S(q)$.
Note that no lower bound on the distance between p and q exists.
So we distinguish the points m such that:

$$\forall \epsilon > 0, \exists (p, q), p \neq q$$
$$|p - m| < \epsilon, |q - m| < \epsilon$$
$$S(p) = S(q)$$

and call them mitter points. Self-intersections components containing at least one mitter point are called local, otherwise we call them global.

A first order expansion indicates that mitter points are necessarily points where the Jacobian $J(S)(m)$ is rank deficient. Let us also remark that a double point is defined up to the equivalence relationship $(p,q) \equiv (q,p)$, i.e. symmetry. Because of these three problems, it is hard to compute self-intersections even in case of a polynomial parametrization of low degree. Indeed the degree of the intermediate equations to be solved increases by the removal of trivial solutions. Also mitter points introduce ill-conditioned solutions and dealing with symmetry introduces an over-cost. Algorithmic approaches similar to the ones developed for the problem of intersection, based on loop detection, separating the case of open and closed components, (see e.g. [5]) can be designed. We also mention approaches via approximate implicitization see e.g. [2] or more recently [12]. In this paper, we propose a pragmatic approach to efficiently computing the selfintersection locus of a procedural parametric surface. Such surfaces result from a sequence of operations specified by the user of a CAGD system (CAGD stands for Computer Aided Geometric Design); we present a sampling algorithm. The reasons of our choice are given below and discussed again in the conclusion. By sampling algorithm we mean the use of a grid of points and triangles which approximate the surface as in the definition given in [10]. The relevance of the approach depends on the sampling rate and the robustness of triangle intersections. The timings and illustrations given in the last section show that our implementation is able to deal with complicate examples, it also indicates that our method will be reliable in a real-word context.

1.1 Motivations

In a modern CAGD software, solids are represented by faces which are trimmed parametrized surfaces. In order to get well-defined solids, these surfaces must be self-intersection free as all the algorithms performing usual CAGD operations could be damaged by a self-intersection. Therefore, it is important to detect self-intersection, then overcome the difficulty or inform the CAGD user. In some situations such that offsetting, drafting, pipe construction, there exist some algorithms which compute self-intersections[3]; but nowadays shape generation algorithms are more and more complex and evolve quickly. So we believe that an algorithm-independent approach to self-intersection is an important issue. We propose a sampling algorithm because sampling is a very general and simple representation of the surfaces considered during the design process. Other representations such as NURBS are used as a standard for exportation of models, but systems can hardly be entirely based on such algebraic representations.

---

[3]see for example [13] and [14]

2 Main ideas

The principle of our algorithm is to decompose the initial domain into subsets where the map S is injective. Then the self-intersection problem is reduced to several simpler problems of intersection. For that purpose, we need sharp sufficient conditions for the injectivity of S restricted to an open set. We simplify this question by considering the injectivity of S composed with a projection, which is a stronger condition. In the following, F denotes a regular map from $\mathbb{R}^2$ to $\mathbb{R}^2$. U is a bounded and simply connected open subset of $\mathbb{R}^2$, $\delta(U)$ is the border of U. The couple (x, y) denotes the two coordinates functions of the map F in a fixed frame.

2.1 Injectivity criteria

We say that a curve C is simple, if it is closed, connected continuous and delimiting a connected open subset of $\mathbb{R}^2$. This concept was studied by C. Jordan during the XIX th century, so it is also called a Jordan curve. If F restricted to C is injective, then the curve $F(C)$ is also simple.

Definition 1. *F is said locally injective over U if its differential is injective at any points of U:*

$$\forall p \in U, |J(F)(p)| \neq 0.$$

In that case, in Topology one says that F is a non-ramified covering. But here we will not use this language, and we will provide complete elementary proofs.

Lemma 2. *If F is locally injective over U we have:*

$$\delta(F(U)) \subset F(\delta(U)).$$

Proof. The closure $\overline{U}$ is compact and F is continuous, so $F(\overline{U}) = \overline{F(U)}$ is compact, moreover $\delta(F(U))$ and $F(\delta(U))$ are subsets of $F(\overline{U})$. We must prove that $\delta(F(U)) \cap F(U)$ is empty. If $\exists N \in \delta(F(U)) \cap F(U)$ then $\exists M \in U$ such that $N = F(M)$. As F is a local diffeomorphism around M, N is in the interior of $F(U)$ so is not in $\delta(F(U))$; we get a contradiction.

Lemma 3. *Let C be a simple curve delimiting an open subset U of $\mathbb{R}^2$. Then, a non empty and strict part of C cannot be the border of an open subset of U.*

Proof. We can, by an homeomorphism, replace the pair (C, U) by the unit circle and unit disk. Assume V is a not empty strict and open subset of U, such that there exists a point M_0 on C not in the border $\delta(V)$. Pick a point M_2 in V, and consider $L := [M_0 M_2] \cap V$. Let M_1 be the point of L nearest to M_0. By construction M_1 is in the border $\delta(V)$, hence $M_1 \neq M_0$ and $M_1 \neq M_2$. As U is a disc, the line $M_0 M_2$ is a chord. So we deduce that $M_1 \in U$ therefore it is not in C. Therefore we get a contradiction.

Theorem 4. *If the border of U is a simple curve, F is locally injective over U and F is injective over $\delta(U)$. Then F is injective over U.*

Proof. F is locally injective over U so $\delta(F(U)) \subset F(\delta(U))$ by lemma 1. As $F_{|\delta(U)}$ is injective, hence $F(\delta(U))$ is a simple curve of $\mathbb{R}^2$. $F(U)$ is open and not empty so $\delta(F(U)) = F(\delta(U))$, by lemma 2. Assume there exist two points A and B such that $F(A) = F(B)$. We denote by v_A (included in U), v_B (included in U) and v (included in $F(U)$), the maximal triple of open connected subsets containing respectively A, B, $F(A)$, and such that F is a diffeomorphism from v_A to v and from v_B to v.
Because this condition is an open condition and the triple is maximal, we get: $\delta(v_A) \subset \delta(U)$ and $\delta(v_B) \subset \delta(U)$. Then lemma 2 implies that $v_A = U = v_B$, and we get a contradiction.

Theorem 5. *If F is locally injective over U and if for all x_0; the level set :*

$$L_{x_0} = \{x(u,v) = x_0\} \cap U$$

is connected. Then F is injective over U.

Proof. The curve $L = L_{x_0}$ included in U is regular: $|J(F)| \neq 0 \Rightarrow \nabla x \neq 0$ and connected by hypothesis. We proceed by contradiction: If there exist two points $A \neq B$ such that $F(A) = F(B)$ with $x(A) = x(B) = x_0$, we can build a continuous regular path $\gamma : [0,1] \to L$ injective, such that $\gamma(0) = A$ and $\gamma(1) = B$. The function $Y = y \circ \gamma$ is also continuous and regular and $Y(0) = Y(1)$. By Rolle's theorem, there exists a local extrema of Y, $t_0 \in]0,1[$; let $M_0 = \gamma(t_0)$.
As $|J(F)(M_0)| \neq 0$, F restricted to an open neighborhood V of M_0, is injective. Because t_0 is a local extrema of Y, there exist two values t_1 and t_2 as close as we want to t_0 such that $t_1 < t_0 < t_2$ with $Y(t_1) = Y(t_2)$. Now, as we have $x(\gamma(t_1)) = x(\gamma(t_2)) = x_0$, setting $M_1 = \gamma(t_1) \in V$, $M_2 = \gamma(t_2) \in V$, we have $F(M_1) = F(M_2)$. This contradicts the injectivity of F over V.

Theorem 6. *If F is locally injective over U, and $x_{|\delta(U)}$ has only two extrema. Then F is injective over U.*

Proof. We consider the level set curve L_{x_0}, it is a smooth curve and does not contain a closed loop ; otherwise a connected component of its image by F would be a compact segment of line and this would contradict the fact that $F(U)$ is open. We deduce that L_{x_0} is made of connected curves segments, and each of them intersects the border of U in two points. Let γ be a parametrization of $\delta(U)$.
Assuming that F is not injective, the previous theorem implies that there exist two components of L_{x_0} for a given x_0, and the function $\gamma \circ x$ takes four times the value x_0 over $\delta(U)$. By Rolle's theorem it has at least three local extrema, and we get a contradiction.

Theorem 7. *If F is locally injective over a rectangular domain R and x is monotonic with respect to u and with respect to v. Then F is injective over U.*

Proof. If x is monotonic it reaches its minimal and maximal value on $\delta(R)$. We can assume that $R = [0,1]^2$, and up to the substitutions $u = 1 - u$ and $v = 1 - v$ we can assume that $\frac{\partial}{\partial u}x > 0$ and $\frac{\partial}{\partial v}x > 0$.
So x strictly increases on the border and we have $x(0,0) < x(1,0) < x(1,1)$ and $x(0,0) < x(0,1) < x(1,1)$. Then there exist only two extrema $(0,0)$ and $(1,1)$ over $\delta(R)$ and F is injective by theorem 6.

Theorem 4 validates the criterion used by the authors of [15] in order to define their mesh self-collision algorithm. Their hierarchical triangulation keeps track of the dot-products of normal vectors with a predefined set of directions. Thus it defines areas where one or many projections can be locally injective and allows the use of an exclusion test based on the self-intersection of the projected border.

In our setting, theorem 7 is more interesting because we plan to use a grid. Theorem 7 says that any rectangular part of our domain is injective if it is monotonic and locally injective. By one hand, using a hierarchical representation of our grid, keeping track of partial derivatives and Jacobian signs associated to several projections, we can define large rectangular part of our grid which are self-intersection free. On the other hand, we can also use the "projected border" criterion for domains with a more complicated shape.

The first part of our algorithm is dedicated to define grid parts where all the local informations needed by our theorems are verified for a predefined set of projections. Then we apply theorem 6 in order to subdivide these parts into regions where S is injective. After that, we build two dimensional bounding box trees encoding each region. Then, we merge all these trees into a larger one encoding the whole grid. Finally all these structures are used to define a sampling algorithm computing self-intersection.

In the sequel, by abuse of language for a fixed F, when F restricted to U is injective, we will simply say that U is injective.

3 A sampling algorithm

Basic assumptions

We consider canonical projections of the surface S, i.e (x,y), (y,z), and (z,x). Our first goal is to define grid parts where these three projections are locally injective and monotonic for each coordinate function. For that purpose, we do not consider local self-intersection (i.e. mitter points), so we assume that all the grid elements are monotonic and *locally injective* for each projection.

We assume that the sample is given by a $n \times n$ grid of points $S_{i,j}$, then given a grid element,

$$Q_{i,j} = \left[\frac{i}{n} \frac{i+1}{n}\right] \times \left[\frac{j}{n} \frac{j+1}{n}\right] \tag{3.1}$$

we define its partial derivatives by a finite difference scheme around the point $S_{i,j}$. The monotonicity code of a grid element is by definition the list of the signs of its normal vector and its partial derivatives.

3.1 Monotonic regions

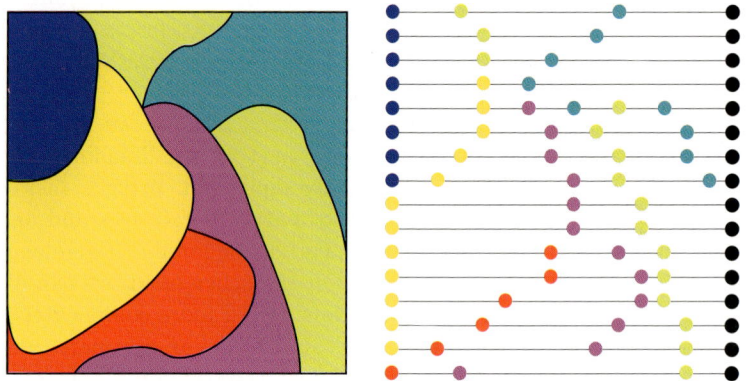

Fig. 1. Exact monotonic regions and marks given by a sample

For each line of the grid, we define the intervals where the monotonicity code is constant: We scan each line and mark positions where there is a variation in the code. We obtain an array of lines; each such (segment of) line is represented by two marks. These sequences of marks define intervals where the monotonicity code is constant. We decided to represent grid parts of constant code by linking, from one line to another, the marks defining intersecting intervals sharing the same code.

Remark 8. This representation by a list of intervals is in some sense suboptimal because it forbids regions which are disconnected along a line. However it is well suited for our purpose as it will appear in the next section.

3.2 Injective regions

Since a region is structured as a list of intervals, its border can be seen as the first and the last interval plus the right and the left bounds of the intervals.

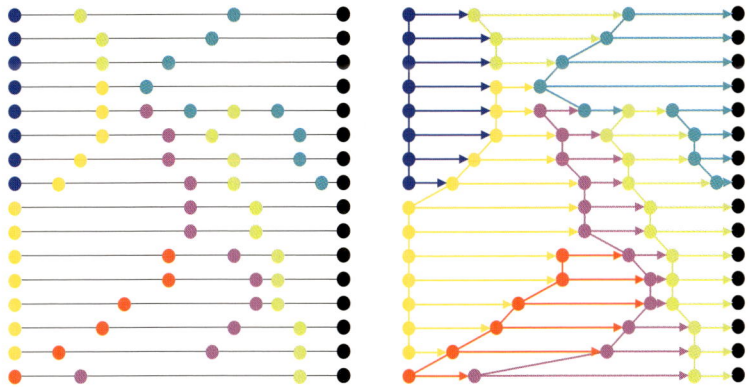

Fig. 2. step two: link the marks

Because the first and the last interval are monotonic local extrema of our function (x, y or z), if they exist along the border, are on the left or the right frontier of our regions.

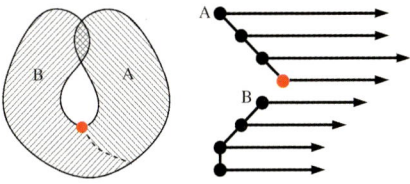

Fig. 3. subdivision of a region (image & list)

Theorem 9. *If i is the position of the first local extrema of x along the border of a region, then the region defined by $1..i$ (i.e the first i intervals), is injective.*

Proof. In order to simplify the notation, suppose for example that the two partial derivatives of x are positive and that the left bound of the ith interval is a local minimum x_0. The function x increases on this interval, therefore x_0 is a global minimum for the region $1..i$. Thus $1..i$ is injective by theorem 3. The argumentation is similar in the other cases.

So, once detected the local extrema of x, y, and z along the border of a region, we obtain injective subpart simply by splitting a list. The difficulty is concentrated in the computation of local extrema, because our curves are defined on a grid. In order to keep down the number of splits, we first select all the extrema for all the coordinate functions and then select the coordinate function, (above we took x) with the smallest number of extrema.

3.3 Region tree

We associate to a region represented by an interval list, as in subsection 3.1, a hierarchy H of rectangular boxes encoding it. This hierarchy is a binary tree where each node is a 2D box, it contains the boxes of its sons. The leafs of this tree provide a partition of the region (see figure 4). We build this hierar-

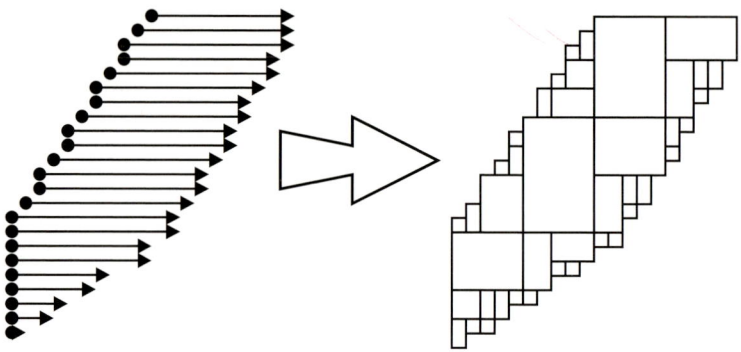

Fig. 4. leafs covering one of our region

chy starting from a box that contains our region, then performing a recursive combination of box and list subdivisions. For that purpose, we first build a binary tree T which will store all the informations needed by the procedure. It contains minima and maxima values of the left and right bounds of intervals for all the sublists obtained by binary splits.

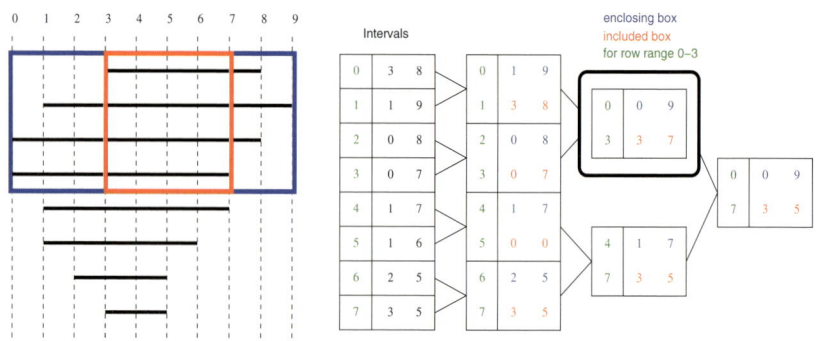

Fig. 5. min/max values associated to binary splits

Each node N of T corresponds to a subregion and provides a box $R_I(N)$ included in it and $R_E(N)$ a box that encloses it (see figure 5). The building time of this tree is linear with respect to the length of the interval list. Let us denote by $U(T)$ and $D(T)$ the left and right sons of T. If B is any rectangle, we denote by $W(B)$ and $E(B)$ the left and right part of B (see figure 6), and by $w(B)$ and $h(B)$ the width and height of the box B. The hierarchy tree H is defined recursively from T using the boxes given by the pseudo-code:

$$\begin{array}{l} B := B \cap R_E(T) \\ \text{If } B = \emptyset \text{ Or } B \subset R_I(T) \text{ Then } H := B. \\ \text{If } w(B) > h(B) \\ \text{Then } H_l := H(W(B), T) \\ \qquad H_r := H(E(B), T) \\ \text{Else } H_l := H(B, U(T)) \\ \qquad H_r := H(B, D(T)) \end{array}$$

where H_l and H_r denote the left and right sons of H.

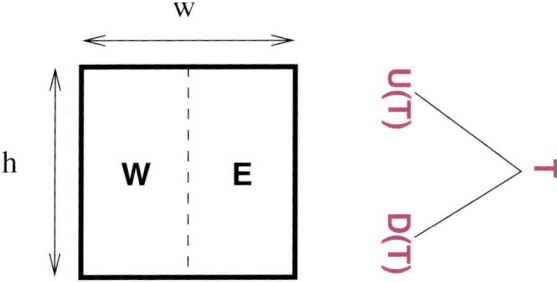

Fig. 6. notations

3.4 Hierarchy of regions

Once we have a hierarchical representation of each regions, we construct a neighborhood graph G on the set of our regions. We aim to get an almost balanced tree such that the leafs of the tree are the elements of G, i.e the regions. Recall that each region is itself represented by a tree. Starting from this graph we define a (new) hierarchy for the whole grid. We call size of a region the number of grid elements it contains. We start from n vertices representing our n regions and reference them into a heap based on their size,

and proceed as follows. While the heap contains more than one vertex,
- get the smallest one s,
- get v, smallest neighbor of s,
- build $g = s \cup v$,
- add it in the graph with neighborhood of s and v,
- suppress v and s from the heap and the graph.

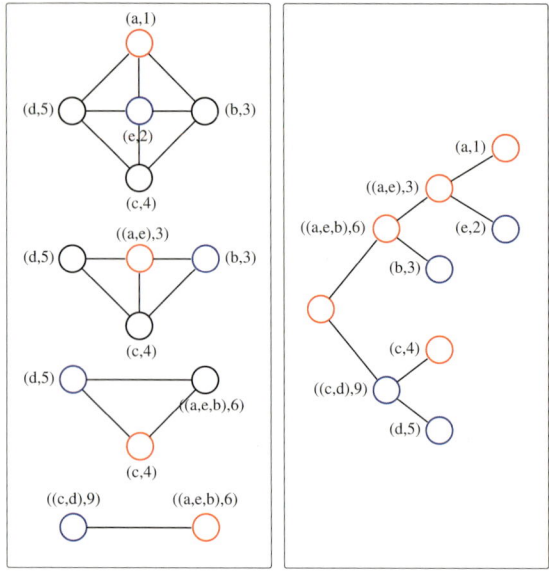

Fig. 7. conversion from graph to hierarchy

Each time we build a new vertex, we build a new hierarchy node. At the end we have a hierarchy for the whole grid, this process is illustrated on an example in figure 7. Nodes are given by a sequence of letters, a weight is associated to each nodes, the left part represents the process, the right part represents the result.

3.5 Hierarchy self-collision

The previous hierarchy (completed with the trees constructed in subsection 3.4) can be converted into a bounding box hierarchy in the image space. This conversion goes as follows: the leafs of the hierarchy are leafs of a region tree defining monotonic rectangular domain. So we can, by comparisons of values of x, y, z on the corners of these domains define an axis aligned bounding box for each leaf. When a node is not a leaf, its image box associated to the represented domain(which is defined by the set of its leafs), is the smallest box containing all the image boxes of its sons.

We define the collision of hierarchy nodes A and B by the set of pair (a, b) such that a (resp. b) is a leaf of A (resp. B), and such that the image box of a, denoted by $B_I(a)$ intersect the image box of b.
We handle collisions by the following pseudo-code for $C(A, B)$:

$$C(A,B) := \begin{vmatrix} \text{- }(A \text{ and } B \text{ are not leaves)} \\ \text{If } B_I(A) \cap B_I(B) = \emptyset \text{ Then } \emptyset. \\ \text{If } |B_I(A)| > |B_I(B)| \\ \text{Then } C(L(A), B) \cup C(R(A), B) \\ \text{Else } C(A, L(B)) \cup C(A, R(B)). \\ \text{- } ... \\ \text{- }(A \text{ and } B \text{ are leaves)} \\ C(A, B) := (A, B) \end{vmatrix}$$

here $|B_I(A)|$ denotes the size of the image box of A, its use prevents from the case where the subdivision of A is not useful ($B_I(A) \subset B_I(B)$ for example). $L(A)$ and $R(A)$ are the left and right sons of A. The special cases where A or B is a leaf is not shown.

From this straightforward use of bounding box trees we can define recursively the self-collision of a hierarchy H from collision and self-collision of its sons. The self-collision of the hierarchy of regions defines a set of leaf pair representing a subset of U^2 containing the self-intersection. Then we refine this location of self-intersection by dynamic updates of the bounding box trees during the preceding function (see figure 8). We can also use a finest strategy using binary search for leaf/leaf collision.

So we obtain a set of grid element pairs containing the self-intersection. These are the main steps of the algorithm, but it is worthwhile to consider several special cases of region/region intersection and hierarchy self-collision. For instance, if we handle a group of regions sharing a locally injective projection we can extract, as in [15], the border of their union in order to prevent the self-collision for the group.

Also, if some regions share the monotonicity of a projection, by the same process explained in section 3.3, we can get a hierarchical representation of their union . We use it to test inclusion of a leaf pair in a rectangular subdomain of the union during the search, or to clip them after a collision without dynamic updates. We will not give technical details about these implementations, although they are not completely trivial. Let us only remark that in case of local self-intersections, which often occur in CAGD, these optimizations do not have an important effect (10 to 20% of total time in our examples).

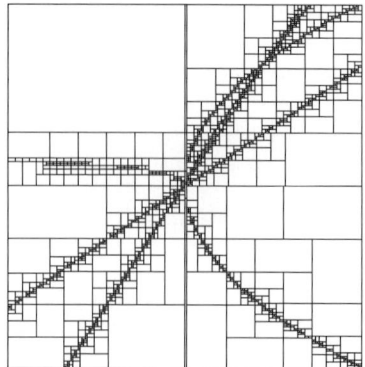 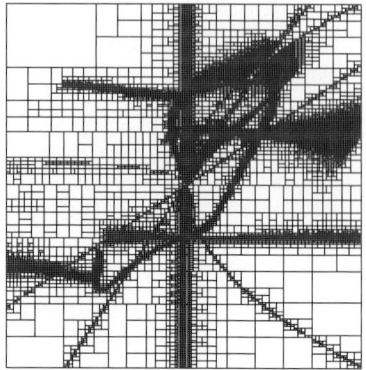

Fig. 8. leaves of bounding box trees before and after collisions

4 Curves reconstruction

Starting from the set of collisions previously built in section 3.5, we define self-intersection curves by using triangles intersections. The principle is to define a set of double curves which are lists of double points and link them with respect to proximity of their end points. We begin with a set of small double-curves (in fact small line segments) coming from triangles intersections. We use double-point instead of image points in the comparisons, it is less ambiguous but we have to compare end-points up to the equivalence relationship $(p,q) \equiv (q,p)$. Our implementation uses data-structures derived from kd-trees in order to define a proximity graph. The curve merging process is then based on the length of double-curves: we always link the smallest one with another curve in the set. This choice insures that if the two curves don't share the same orientation, the list reversal algorithm is done on the smallest curve.

Let us remark that beside the robustness problem due to triangle intersections, this curve retrieval process has currently one big problem: as it is based on proximity, it is unable to produce a topological description of the self-intersection. This point is illustrated in figure 9, the black curve crosses a singular point while the blue one is singular. However we believe that this kind of problems can be generally solved by extra evaluations of the surface derivatives during the linking process.

5 Experimentations

Figures (12,13,14,15) show some examples of self-intersections computed by our algorithm. Each figure shows several views of the surface with its self-intersection curves and a graphical representation of the time spent by our

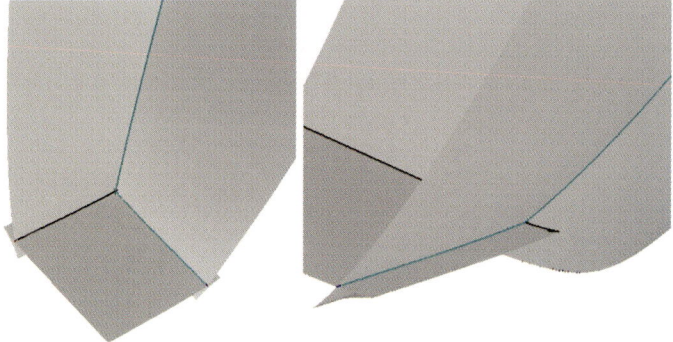

Fig. 9. offset of a rounded corner with a singular point

algorithm. These timings are given for three steps: sampling (red curve), building (green curve) which correspond to hierarchy building, and search (blue curve) which correspond to collisions detection and curves reconstruction, these curves depend on n, assuming a $n \times n$ grid. On some example the bounding volumes hierarchy leaves are shown (blue boxes).

As one can see on the timings table, our algorithm tends to be as fast as the sampling process. Therefore, we think we have reached a limit of sampling method efficiency. The key of efficiency is monotonicity because it allows a fast construction of bounding box trees. For that reason, we don't believe that oriented bounding boxes trees ([4]) could be interesting in our context. Our approach can be used with n-dop hierarchies ([7]), just by taking another monotonicity code definition.

6 Conclusion

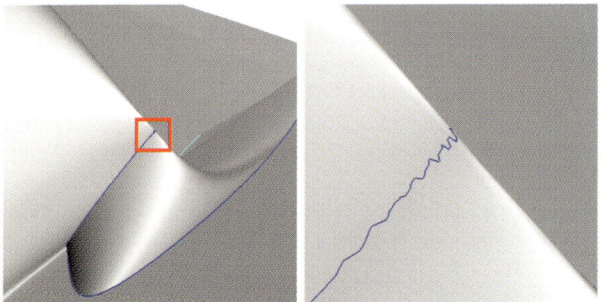

robustness problem around a mitter point.

In this paper we presented a fast algorithm for detection of self-intersection. It have been tested on real CAGD examples by **think**[3] as a plug-in of

thinkdesign and we have a good feedback about its performance and reliability. So we think it can be useful for CAGD users who wants to verify the correctness of their models. Minors modifications that will not increase too much its computation time, can lead to improvements.

Let us remark that most of the grid elements will lie in monotonic regions of S. If our algorithm misses a self-intersection curve, it must lie completely on grid elements that are across the exact boundary of regions. Since self-intersection is an unwanted phenomenon, this case must occur accidentally. Moreover, we could easily improve this point by computing more precisely the positions of monotonicity changes by Newton method. Therefore we claim that the self-collision detection algorithm based on bounding box trees is reliable.

From our point of view, the principal part of our work to be improved, is the use of triangle intersections: our triangles intersections could be replaced by turning points computation using a Newton-like method.

As noted by [3], for surface intersection, it is worthwhile and more reliable to express a differential information in order to detect existence of loops. In [3], the authors propose to examine the gradient field of a distance function from F to S. However, in their setting, their distance function is well defined only if the surfaces are self-intersection free and it relies on the projection of a dense set of points. As a consequence our data-structures of regions (bounding box trees on injective regions), are well suited to use their framework[4]: the distance function is well defined and projections can be handled by our trees.

We will return to these questions in a future work.

We thank the referees for their remarks and suggestions.

References

1. L. Andersson, T. Peters, N. Stewart. Self-intersection of composite curves and surfaces. *Comp. Aided Geom. Design*, 15:507–527, 1998.
2. T. Dokken. Approximate implicitization for surface intersection and self-intersection. In *ECITTT Euroconference on CAE Integration - tools, Trends and Technologies*, 2001.
3. F. Wolter G. Kriezis, N. Patrikalakis. Topological and differential equation methods for surface intersections. *Computer-aided Design*, 24(1):41–55, 1990.
4. S. Gottschalk, M. C. Lin, D. Manocha. OBBTree: A hierarchical structure for rapid interference detection. *Computer Graphics*, 30(Annual Conference Series):171–180, 1996.
5. M. E. Hohmeyer. A surface intersection algorithm based on loop detection. In *ACM Symposium on Solid Modeling Foundations and CAD/CAM Applications*, pages 197–207, 91.
6. M. Hosaka. *Modeling of Curves and Surfaces in CAD/CAM*. Springer-Verlag, New York, NY., 1992.

---

[4]note that injectivity of a projection is also needed in [9]

7. J. T. Klosowski, M. Held, J. S. B. Mitchell, H. Sowizral, K. Zikan. Efficient collision detection using bounding volume hierarchies of k-DOPs. *IEEE Transactions on Visualization and Computer Graphics*, 4(1):21–36, /1998.
8. D. Lasser. Self-intersections of parametric surfaces. In *Proceedings of the International Conference on Engineering Graphics and Descriptive Geometry, Vienna*, Volume 1, pp. 322–321, 1988.
9. Y. Ma, Y.-S. Lee. Detection of loops and singularities of surface intersections. *Computer-aided Design*, 30(14):1059–1067, 1998.
10. N. M. Patrikalakis. Surface-to-surface intersections. *IEEE Computer Graphics and Applications*, 13(1):89–95, January 1993.
11. T. W. Sederberg, R. J. Meyers. Loop detection in surface patch intersections. *Comp. Aided Geom. Design*, 5(2):161–171, 1988.
12. J. B. Thomassen. Self-intersection problems and approximate implicitization. In *Computational Methods for Algebraic Spline Surfaces, Springer*, pages 155–170, 2005.
13. W. Cho, T. Maekawa, N. M. Patrikalakis. Computation of self-intersections of offsets of bezier surface patches. *Journal of Mechanical Design, ASME Transactions, 119(2):275-283*, 1997.
14. T. Sakkalis, T. Maekawa, N. M. Patrikalakis, G. Yu. Analysis and applications of pipe surfaces. *Comp. Aided Geom. Design*, 15:437–458, 1998.
15. P. Volino, N. Magnenat Thalmann. Collision and self-collision detection: Efficient and robust solutions for highly deformable surfaces. In Dimitri Terzopoulos and Daniel Thalmann, editors, *Computer Animation and Simulation '95*, pages 55–65. Springer-Verlag, 1995.

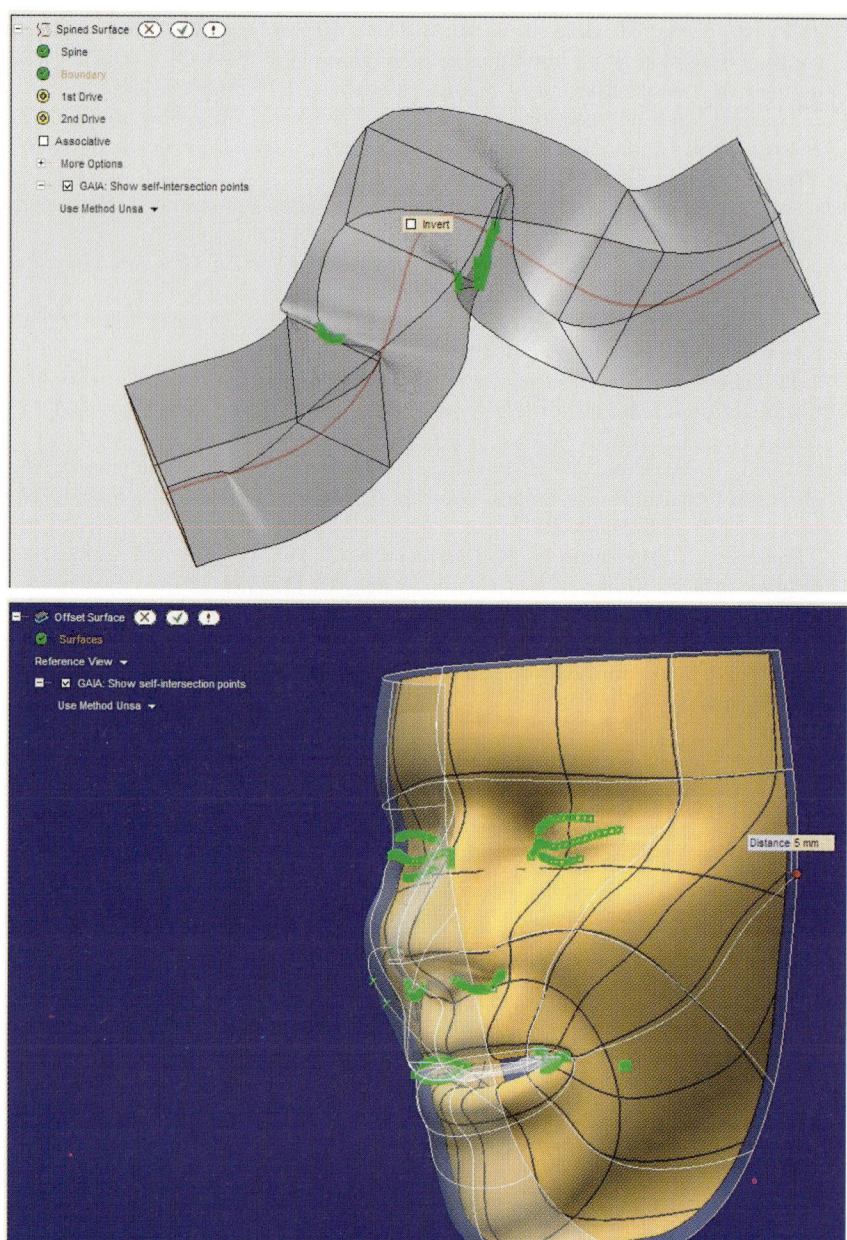

Fig. 10. self-intersection detected in real-time during sweep and offset operations (screen-shots made by think3 for the GAIA II project)

A sampling algorithm computing self-intersections of parametric surfaces   201

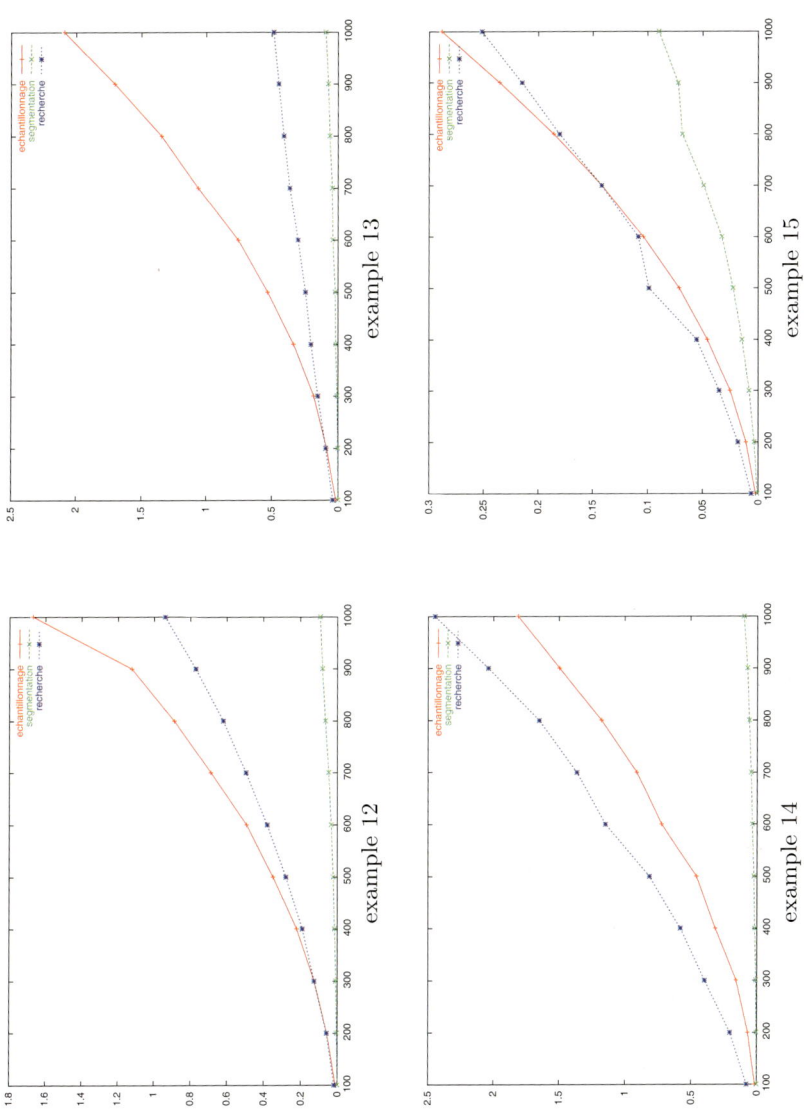

Fig. 11. timings of examples

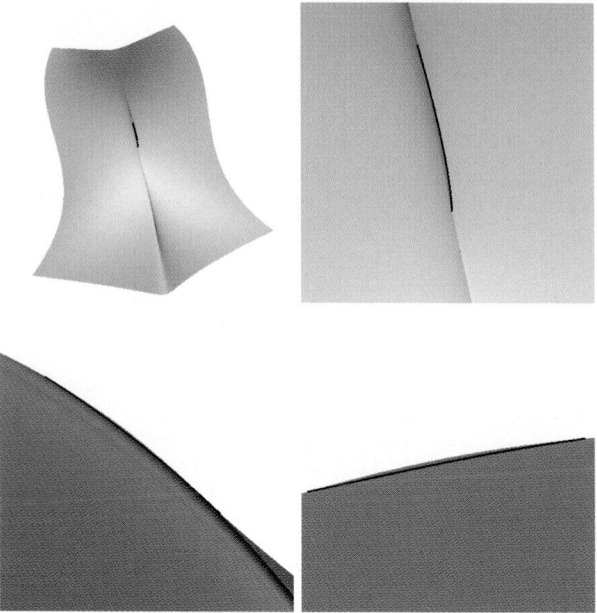

Fig. 12. this example is a bi-cubic patch with a small local self-intersection.

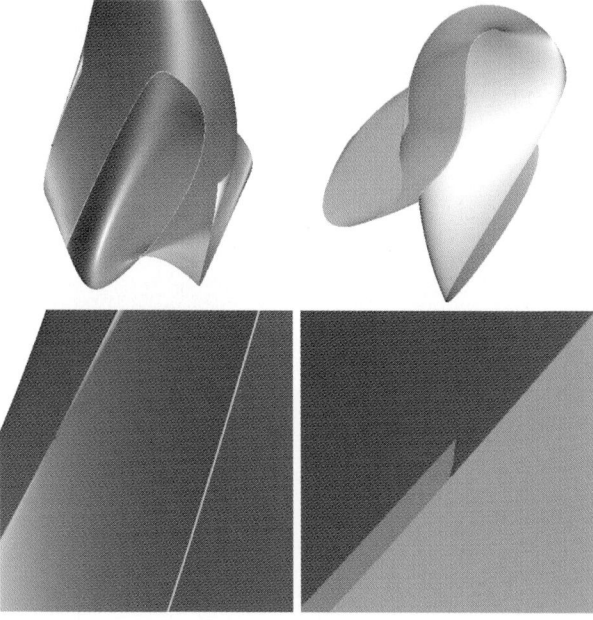

Fig. 13. Bezier surface of degree $(4, 4)$ with two local self-intersection.

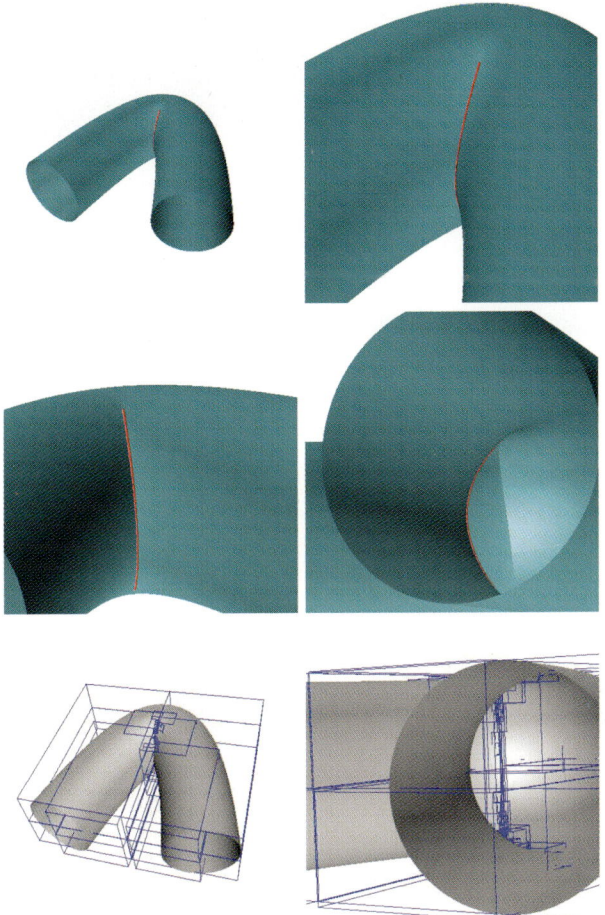

Fig. 14. a pipe, in this example computation time is near to sampling time.

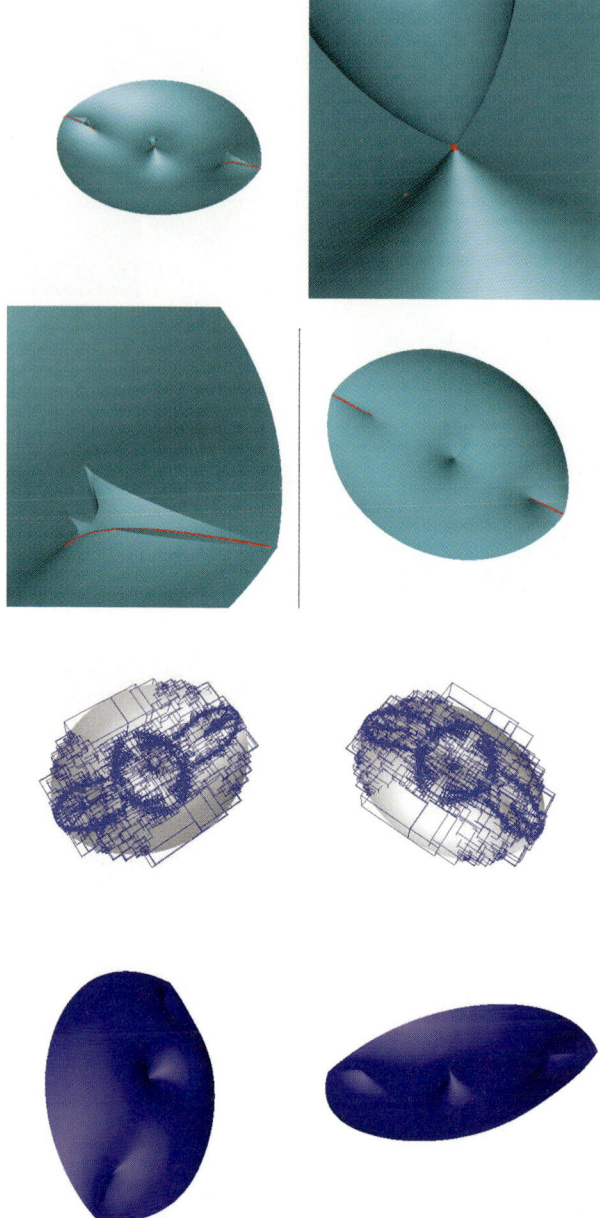

Fig. 15. an offset of the blue surface below, with an isolated singularity.

Elimination in generically rigid 3D geometric constraint systems

Jörg Peters, Meera Sitharam, Yong Zhou, and JianHua Fan

University of Florida jorg@cise.ufl.edu

Summary. Modern geometric constraint solvers use combinatorial graph algorithms to recursively decompose the system of polynomial constraint equations into generically rigid subsystems and then solve the overall system by solving subsystems, from the leave nodes up, to be able to access any and all solutions. Since the overall algebraic complexity of the solution task is dominated by the size of the largest subsystem, such graph algorithms attempt to minimize the fan-in at each recombination stage.

Recently, we found that, especially for 3D geometric constraint systems, a further graph-theoretic optimization of each rigid subsystem is both possible, and often necessary to solve wellconstrained systems: a minimum spanning tree characterizes what partial eliminations should be performed before a generic algebraic or numeric solver is called. The weights and therefore the elimination hierarchy defined by this minimum spanning tree computation depend crucially on the representation of the constraints. This paper presents a simple representation that turns many previously untractable systems into easy exercises. We trace a solution family for varying constraint data.

1 Introduction and motivation

Specifying geometry via constraints is an elegant and succinct approach to defining geometric composites in applications such as computer aided design, robotics, molecular modeling and teaching geometry (see e.g. [3, 22, 21]). However, being able to navigate to and access *all* valid configurations satisfying the constraints is a difficult, ongoing challenge, especially in the practically relevant 3D case. This paper develops an automated strategy for resolving generically rigid configurations by developing an interesting connection between the algebraic complexity of the underlying polynomial system and the topology of its set of its subsystems.

A *geometric constraint system* relates a finite set of geometric objects by a finite set of constraints. Constraints are represented as algebraic equations and inequalities whose variables are the coordinates of the participating geometric objects. For example, a distance constraint of d between two points (x_1, y_1) and (x_2, y_2) in 2D is written as $(x_2 - x_1)^2 + (y_2 - y_1)^2 = d^2$. The *solution or realization* of a geometric constraint system is the (set of) *real zero(es)* of the algebraic system, each representing a valid choice of position, orientation and any other parameters of the geometric elements in the Euclidean plane, on the sphere, or in Euclidean 3D space. Wellconstrained systems have a finite but potentially very large number of zero-dimensional solutions. Underconstrained systems have infinitely many solutions and overconstrained systems have no solution unless they are consistently overconstrained. Wellconstrained or consistently overconstrained systems are called *rigid*.

Modern geometric constraint solvers recursively decompose, solve and recombine the polynomial geometric constraint system according to an optimized partial ordering, called DR-plan. As graph algorithms, they address *generically rigid* systems, i.e. systems that are rigid except possibly for polynomial dependencies arising from specific geometric inputs. A subgraph in a DR-plan corresponding to generically rigid subsystems is called *cluster* (see Section 2 for the formal definition). *Resolving a cluster* $C = \cup_{i=1}^{n} C_i$ means computing position and orientation of all C_i in the common coordinate system of C so that all constraints are satisfied.

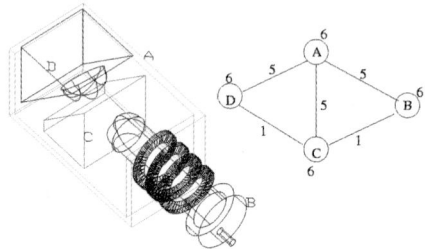

Fig. 1. An underconstrained 3D example and its dof constraint graph. Each object (housing A, screw B, clipped prism C and prism D) has 6 dofs. Incidence of B, C, D with A removes 5 dofs each. Then each of the pairs AB, AC, AD is underconstrained with one extra degree of freedom and the remaining two pairs BC and CD are connected by distance constraints that remove 1 dof each, leaving it with a density of -7.

This paper develops an automated strategy for resolving generically rigid clusters by further developing an interesting connection between the algebraic complexity of a cluster's polynomial system and the topology of its set of child clusters. It shows how a careful formulation of the problems can reduce the algebraic complexity so that many previously untractable systems turn into easy exercises.

Section 2 defines key terms, such as constraint graph, degree of freedom analysis and generic rigidity. And it outlines a general graph-based algorithm for partial elimination within one cluster. Section 3 defines the representation and the weights that drive the partial elimination. Section 4 discusses the solution of the remaining, active system and Section 5 illustrates the findings by tracing the solutions of a 3D geometric constraint system for a varying parameter.

2 Graph structures for elimination

The decomposition and ordering of elimination of the constraint system is optimized using three graphs: the constraint graph, the DR-plan and the overlap graph. The constraint graph represents the overall system of constraints, the DR-plan is a directed acyclic graph that represents the decomposition (and later recombination) of the constraint system into clusters. The overlap graph is the focus of this paper. It guides the partial resolution of the cluster constraints until only a small active system of active constraints remains to be solved by a general purpose solver.

Rather than focussing directly on elimination within a cluster (addressed in the subsection on overlap graphs, last in this section), this section sketches the larger picture including the DR-plan to show that, once we can solve clusters, we can solve the whole constraint system. Moreover, there are some subtle issues concerning well-posed constraint systems that need to be discussed for completeness.

Constraint Graph, density, cluster and rigidity: A geometric constraint graph $G = (V, E, w)$ corresponding to geometric constraint system is a weighted graph with n vertices V representing geometric objects and m edges E representing constraints. The weight $w(v)$ of a vertex v counts the degrees of freedom (*dofs*) of the object represented by v and the weight $w(e)$ of an edge e counts the dofs removed by the constraint represented by e. Figure 1 illustrates a small 3D constraint system and its dof constraint graph. A subgraph $A \subseteq G$ that satisfies

$$d(A) := \sum_{e \in A} w(e) - \sum_{v \in A} w(v) \geq -d_0$$

is called *dense* and $d(A)$ is called *density* of A. Here d_0 is a constant, typically $\binom{D+1}{2}$, i.e. the dofs of a rigid body in D dimensions. For 2D Euclidean geometry, we expect $d_0 = 3$ and for 3D geometry $d_0 = 6$. For a rigid body fixed with respect to a global coordinate system, $d_0 = 0$.

The following purely combinatorial notions for computing a good decomposition are based on density.

A *dof-rigid cluster*, short cluster, is a constraint graph all of whose subgraphs, including itself, either have density at most $-d_0$ (wellconstrained) or can be replaced by well-constrained subgraphs so that G remains dense (well-overconstrained).

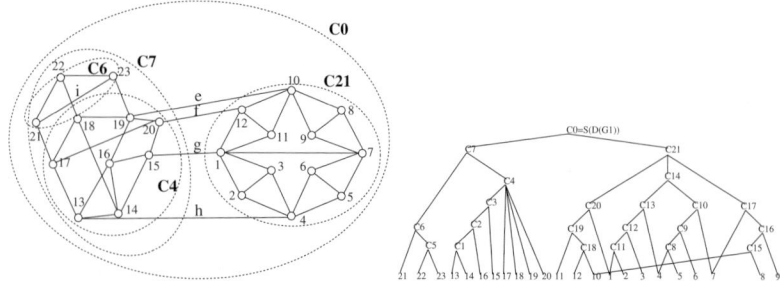

Fig. 2. Fan-in minimizing DR-plan (*right*) of a constraint graph (*left*) ; all vertex weights are 2, all edge weights 1.

A dense graph is *minimal* if it has no dense proper subgraph. Minimal dense subgraphs are clusters, but a dense graph that is not minimal can be underconstrained, i.e. not a cluster: the apparently correct density can be the result from summation with another subgraph of density greater than $-d_0$.

A constraint system is *generically rigid* if it is rigid (does not flex or has only finitely many non-congruent, isolated solutions) for all nondegenerate choices of coefficients of the system. A generically rigid system yields a cluster, but the converse is not always the case. In 3D, there are wellconstrained, generically non-rigid clusters due to the presence of generic 'hidden' *constraint dependencies* such as the 'banana' and 'hinge' configurations [6, 16, 17, 15]. These can result in overconstraints that are not detectable by a dof count.

Since to date, no tractable, combinatorial method of determining generic rigidity has been proven (see conjectures in [6, 15, 30]), we restrict the scope of our approach to the large (and possibly complete) class of constraint graphs whose rigidity is fully verified by the DR-plan [30] (2003 version) based on [12, 15, 29, 28]. This verification imposes a solving priority on the DR-plan and some dependences between clusters that do not contain one another but covers all known constraint graph dependencies. Therefore, in the following, we may assume that all hidden or explicit overconstraints have been removed.

DR-plan, covering sets, trivial subgraphs, completeness: The DR-plan of a constraint graph G is a directed acyclic graph whose nodes represent clusters in G, and whose edges represent containment. The leaves or sinks of the DR-plan are all the primitive clusters of G and the roots or sources are all the maximal clusters of G. For rigid graphs, the DR-plans have a single source but there could be many DR-plans for G. The DR-planner is a graph algorithm that generates the decomposition-recombination plan by working bottom up: at stage i, the DR-plan picks a wellconstrained cluster C_i in the current constraint graph G_i, and uses an abstract, unevaluated simplification of C_i to create a transformed constraint graph G_{i+1}. Since the complexity of finding real zeroes of sparse polynomial systems is exponential in the number of variables, the size of the largest subsystem in a DR-plan dominates the complexity of constraint solving, and DR-planners use a combinatorial degree of freedom (*dof*) analysis to minimize the number of child clusters [4, 11], or, equivalently, the fan-in, to isolate clusters (Figure 2).

Trivial subgraphs are subgraphs that correspond (after resolving incidence constraints) to a single 2D or 3D point or to a line segment in 3D. Whenever two rigid clusters overlap on a non-trivial subgraph, the cluster induced by their union is rigid. A *covering set* S of a cluster C is a set of child clusters $C_i \neq C$ whose union covers all the vertices of C. A covering set S is a *complete set of maximal clusters* if one of the following two conditions holds. Either it consists of only two clusters whose intersection is a non-trivial subgraph, or it covers all edges in C and every C_i is *maximal*, i.e. the only proper cluster of C that contains C_i is one that intersects every other cluster in S on a non-trivial subgraph. Note that S is not unique and we will use later on that there can be several covering sets of maximal clusters within S.

Even for a wellconstrained cluster C, it is still a nontrivial task to pick a guaranteed stable, independent system of polynomial equations corresponding to C. Moreover, shared objects between clusters may appear to have inconsistent coordinates due to numerical roundoff (see [19] for a description of and solution to this problem, which is not our focus here.)

Overlap Graph, spanning tree and active constraints. Figures 3 and 4 illustrate the concepts to be discussed. The *overlap graph* of a subset S of child clusters of a cluster C is an undirected graph. Its vertices are the clusters in S and there is an edge (i, j) with weight $w(k)$ if child clusters C_i, C_j overlap in a trivial subgraph with k distinct points; $w(k)$ counts the dofs after one cluster is expressed in the coordinate system of the other, e.g. $k = 1$ for a rotation about a common edge between C_i and C_j.

Any spanning tree T of the overlap graph of the clusters in S induces a system of equations for resolving C. Since C is assumed to be wellconstrained, the number of variables and equations equals the total edge weight of the the spanning tree. Most sparse polynomial system solvers such as [10, 31] (geometric constraint systems are sparse) take time exponential in the number of variables. Hence the overwhelming factor in the algebraic complexity of the system is the number of variables which is counted by the sum of weights. Considering all covering sets S of C, we therefore select all spanning trees that minimize this sum.

The ordering expressed in each spanning tree defines a partial elimination of constraints. The remaining constraints (non-tree edges in S) are called *active* and form the active system of equations that must be solved. Active constraints represent, for example distance and point-matching of cluster pairs in S. While the main goal is to reduce the number of variables, the polynomial degree and separation of variables of the active system is a secondary consideration since the time complexity is polynomial in these parameters. The choice of a root of S strongly influences degree and separation as we will see in the next section.

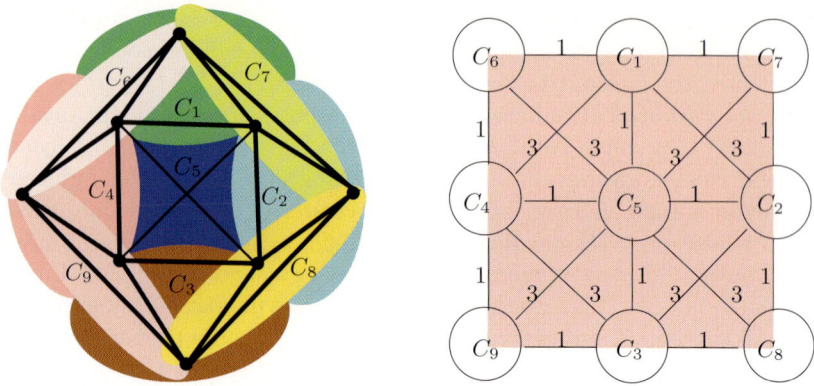

Fig. 3. (*left*) Constraint graph of problem quad: given are 8 vertices (indicated as •) and 18 distance constraints (edges). The graph structure is that of a square-base pyramid (see also Figure 5, inset) with the apex split into four vertices and the triangular faces folded down. The vertices are not labeled since they have different labels (and different local coordinates) in different clusters C_i. Incidence of these different instances of a vertex has to be explicity enforced. The edges are not labeled with distances, since geometry does not influence the construction of the overlap graph. In Section 5, specific edge lengths are assigned. (*right*) The corresponding weighted overlap graph (all remaining edges of the complete graph are of weight 6 and are omitted).

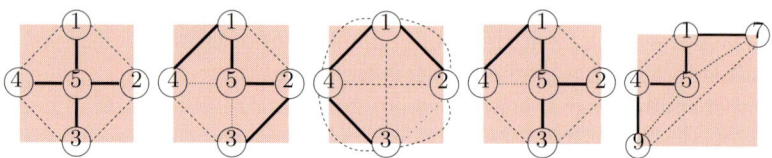

Fig. 4. Five spanning trees (*solid* edges) of covering sets of quad (same layout as in Figure 3(*right*) ; label i stands for C_i). The sum of weights are from left to right 4,8,9,6 and 4. These numbers match the number of *dashed* distance constraints (= 1 constraint) between clusters and *dotted* overlaps that are not tree edges (sharing of nodes = 3 constraints). When optimally rooted (with root C_5), the sum of the depths of the first, leftmost tree is 4, and for the rightmost, it is 6. Therefore the first tree is preferred for the partial elimination.

3 Partial elimination

Based on the characterization in Section 2, we can optimize the partial elimination of a cluster C by the following algorithm

(1) Determine $\mathcal{S}_C$, the set of all minimum weight spanning trees for all covering sets S of C (first and last tree in Figure 4) and

(2) minimize the sum of the depths of all nodes for all choices of rooted trees in $\mathcal{S}_C$.

In general, due to the large number of possible covering sets, the above algorithm requires solving an NP-complete problem. Moreover, keeping track of the spanning trees with the low complexity requires complex data structures [20]. Fortunately, in practical applications, the number of child clusters is no more than ten and even an exhaustive search through all covering sets S is an efficient option.

This section now explains the partial elimination implied by the spanning tree and characterizes the degree of the resulting active equations. Once we solve the active system, recombining clusters according to the DR-plan translates solutions of the clusters into solutions of the overall constraint system. The goal of partial elimination is to express, at each level of the spanning tree, all child cluster in the common coordinate system of the current parent cluster. For a, an instance of a point in the cluster C_i, we write

$$\mathbf{x}_{a,C_i} := \begin{bmatrix} x_{a,C_i} \\ y_{a,C_i} \\ z_{a,C_i} \end{bmatrix} \in \mathbb{R}^3 \text{ the coordinates of } a \text{ in } C_i\text{'s local coordinate system}$$

$$\mathbf{x}_{a,C_j,C_i} := \begin{bmatrix} x_{a,C_j,C_i} \\ y_{a,C_j,C_i} \\ z_{a,C_j,C_i} \end{bmatrix} \in \mathbb{R}^3 \text{ the coordinates of } a \text{ in } C_j\text{'s coordinate system.}$$

When, as is usually the case, two clusters do not join to form a composite rigid cluster but can moved with respect to one another, the coordinates $\mathbf{x}_{a,C_j,C_i}$ are parametrized in these dofs and the dofs will appear as variables in the active system. These dofs depend, for example, on rotation angles. Once the active system is solved, the parametrized coordinates are resolved to a final position $\mathbf{x}_{a,C}$. We express the transformation experienced by a child node, in terms of the following transformations:

T_a translation that maps a to the origin. (T_a^{-1} maps the origin to a.)
R_{ab} rotation that maps $b - a$ to the x-axis.
M_{bc} the matrix $[b, c, b \times c] \in \mathbb{R}^{3 \times 3}$ whose columns span a coordinate system in $\mathbb{R}^3$.
T undetermined translation (three dofs).
R undetermined rotation about the x-axis (1 dof).
Q undetermined unit quaternion (orientation, 3 dofs).

If C_j and C_i overlap in k points, mapping each point, say $\mathbf{x}_{a,C_i}$ to $\mathbf{x}_{a,C_j,C_i}$, as required by the overlap leaves degrees of freedom in terms of the undetermined maps T, R and Q as follows.

k	overlap points	map
3	$(a,b,c) \to (a',b',c')$	$T_{a'}^{-1} M_{b'-a',c'-a'} M_{b-a,c-a}^{-1} T_a$
2	$(a,b) \to (a',b')$	$T_{a'}^{-1} R_{b'-a'}^{-1} R R_{b-a} T_a$
1	$(a) \to (a')$	$T_{a'}^{-1} Q T_a$
0	none	QT

Here $(a, b) \to (a', b')$ means that $\mathbf{x}_{a,C_j,C_i} = \mathbf{x}_{a',C_j}$ and $\mathbf{x}_{b,C_j,C_i} = \mathbf{x}_{b',C_j}$, i.e. a and b are mapped to their two counterparts a' and b', respectively. If there is no overlap, i.e. $k = 0$, then the orientation Q and translation T are completely free.

We assume for now that the overlap points are in general position. According to Section 2, the weight of each edge in the overlap graph corresponds to the dofs of the overlap of the pair of clusters connected by the edge. Since we need zero parameters to describe the position and orientation of a cluster with respect to another if the two share three points (rigid joint), one if they share an edge, three if they share a point and six if they share no point,

$$w(0) = 6, \quad w(1) = 3, \quad w(2) = 1, \quad w(3) = 0.$$

The challenge is now to formulate the active constraints as low degree polynomial constraints whose dofs match the weight $w(k)$. Choosing the rotation angles as dofs matches the weight but does not lead to polynomial equations. With

$$R := \begin{bmatrix} 1 & 0 & 0 \\ 0 & c & -s \\ 0 & s & c \end{bmatrix}, \quad s^2 + c^2 = 1, \text{ and}$$

$$Q := \begin{bmatrix} 1 - 2q_2^2 - 2q_3^2 & 2(q_1q_2 + q_0q_3) & 2(q_1q_3 - q_0q_2) \\ 2(q_1q_2 - q_0q_3) & 1 - 2q_1^2 - 2q_3^2 & 2(q_2q_3 + q_0q_1) \\ 2(q_1q_3 + q_0q_2) & 2(q_2q_3 - q_0q_1) & 1 - 2q_1^2 - 2q_2^2 \end{bmatrix}, \quad \sum_{i=0}^{3} q_i^2 = 1,$$

we do obtain quadratic polynomial equations in the dofs c,s and q_i, $i = 0,1,2,3$. However, the weights are too high: we have two variables for $k = 2$ and four variables for $k = 1$ instead of one and three. A simple, but very effective insight is that we can parametrize the variables c, s, q_j by stereographic projection:

$$c := \frac{1 - t_0^2}{1 + t_0^2}, \quad s := \frac{2t_0}{1 + t_0^2}, \quad q_0 := \frac{1 - \sum_{i=1}^{3} t_i^2}{1 + \sum_{i=1}^{3} t_i^2}, \quad q_j := \frac{2t_j}{1 + \sum_{i=1}^{3} t_i^2}, j = 1, 2, 3.$$

This yields

overlap points k	dofs $w(k)$	(rational) degree in the t_j of the transformation
3	0	0
2	1	2
1	3	4
0	6	4 (1 for T)

The partial elimination now proceeds as follows. We traverse the selected, minimal, rooted spanning tree in depth first order. At each node, we express the child nodes (and all their children which are already expressed in the child's coordinate system) in the parent's coordinate system. The constraints obtained after this partial elimination, are called active. Since we introduce new variables with each step of the partial elimination, the *coordinate degree* of the numerator and of the denominator of the constraint systems is 2 throughout the elimination. (The coordinate degree is an n-tuple listing the degrees for each variable, e.g. (2,2) for x^2y^2.) In our examples, the active system consists of point-matching constraints and distance constraints. After clearing the denominator, the coordinate degree of a point-matching constraint is 2 and that of a distance constraint is 4.

Note that the total degree increases with the depth of the spanning tree. Moreover, since all descendant clusters are transformed with the child cluster, the equations are more separated and sparse if the spanning tree is shallow; and that means less work.

4 Solving the active system

We are interested in real solutions to the constraint system and currently use a recursive subdivision solver for tensor-product Bézier polynomials [14] which allows us to focus on a finite domain and real solutions. In fact, we encountered the noteworthy problem when we attempted to solve the equations of the example quad (see Figure 3) with an algebraic solvers, the online version of Synaps [1]. Since algebraic solvers typically consider both real and complex solutions, we did not receive an answer to quad, whose generically rigid constraints are rigid over the reals (isolated roots) but algebraically dependent over the complex numbers, yielding a 1-parameter family of complex solutions. (We did eventually get our results confirmed by Synaps after substituting one correct real value for one of the parameters).

When using a subdivision-based solver on the standard Bézier domain $[0..1]^N$, there is a price to pay for the stereographic parameterization. For numerical stability, we restrict the parametrization to $t_i \in [-1..1]$ and create another active constraint system for $1/t_i \in [-1..1]$. Change of variables then maps each finite domain to the Bézier domain $[0..1]$. This yields up to 2^N separate constraint systems when there are N variables. However, since every subdivision step generates in principle an exponential number of subproblems (to be efficiently pruned), the cost of starting with an exponential number of systems is negligible and preferable to a higher number of variables and equations. For the particular solver, we also needed to filter out duplicate roots obtained on the multiply covered boundaries of the subdivision domains.

5 Solutions to families of constraint problems

As our main example, we consider the constraint system shown in Figure 3. The eight nodes of quad can be viewed as a Stewart mechanism with a quadrilateral fixed base and a quadrilateral work platform. (Below, we will expand and shrink the work platform according to a parameter r, while in typical engineering applications the work platform is moved by adjusting the edge lengths between the two platforms.) Evidently, choosing the minimal covering set (with four clusters, Figure 4, center) does not yield the best partial elimination since the sum of dofs is 9. The minimal dof sum is 4, and the leftmost tree, with root C_5 is preferred to the rightmost tree, because its depth is 1. That is, the algorithm *automatically* generates the reduced, active system that we might have derived, without proof of minimality, by intuitive selection: fix cluster C_5 as the quadrilateral base platform of the Stewart mechanism and parametrize the four points of the moving platform of the Stewart mechanism each by one parameter. There are four overlap constraints, between C_5 and $C_j, j = 1, 2, 3, 4$, that imply and hence allow to automatically discard, the overlaps between C_i and $C_j, j := i \mod 4+1$. The coordinates of the root cluster C_5 are

$$\mathbf{x}_{1,C_5} := \begin{bmatrix} 1 \\ 1 \\ 0 \end{bmatrix}, \quad \mathbf{x}_{2,C_5} := \begin{bmatrix} -1 \\ 1 \\ 0 \end{bmatrix}, \quad \mathbf{x}_{3,C_5} := \begin{bmatrix} -1 \\ -1 \\ 0 \end{bmatrix}, \quad \mathbf{x}_{4,C_5} := \begin{bmatrix} 1 \\ -1 \\ 0 \end{bmatrix},$$

a 2-unit square. The local coordinates of each of the cluster C_i, $i = 1, 2, 3, 4$ are

$$\mathbf{x}_{1,C_i} := \begin{bmatrix} -1 \\ 1 \\ 0 \end{bmatrix}, \quad \mathbf{x}_{2,C_i} := \begin{bmatrix} 1 \\ 1 \\ 0 \end{bmatrix}, \quad \mathbf{x}_{3,C_i} := \begin{bmatrix} 0 \\ 4 \\ 0 \end{bmatrix}.$$

The four distance constraints are

$$\|\mathbf{x}_{3,C_i} - \mathbf{x}_{3,C_j}\| = r. \quad j := i \mod 4 + 1.$$

Figure 5 illustrates all possible configurations for different choices of r (with $r = 0$ being the minimum and $r = 2\sqrt{10}$ the maximal distance possible for real solutions). The time needed to solve is in the range of seconds.

While the active system corresponding to a similar five-sided platform is not solvable by Maple, Maple `solve` returns an answer to the active constraints of `quad`: three 1-parameter families of solutions and one singleton. Painstaking examination confirms correctness of the output of the subdivision solver: there are only isolated, 0-dimensional real solutions.

References

1. B. Mourrain et al. Synaps Library, web interface *http://www-sop.inria.fr/galaad/logiciels/synaps/html/index.html*
2. R. Latham and A. Middleditch. Connectivity analysis: a tool for processing geometric constraints. *Computer Aided Design*, 28:917–928, 1996.
3. C. M. Hoffmann, A. Lomonosov and M. Sitharam. Geometric constraint decomposition. In Bruderlin and Roller Ed.s, editor, *Geometric Constraint Solving*. Springer-Verlag, 1998.
4. C. M. Hoffmann, A. Lomonosov and M. Sitharam. Decomposition of geometric constraints systems, part i: performance measures. *Journal of Symbolic Computation*, 31(4), 2001.
5. C. M. Hoffmann, A. Lomonosov and M. Sitharam. Decomposition of geometric constraints systems, part ii: new algorithms. *Journal of Symbolic Computation*, 31(4), 2001.

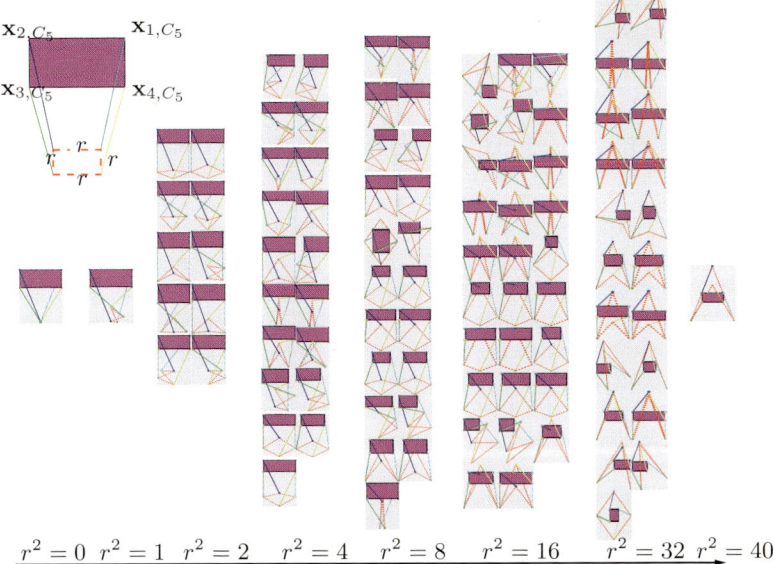

Fig. 5. All possible configurations of quad parametrized by r, the value of the four distance constraints displayed as thick *red* dashed lines. The gallery contains both macro information (number of non-isomorphic solutions) and micro information (geometry of the realization). For example, (cf. *upper left* inset). $r = 0$ corresponds to a square-based pyramid (with the solid (*purple*) rectangle representing cluster C_5; for each of C_1, C_2, C_3 and C_4 only one edge is displayed.) $r^2 = 40$ corresponds to a saddle with two opposing triangle faces flipped up and two flipped down. Find the planar configuration!

6. J. E. Graver, B. Servatius and H. Servatius. *Combinatorial Rigidity*. Graduate Studies in Math., AMS, 1993.
7. I. Fudos and C. M. Hoffmann. A graph-constructive approach to solving systems of geometric constraints. *ACM Transactions on Graphics*, 16:179–216, 1997.
8. A. Middleditch and C. Reade. A kernel for geometric features. In *ACM/SIGGRAPH Symposium on Solid Modeling Foundations and CAD/CAM Applications*. ACM press, 1997.
9. D. Cox, J. Little and D. O'Shea. *Using algebraic geometry*. Springer, 1998.
10. I. Emiris and J. Canny. A practical method for the sparse resultant. In *International Conference on Symbolic and Algebraic Computation, Proceedings of the 1993 international symposium on Symbolic and algebraic computation*, pages 183–192, 1993.
11. C Hoffman, M Sitharam and B Yuan. Making constraint solvers more useable: the overconstraint problem. *CAD*, 36(4), 377-399, 2004.

12. A. Lomonosov and M. Sitharam. Graph algorithms for geometric constraint solving. In *submitted*, 2004.
13. S. Ait-Aoudia, R. Jegou and D. Michelucci. Reduction of constraint systems. In *Compugraphics*, pages 83–92, 1993.
14. J Gaukel. Effiziente Lösung polynomialer und nichtpolynomialer Gleichungssysteme mit Hilfe von Subdivisionsalgorithmen, PhD thesis, Mathematics, TU Darmstadt, Germany, 2003.
15. M Sitharam and Y Zhou. A tractable, approximate, combinatorial 3d rigidity characterization. *proceedings of ADG 2004*, 2004.
16. H. Crapo. Structural rigidity. *Structural Topology*, 1:26–45, 1979.
17. H. Crapo. The tetrahedral-octahedral truss. *Structural Topology*, 7:52–61, 1982.
18. M Sitharam, J. Peters and Y Zhou. Solving minimal, wellconstrained, 3D geometric constraint systems: combinatorial optimization of algebraic complexity *submitted, http://www.cise.ufl.edu/~sitharam*, 2004.
19. M Sitharam, A. Arbree, Y Zhou and N Kohareswaran. Solution management and navigation for 3d geometric constraint systems. *to appear, ACM TOG*, 2005.
20. D. Eppstein. Representing all minimum spanning trees with applications to counting and generation. Technical Report 95-50, Univ. of California, Irvine, Dept. of Information & Computer Science, Irvine, CA, 92697-3425, USA, 1995.
21. I. Fudos. *Geometric Constraint Solving*. PhD thesis, Purdue University, Dept of Computer Science, 1995.
22. G. Kramer. *Solving Geometric Constraint Systems*. MIT Press, 1992.
23. G. Laman. On graphs and rigidity of plane skeletal structures. *J. Engrg. Math.*, 4:331–340, 1970.
24. J. Owen. www.d-cubed.co.uk/. In *D-cubed commercial geometric constraint solving software*.
25. J. Owen. Algebraic solution for geometry from dimensional constraints. In *ACM Symp. Found. of Solid Modeling*, pages 397–407, Austin, Tex, 1991.
26. J. Owen. Constraints on simple geometry in two and three dimensions. In *Third SIAM Conference on Geometric Design*. SIAM, November 1993. To appear in Int J of Computational Geometry and Applications.
27. J.A. Pabon. Modeling method for sorting dependencies among geometric entities. In *US States Patent 5,251,290*, Oct 1993.
28. M. Sitharam. Frontier, an opensource 3d geometric constraint solver: algorithms and architecture. *monograph, in preparation*, 2004.
29. M. Sitharam. Graph based geometric constraint solving: problems, progress and directions. In Dutta and Janardhan and Smid, editor, *AMS-DIMACS volume on Computer Aided Design*, 2004.
30. M. Sitharam. Frontier, opensource gnu geometric constraint solver: Version 1 (2001) for general 2d systems; version 2 (2002) for 2d and some 3d systems; version 3 (2003) for general 2d and 3d systems. In *http://www.cise.ufl.edu/~sitharam, http://www.gnu.org*, 2004.
31. B. Huber and B. Sturmfels. A polyhedral method for solving sparse polynomial system. *Math. Comp.*, 64:1541–1555, 1995.
32. W. Whiteley. Rigidity and scene analysis. In *Handbook of Discrete and Computational Geometry*, pages 893–916. CRC Press, 1997.

Minkowski decomposition of convex lattice polygons

Ioannis Z. Emiris and Elias P. Tsigaridas

Department of Informatics and Telecommunications
National University of Athens, HELLAS {emiris,et}@di.uoa.gr

Summary. A relatively recent area of study in geometric modeling concerns toric Bézier patches. In this line of work, several questions reduce to testing whether a given convex lattice polygon can be decomposed into a Minkowski sum of two such polygons and, if so, to finding one or all such decompositions. Other motivations for this problem include sparse resultant computation, especially for the implicitization of parametric surfaces, and factorization of bivariate polynomials. Particularly relevant for geometric modeling are decompositions where at least one summand has a small number of edges. We study the complexity of Minkowski decomposition and propose efficient algorithms for the case of constant-size summands. We have implemented these algorithms and illustrate them by various experiments with random lattice polygons and on all convex lattice polygons with zero or one interior lattice points. We also express the general problem by means of standard and well-studied problems in combinatorial optimization. This leads to an improvement in asymptotic complexity and, eventually, to efficient randomized algorithms and implementations.

1 Introduction

In this paper we study the decomposition of convex polygons with integral vertices (also called lattice polygons) under the Minkowski sum, which is defined as follows:

Definition 1. *For any two subsets A and B in $\mathbb{Z}^2$, their* Minkowski sum *is $A \oplus B = \{a+b | a \in A, b \in B\}$. We call A and B the* summands *of $A \oplus B$.*

The definition of the Minkowski sum can be generalized to arbitrary dimension.
 The decomposition problem has a great interest on its own. The recent work on toric Bézier patches (e.g [7, 12, 13]), in geometric modeling, motivates several questions around this problem, mainly testing whether a given lattice polygon can be written as a Minkowski sum of two such polygons and,

if so, finding one or all such decompositions. Another application in implicitization is the construction of matrices for the sparse resultant of 3 bivariate polynomials, cf [13, sec.10.3] or [23].

One important application of general Minkowski decomposition is bivariate (and, eventually, multivariate) polynomial factorization. This is so because, given a bivariate (multivariate) polynomial, we can associate with it its Newton polytope. As first observed by Ostrowski ([15]), if the polynomial factors, then its Newton polytope decomposes.

First, we focus on Minkowski decompositions where at least one of the summands is of constant size, namely it is a line segment, a triangle or a quadrangle. These are particularly relevant when manipulating toric Bézier patches with depth. In [13], the authors "extend blossoming, degree elevation and implicitization techniques to arbitrary toric Bézier patches. [...] The key idea to each of these algorithms is to employ decompositions based on the Minkowski sum". They add [13, Sec. 10.1] that "This approach to evaluation, blossoming, and dual functionals works for any toric Bézier patch whose lattice polygon decomposes into the Minkowski sum of line segments and unit triangles". A key step in the algorithm of [23] (cf [13, Sec. 10.3]) for constructing resultant matrices for implicitization is to "decompose [Newton polygon] A into a Minkowski sum of simpler lattice polygons, typically line segments and triangles".

We estimate the hardness, from an asymptotic complexity viewpoint, and propose efficient algorithms for the case of constant-size summands. We relate Minkowski decomposition to the k−SUM problem, where an algorithm with time complexity $O(n^{\lceil k/2 \rceil})$ or $O(n^{\lceil k/2 \rceil} \lg n)$ exists but there are no matching lower bounds. We have implemented these algorithms and illustrated them on all lattice polygons with zero and one interior lattice points. Moreover, we performed experiments on various data sets against the algorithm of Gao and Lauder ([5]), which solves the general problem of Minkowski decomposition.

The decision problem of whether a lattice polygon admits a Minkowski decomposition is NP-complete [5]. In the same paper, a pseudo-polynomial algorithm is given with complexity in $O((nDE)^3)$, where n is the number of edges in the polygon and DE is their maximum integer length. Note that DE is exponential with respect to the bit size of the input, which is $O(n \lg (DE))$. We express the general problem by means of standard and well-studied problems in combinatorial optimization, such as the SUBSET-SUM problem. This leads to an algorithm that improves the above bound by a factor of nD. Our approach also leads immediately to approximation algorithms, to practical methods amenable to fast implementations and to a probabilistic algorithm. The implementation goes beyond the scope of the present paper.

Our paper is organized as follows. The next section defines the problem and overviews relevant work. Section 3 presents our approach for Minkowski decomposition of a lattice polygon to two summands, where at least one has a given constant number of edges. Section 4 presents the implementation of our algorithms, experiments with various random polygons and the decomposition of all, up to unimodal transformations, lattice polygons with zero and one interior lattice points. In Section 5 we propose an algorithm for general decomposition of a lattice polygon, that has better time complexity than the one known so far. The last section presents our future research aims on the problem of Minkowski decomposition.

In what follows $O(\cdot)$ (resp. $O_B(\cdot)$) indicates arithmetic (resp. bit) complexity.

2 Definitions and previous work

The general problem that we deal with is:

Problem 2. MINKOWSKI-DECOMPOSITION Given a lattice polygon Q, with n vertices, decide if it is *decomposable*, that is if there are lattice polygons A and B such that $A \oplus B = Q$, where $\oplus$ denotes the Minkowski sum.

We are given a lattice polygon Q, with vertices $v_0, v_1, \ldots, v_{n-1}$, where $v_j \in \mathbb{Z}^2, 0 \leq j \leq n-1$. For every edge of the polygon we associate the vector $u_1 = (v_1 - v_0), \ldots, u_n = (v_0 - v_{n-1})$. The polygon is completely characterized by this sequence of vectors and the initial vertex v_0. In what follows edge means its vector.

Definition 3. Let $u = (a,b)$ be a vector and $d = \gcd(a,b)$. The primitive vector *of u is $e = (a/d, b/d)$.*

We denote the sequence of all vectors u_i as $\mathcal{U}$ and we call it the edge sequence. For every vector $u_i = (a_i, b_i)$ of Q we associate the primitive vector e_i, $1 \leq i \leq n$. We call the sequence of all primitive vectors, primitive edge sequence and denote it by $\mathcal{E}$. Additionally we call $\mathcal{A}$ the set of all possible vectors

$$\mathcal{A} = \{k_i e_i | 1 \leq i \leq n, 1 \leq k_i \leq d_i\}$$

where $d_i = \gcd(a_i, b_i)$. Let

$$D = \max\{d_1, \ldots, d_n\},$$
$$E = \max\{e_{1x}, e_{1y}, \ldots, e_{nx}, e_{ny}\},$$

where (e_{ix}, e_{iy}) are the coordinates of the primitive vector e_i. Moreover let g be the time needed for the computation of the gcd of two numbers of magnitude DE. Using the HALF-GCD algorithm ([22]) the gcd has bit complexity

$$g := O_B(\lg(DE) \lg^2 \lg(DE) \lg \lg \lg(DE))$$

The cost for computing $\mathcal{A}$ is

$$O_B\left(ng + nD\,\mathsf{M}(\max\{D,E\})\right) = O_B(nD\mathsf{M}(\max\{D,E\})),$$

where $M(\tau)$ is the time needed for the multiplication of two numbers of length τ. If we use FFT ([22]) the bit complexity of the multiplication is:

$$M(\tau) = \tau \lg \tau \lg \lg \tau \tag{2.1}$$

In terms of arithmetic complexity the cost of computing $\mathcal{A}$ is $O(nD)$. However, when the computation of $\mathcal{A}$ is needed, its cost is dominated by other steps in the algorithms that we derive.

Lemma 4. *Consider a lattice polygon Q, such that $Q = A \oplus B$. Every edge of Q is determined uniquely as the Minkowski sum of an edge of A and a vertex of B, or as the sum of a vertex of A and an edge of B, or as a sum of two parallel edges of A and B.*

Hence, the set of the normals of Q is the union of the sets of the normals of A and B.

Using Lemma 4, it is easy to show ([5]):

Lemma 5. *A (lattice) polygon is a summand of Q if and only if its edge sequence is of the form $\{k_j e_j\}_{j \in J}$, where $J \subseteq \{1, \ldots n\}$, $0 \leq k_j \leq d_j$, $k_j \in \mathbb{Z}$ and $\sum_{j \in J} k_j e_j = (0,0)$ (the sum of the vectors that correspond to its edges is zero).*

Theorem 6. *([5]) The decision problem of whether a lattice polygon has a Minkowski decomposition is NP-complete. There is an algorithm that decides if a polygon is decomposable, which has complexity $O(nDT)$, where T is the number of interior lattice points of the polygon. However, $T = O((nDE)^2)$ hence the complexity of the algorithm is $O(n^3 D^3 E^2)$.*

Note that, if a polygon is decomposable, there is possibly an exponential number of decompositions. The algorithm is pseudo-polynomial because its running time is polynomial in the length of the sides of the polygon rather than the logarithm of the lengths. In Section 5 we will present an algorithm that improves the time complexity by a factor of nD.

The bound $T = O(n^2 D^2 E^2)$ ([6], [8, Chap. 7]) is tight. One way that it can be achieved is as follows. Consider the lattice polygon of Fig. 1, where its edge sequence is

$$s_1 = (1, DE), s_2 = (2, DE), \ldots, s_n = (n, DE), (0, -nDE), (-\frac{n(n+1)}{2}, 0)$$

The area of the polygon is $\Theta(n^3 DE)$. If we assume that $n = \Theta(DE)$, then its area is $\Theta(n^2 D^2 E^2)$. The number of interior lattice points is asymptotically greater than the number of boundary lattice points. Also notice that #(Boundary points) $= O(n^2)$. Now, using Pick's formula

$$\text{Area} = \#(\text{Interior points}) + \frac{\#(\text{Boundary points})}{2} - 1$$

we can deduce that the number of interior lattice points is asymptotically $\Theta((nDE)^2)$.

Fig. 1. A polygon with area $O((nDE)^2)$.

3 Constant-size summands

In this section we focus on the problem of decomposing a lattice polygon to two summands, where at least one has a given fixed number of edges. Remember that the input is a point sequence of cardinality n. We are dealing with two different problems:

Problem 7. Decision k-SUMMAND
Given a lattice polygon, decide whether there is a Minkowski decomposition to two summands, such that at least one of them has k edges.

Problem 8. Enumeration k-SUMMAND
Given a lattice polygon, enumerate all Minkowski decompositions of it, to two summands, where at least one of them has k edges.

In what follows, we examine in detail the cases where one summand is a segment (2-SUMMAND), a triangle (3-SUMMAND) or a quadrilateral (4-SUMMAND). The latter is generalisable to any fixed-size summand. We deal with both decision and enumeration problems. When we do not mention whether it is a decision or enumeration problem, it is clear from the context to which one we refer to.

The decision k-SUMMAND problem can be solved using the k-SUM problem. The latter is defined as follows:

Problem 9. k-SUM
Given a set of m integers and a goal sum S, decide whether there are k of them that add up to S.

The best known algorithm for the k-SUM problem has time and space complexity ([20, 21]) $O(m^{\lceil k/2 \rceil} \lg m)$ and $O(m^{\lceil k/2 \rceil})$, respectively. When k is odd the time complexity can be improved to $O(m^{\lceil k/2 \rceil})$. However, the derivation of a non-trivial lower bound for the k-SUM problem in the algebraic decision tree model or in the algebraic computation tree model is a major open problem. The only known result is due to Erickson ([3]), who proved an $\Omega(m^{\lceil k/2 \rceil})$ lower bound in a certain restricted variant of the linear decision tree model.

Theorem 10. *An instance of the k-SUMMAND problem can be transformed to an instance of k-SUM, such that the instance of k-SUMMAND has a solution if and only if the corresponding instance of k-SUM has a solution.*

Proof. Consider a lattice polygon Q, with n vertices. We compute $\mathcal{A}$, in $O(nD)$. Every vector in $\mathcal{A}$ is of the form $ke_i = k(e_{ix}, e_{iy})$, where $1 \le i \le n$ and $1 \le k \le d_i$. For every vector in $\mathcal{A}$ we associate the number $\alpha_{ik} = k(e_{ix} + Le_{iy})$, where $L = (k+1)DE$.

This new set of α_{ik}'s has at most nD elements. Let the target be $S = 0$. If we find k elements of this set, such that all of them correspond to different primitive vectors and that sum up to zero, then a k−SUMMAND exists.

Notice that the size of the instance of k−SUM is $O(nD)$. □

The above transformation allows us to solve the k−SUMMAND problem using the straight-forward algorithms of the k−SUM and as a consequence provides us with upper bounds for both time and space complexity. For $k = 2, 3, 4$ we have:

Decision 2−SUMMAND can be solved in $O(nD \lg(nD))$ time and $O(nD)$ space.

Decision 3−SUMMAND can be solved in $O(n^2 D^2)$ time and $O(nD)$ space.

Decision 4−SUMMAND can be solved in $O(n^2 D^2 \lg(nD))$ time and $O(nD)$ space.

As for the general case, the decision k−SUMMAND problem can be solved in $O\left((nD)^{\lceil k/2 \rceil} \lg(nD)\right)$ and $O\left((nD)^{\lceil k/2 \rceil}\right)$ time, for k even and odd respectively and $O\left((nD)^{\lceil k/2 \rceil}\right)$ space.

We improve almost all the bounds in the subsequent sections.

Following [4], we give the following definition:

Definition 11. *Given two problems* PR_1 *and* PR_2 *we say that* PR_1 *is* $f(n)$−*solvable using* PR_2 *if and only if every instance of* PR_1 *of size* n *can be solved using a constant number of instances of* PR_2 *of at most linear size and* $O(f(n))$ *additional time. We denote this by*

$$\mathrm{PR}_1 \lll_{f(n)} \mathrm{PR}_2$$

Note that reduction implies, that when $f(.)$ is sufficiently small, lower bounds for the time complexity of PR_1 carry over to PR_2 and upper bounds for PR_2 hold for PR_1.

In order to prove lower bounds for the k−SUMMAND problem we use the following:

Theorem 12. k−SUM $\lll_{n \lg n}$ k−SUMMAND

Proof. Consider the sequence $\{a_i\}_{1 \le i \le n}$, where $a_i \in \mathbb{Z}$. We assume that the sequence is sorted; if it is not, then we sort it in $O(n \lg n)$ time. Let $M = \max_i |a_i|$ and $L = (k+1)M$. We form the sequence $\{s_i = a_i + L\}_{1 \le i \le n}$, where $0 \le s_1 \le \cdots \le s_n$. Next we consider the edge sequence (see Figure 2):

$$(s_1, 1), (s_2, 1), \ldots, (s_n, 1), (0, -n), (-kL, -1), \left(-\sum_{i=1}^{n} a_i - (n-k)L, 1\right)$$

This sequence is an edge sequence of a lattice polygon, since both the sum of the ordinates and the sum of the abscissae of the vectors equal zero, and the angles of the edges are sorted, in clockwise order.

This polygon has a k−SUMMAND, if and only if there are k numbers in the sequence $\{a_i\}$ that sum up to zero, since in this case the edge sequence of the k−SUMMAND will be of the form

$$(s_{i_1}, 1), (s_{i_2}, 1), \ldots, (s_{i_k}, 1), (0, -(k-1)), (-kL, -1)$$

where $i_j \in J$ and J is a subset of $\{1, \ldots, n\}$ of cardinality k. Proving the forward direction is easy. The reverse can be proven by considering the cases of summand edges and checking whether the y−coordinates sum to zero. □

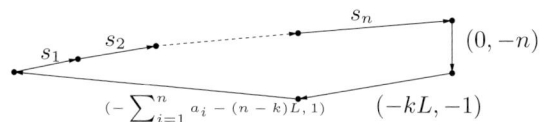

Fig. 2. Reduction of a k−SUM to a k−SUMMAND problem.

The previous reduction indicates that the k−SUMMAND problem is at least as hard as the k−SUM problem, maybe harder. Actually, this is the case when $D > 1$.

We consider as direction of a vector, the rational tangent of the angle between the positive x−semi-axis and the vector in a counter-clockwise orientation. For the algorithms that we will present direction and angle have the same meaning. The direction (essentially tangent) is represented by a pair of integer numbers, and each of them has magnitude at most DE. We can compare two directions in $O_B(\mathsf{M}(\lg(DE)))$ bit complexity, where $\mathsf{M}(\tau)$ is the time needed for the multiplication of two numbers of length τ (see Eq. (2.1)).

In what follows, we measure the algorithms' complexity using the arithmetic model (real RAM [17]). However, we can deduce the bit complexity if we multiply the derived complexities by either $\mathsf{M}(\lg(DE))$, if the comparison of directions is needed, or $\lg(DE)$, if the comparison of coordinates is needed.

Furthermore, we assume that v_0 is the bottom-left vertex, this means that v_1 is the vertex with the smallest direction. This is without loss of generality, since we can find this vertex in time $O(n)$. The key observation is that vectors in both $\mathcal{U}$ and $\mathcal{E}$ sequences are sorted in increasing order with respect to direction, for any lattice polygon. Taking this into account we deduce algorithms for the $\{2, 3, 4\}$−SUMMAND problem.

3.1 Line summand

Note that a 2−SUMMAND exists if and only if there are at least two parallel edges. In order to decide if a line summand exists, we compute the vectors that correspond to the edges of the lattice polygon, that is the sequence $\mathcal{U}$, in time $O(n)$.

Since $\mathcal{U}$ is sorted with respect to direction, we split it to two sequences, one that has directions in $[0, \pi)$, say $\mathcal{U}_1$ and one that has directions in $[\pi, 2\pi)$, say $\mathcal{U}_2$. We can do this in $O(n)$ time. We consider indices i and j that traverse $\mathcal{U}_1$ and $\mathcal{U}_2$, respectively. This means that i starts from the minimum direction of $\mathcal{U}_1$ and goes towards the maximum direction in $\mathcal{U}_1$ and the same for j in $\mathcal{U}_2$. If the direction of $\mathcal{U}_1[i]$ is smaller (resp. greater) than $\delta - \pi$ where δ is the direction of $\mathcal{U}_2[j]$, we advance i (resp. j). If the direction of $\mathcal{U}_1[i]$ is smaller than the direction of $\mathcal{U}_2[j]$ by π, then a line summand exists. Both the time and space complexity are $O(n)$.

If we are interested in the enumeration 2–SUMMAND problem then we have to find all the vectors with directions differing by π and for every such pair, say with indices i and j, we compute the corresponding primitive vectors, say e_i and e_j, and we output d pairs of vectors, (ke_i, ke_j), where $1 \leq k \leq d$ and $d = \min\{d_i, d_j\}$. Then we advance both indices i and j and continue the algorithm. This algorithm has time complexity $O(n+t)$, where t is the number of all possible line summands, which is at most $\frac{nD}{2}$.

The previous discussion leads to the following theorem:

Theorem 13. *There is an algorithm for the decision 2–SUMMAND problem that has time complexity $O(n)$. There is an algorithm for the enumeration 2–SUMMAND problem that has time complexity $O(n+t)$. The space complexity for both algorithms is $O(n)$.*

Both algorithms are optimal.

3.2 Triangle summand

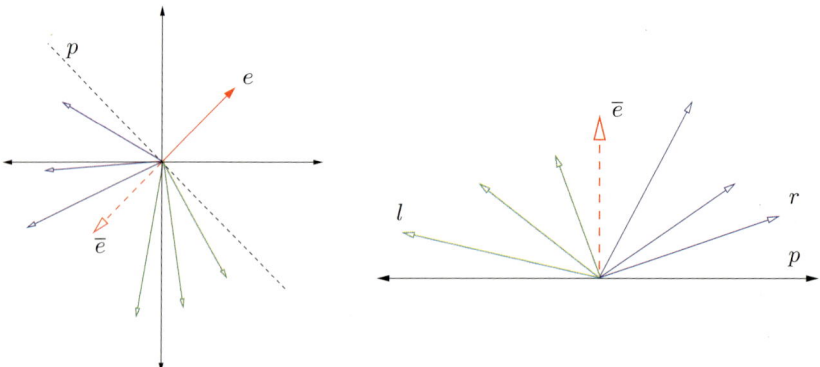

Fig. 3. Computing the triangle summands

In order to solve the decision problem for the 3–SUMMAND, first we compute the primitive edge sequence $\mathcal{E}$ and sequence $\mathcal{A}$, in time $O(nD)$. Note that

$|\mathcal{A}| = O(nD)$. Since $\mathcal{A}$ contains scalar multiples of the vectors in $\mathcal{E}$, we can assume that it is sorted in increasing order, first with respect to direction and then with respect to x and y coordinates.

If a triangle summand exists then for some primitive vector $e \in \mathcal{E}$ there are two indices r and l, where $1 \leq r, l \leq |\mathcal{A}|$, such that the direction of the vector $w = \mathcal{A}[r] + \mathcal{A}[l]$ is opposite to that of e.

Consider the case of the left half of Figure 3. Vector $e = (e_x, e_y)$ is a primitive vector and the dotted vector $\bar{e} = (\bar{e}_x, \bar{e}_y) = (-e_x, -e_y)$ is its opposite. We consider an axis perpendicular to e, this is line p in the figure, and only the vectors from $\mathcal{A}$ that lie in the same half-plane as $\bar{e}$ does. We can find these vectors in time $O(nD)$, since $\mathcal{A}$ is sorted with respect to direction. We denote this sequence of vectors also by $\mathcal{A}$. Note that this sequence is also sorted with respect to direction.

With a suitable axis rotation, the case of the left half of Figure 3 is equivalent to the one of the right half. From now on we will refer to the right half since it is more intuitive. All vectors, except $\bar{e}$, are elements of $\mathcal{A}$.

In order to find if a triangle summand exists, we start with indices $r = 1$ and $l = |\mathcal{A}|$, assuming that $\mathcal{A}$ is sorted from right to left as in Figure 3 (right). Then, we examine all vectors of $\mathcal{A}$ trying to find values for the indices r and l such that the direction of $w = \mathcal{A}[r] + \mathcal{A}[l]$ is equal to the direction $\bar{e}$. If this happens, then we check if $-\frac{w_x}{e_x}$ and $-\frac{w_y}{e_y}$ are the same integer between 1 and d. If this is the case then a triangle summand exists, otherwise we advance both r and l. If the direction of w is smaller, respectively larger, than the direction of $\bar{e}$, we advance r, respectively reduce l, by 1.

We traverse $\mathcal{A}$ in time $O(nD)$ and since we have to do this for every vector in the primitive edge sequence, the total time for the decision algorithm is $O(n^2 D)$ and its space complexity is $O(nD)$.

If we are interested in the enumeration problem we advance both indices r and l, when we find direction equality, and so we enumerate all the possible triangle summands. The total time for the algorithm is $O(n^2 D + t)$, where t is the number of all possible triangle summands.

The previous discussion leads to the following theorem:

Theorem 14. *There is an algorithm for the decision 3–SUMMAND problem that has time complexity $O(n^2 D)$. There is an algorithm for the enumeration 3–SUMMAND problem that has time complexity $O(n^2 D + t)$. The space complexity for both algorithms is $O(nD)$.*

There is an alternative algorithm for the decision problem that runs in $O(n^3)$ arithmetic complexity or $O_B(n^3 g)$ bit complexity and space complexity $O(n \lg(DE))$. First we compute the primitive edge sequence $\mathcal{E}$, in time $O_B(ng)$ bit complexity. If a triangle summand exists then at least one of the $O(n^3)$ systems of Diophantine equations and inequalities

$$a_i e_{ix} + a_j e_{jx} + a_k e_{kx} = 0$$
$$a_i e_{iy} + a_j e_{jy} + a_k e_{ky} = 0$$
$$1 \le a_i \le d_i, 1 \le a_j \le d_j, 1 \le a_k \le d_k$$

where $1 \le i < j < k \le n$, must have an integer solution. As for the bit complexity of the solution of the above system, it is dominated by the computation of $\gcd(e_{ix}, e_{jx}, e_{kx})$ and $\gcd(e_{iy}, e_{jy}, e_{ky})$ and so it is $O_B(g)$.

As for the enumeration problem, we must solve all these systems and thus the algorithm has $O(n^3 + t)$ arithmetic complexity or $O_B(n^3 g + t)$ bit complexity.

From the previous discussion follows that the decision 3−SUMMAND can be solved in polynomial time arithmetic complexity. Typically, n is large compared to D, so Th. 14 is preferable, hence we do not extend this approach further.

3.3 Quadrangle summand

In order to deduce an algorithm for the decision 4−SUMMAND problem, we compute the primitive edge sequence $\mathcal{E}$ and then the sequence $\mathcal{A}$ in time $O(nD)$. We compute the sequence of all vectors that are sums of two distinct vector of $\mathcal{A}$ in time $O(n^2 D^2)$. We call this sequence $\mathcal{A}_2$. We sort $\mathcal{A}_2$, first with respect to the x−coordinate and then with respect to the y−coordinate, in time bounded by $O(n^2 D^2 \lg(nD))$.

For every vector in $\mathcal{A}_2$, we search $\mathcal{A}_2$ for a vector with opposite x and y coordinates. We can do the search in $O(\lg(nD))$ time. Thus the total time of this decision algorithm is $O(n^2 D^2 \lg(nD))$ and its space complexity is $O(n^2 D^2)$.

If we want to enumerate all the possible quadrangle summands we perform the search for every vector in $\mathcal{A}_2$. Thus the complexity for the enumeration algorithm is $O(n^2 D^2 \lg(nD) + t)$, where t is the number of all possible quadrangle summands.

In practice we can eliminate the logarithmic factors since we can use a hash structure in order to keep the elements of $\mathcal{A}_2$. If we want to reduce the space requirements, we can use a special data structure that produces (in increasing or decreasing order) all possible sums of two vectors (see [20]) which has space complexity $O(nD)$ and access time $O(\lg(nD))$.

The previous discussion leads to the following theorem:

Theorem 15. *There is an algorithm for the decision* 4−SUMMAND *problem that has time complexity* $O(n^2 D^2 \lg(nD))$. *There is an algorithm for enumeration* 4−SUMMAND *problem that has time complexity* $O(n^2 D^2 \lg(nD) + t)$. *The space complexity for both algorithms is* $O(nD)$.

3.4 Summand with k edges

For the general k−SUMMAND problem we have to distinguish between two cases, when k is odd or even. As in the previous sections first we discuss the decision problem.

In both cases, first we compute the sequences $\mathcal{E}$ and $\mathcal{A}$ and then we compute all the possible sums of $\lfloor \frac{k}{2} \rfloor$ vectors of $\mathcal{A}$. Since the size of $\mathcal{A}$ is $O(nD)$, this computation can be done in $O((nD)^{\lfloor \frac{k}{2} \rfloor})$ and the space required is of the same order. We call this sequence $\mathcal{A}_{\frac{k}{2}}$.

If k is odd then we sort $\mathcal{A}_{\frac{k}{2}}$, first with respect to direction, then with respect to x−coordinate and finally with respect to y−coordinate. This can be done in $O((nD)^{\lfloor \frac{k}{2} \rfloor} \lg (nD))$. After this we proceed as in the 3−SUMMAND case. For every primitive vector $e \in \mathcal{E}$, we traverse $\mathcal{A}_{\frac{k}{2}}$ with two pointers: One that goes from left to right and another from right to left, in order to find two vectors of $\mathcal{A}_{\frac{k}{2}}$ such that the direction of their sum is opposite to the direction of e. The time complexity of this algorithm is $O(n^{\lceil \frac{k}{2} \rceil} D^{\lfloor \frac{k}{2} \rfloor} + (nD)^{\lfloor \frac{k}{2} \rfloor} \lg (nD))$ or $O(n^{\lceil \frac{k}{2} \rceil} D^{\lfloor \frac{k}{2} \rfloor})$ if we assume that $n > \lg (nD)$.

If k is even then we proceed as in the 4−SUMMAND case, that is we sort $\mathcal{A}_{\frac{k}{2}}$, first with respect to the x−coordinate and then with respect to the y−coordinate. This can be done in $O(n^{\lceil \frac{k}{2} \rceil} D^{\lfloor \frac{k}{2} \rfloor} \lg (nD))$. Note that since k is even $\lceil \frac{k}{2} \rceil = \lfloor \frac{k}{2} \rfloor$. Finally, for every vector of $\mathcal{A}_{\frac{k}{2}}$, we search $\mathcal{A}_{\frac{k}{2}}$ for a vector with opposite x and y coordinates. The search can be performed in $O(\lg (nD))$ time. Thus the total time for the algorithm is $O(n^{\lceil \frac{k}{2} \rceil} D^{\lfloor \frac{k}{2} \rfloor} \lg (nD))$.

As for the enumeration problem, in both cases, we continue the search when a k−SUMMAND is found.

The previous discussion leads to the following theorem:

Theorem 16. *There is an algorithm for the decision k−SUMMAND problem that has time complexity $O(n^{\lceil \frac{k}{2} \rceil} D^{\lfloor \frac{k}{2} \rfloor} \lambda)$, where $\lambda = 1$ if k is odd and $\lambda = \lg (nD)$ if k is even.*

There is an algorithm for the enumeration k−SUMMAND problem that has time complexity $O(n^{\lceil \frac{k}{2} \rceil} D^{\lfloor \frac{k}{2} \rfloor} \lambda + t)$, where t is the number of all possible decompositions to two summands, where at least one of them has k edges.

The space complexity for both algorithms is $O((nD)^{\lfloor \frac{k}{2} \rfloor})$.

4 Implementation and application to polygons with zero and one lattice interior point

This section sketches our implementations of the above algorithms, their application to computing all Minkowski decompositions of all polygons with one lattice interior point as well as polygons without interior lattice points and experiments with various datasets.

We implemented our algorithms in C++ and we used the geometric library CGAL ([1]). CGAL has classes that refer to points, vectors and polygons and operations on them. Additionally, there is a class for the direction (in $\mathbb{Q}$) of a vector and comparisons between them. Our code is freely available at http://www.di.uoa.gr/~et.

We performed all experiments on a 2.6GHz Pentium, with 1GB memory, running Linux, with kernel version 2.6.10. We compiled the programs with g++, v. 3.3.5, with options -O3 -DNDEBUG.

4.1 Experiments with random lattice polygons

We performed various experiments so as to check the efficiency of our algorithms for the decision $\{2, 3, 4\}$–SUMMAND problems. We refer to these algorithms as ET(s), ET(t) and ET(q). We also implemented in CGAL the algorithm of Gao and Lauder ([5]), which decides the general problem of Minkowski decomposition. We refer to this algorithm as GL. The running times of the experiments are presented in Table 1. All the times are in msec.

Columns A_k, B_k, C_k and D_k, where $k \in \{10, 20, 30, 40, 50, 60, 70\}$, refer to 500 lattice polygons with k edges, sampled in $[0, 3000] \times [0, 3000]$. The polygons in B_k, C_k and D_k, were constructed such that they have at least one segment, one triangle, one quad summand, respectively. Columns E_k, refers to 500 lattice polygons that are the convex hull of 50 random lattice points in $[0, k] \times [0, k]$.

In all cases our algorithms are considerably faster. This is the case because the ET algorithms are dedicated for constant-size summands and solve a polynomial problem, while GL is an algorithm for the general problem which is NP-complete. Special notice must be paid to the running times of ET(s), which are more or less the same on all data sets. The reason is that the complexity of ET(s) depends linearly only on the number of edges of the polygon. Additionally, most of the time of the GL algorithm is spent for the computation of the integer points of the tested lattice polygon. As a consequence the running times of GL for the data sets A_k, B_k, C_k, D_k, where the polygons have a large number of lattice points, are not satisfactory. However in E_k, where the polygons have a small number of lattice points, the running times of GL are quite competitive.

Even though current experiments show the superiority of the special purpose algorithms, a more careful implementation and a more detailed experimental analysis is needed.

4.2 Lattice polygons with one lattice interior point

There are only 16 lattice polygons with one lattice interior point, modulo unimodular transformations, as proven in [18] (see also [19]). We compute all decompositions into Minkowski summands. The results are in Figure 4 and 5. These polygons are of particular interest for toric Bézier patches ([7, 12, 13]).

	A_{10}	A_{20}	A_{30}	A_{40}	A_{50}	A_{60}	A_{70}
ET(s)	0.007	0.01	0.02	0.03	0.04	0.04	0.05
ET(t)	1.1	5.1	9.6	16.5	30.1	40.3	56.2
ET(q)	1.6	6.2	11.6	19.1	34.4	46.1	65.1
GL	11150	15270	22050	23995	23370	26205	27315

	B_{10}	B_{20}	B_{30}	B_{40}	B_{50}	B_{60}	B_{70}
ET(s)	0.004	0.006	0.008	0.01	0.01	0.02	0.02
ET(t)	4.3	7.2	9.7	17.3	25.5	39.3	49.9
ET(q)	3.3	9.2	12.1	19.2	29.7	44.8	57.4
GL	27330	50105	37930	53635	46345	54205	36475

	C_{10}	C_{20}	C_{30}	C_{40}	C_{50}	C_{60}	C_{70}
ET(s)	0.003	0.006	0.008	0.01	0.01	0.02	0.02
ET(t)	1.8	3.6	10.4	16.3	27.8	37.5	53.7
ET(q)	2.6	5.3	12.7	18.2	33.0	43.2	62.1
GL	25630	27065	52810	37215	84510	86555	51465

	D_{10}	D_{20}	D_{30}	D_{40}	D_{50}	D_{60}	D_{70}
ET(s)	0.003	0.006	0.008	0.01	0.01	0.02	0.02
ET(t)	1.6	5.2	9.4	19.3	28.3	43.2	54.3
ET(q)	1.9	5.5	11.2	22.4	33.5	49.5	63.1
GL	32950	78840	72230	71240	75805	64690	73335

	E_{10}	E_{20}	E_{30}	E_{40}	E_{50}	E_{60}	E_{70}
ET(s)	0.002	0.003	0.003	0.003	0.004	0.004	0.004
ET(t)	0.1	0.4	0.3	0.4	0.4	0.5	0.5
ET(q)	0.1	0.3	0.4	0.4	0.5	0.6	0.6
GL	0.1	0.2	0.4	0.7	1.2	1.6	2.2

Table 1. Experimental results

4.3 Lattice polygons without interior lattice points

We have computed all the decompositions of lattice polygons with zero interior lattice points and area less than or equal to 3. All possible decompositions are in Figure 6.

Using the enumeration algorithms for $\{2, 3, 4\}$–SUMMAND of the previous section we can decompose all the polygons, up to unimodular transformations, without interior lattice points. All the decompositions are presented in Figure 7.

First we need to define all such polygons and so we state the following theorem:

Theorem 17. *[19] Any lattice polygon without interior lattice points is unimodular equivalent to a polygon $T_{m,n}$ with vertices $\{(0,0), (0,1), (m+n,0), (n,1)\}$, where $m, n \geq 0$, or to the triangle Δ_2 with vertices $\{(0,0), (2,0), (0,2)\}$.*

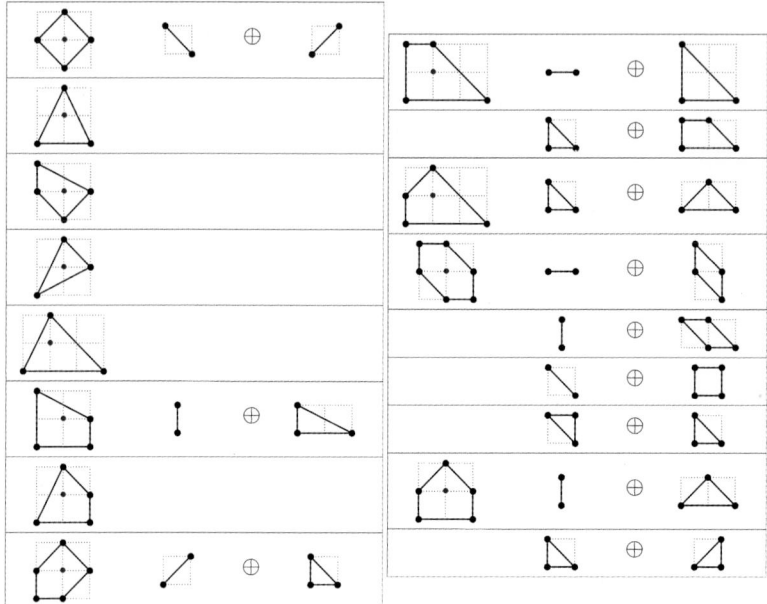

Fig. 4. Minkowski decomposition of lattice polygons, with one interior lattice point (continued in next figure).

The edge sequence of Δ_2 is $\{2(1,0), 2(-1,1), 2(0,-1)\}$. It is easy to see that Δ_2 admits a Minkowski decomposition, to two equal triangle summands. If Δ_1 is the triangle with vertices $\{(0,0),(1,0),(0,1)\}$, then $\Delta_2 = \Delta_1 \oplus \Delta_1 = 2\Delta_1$. Notice that Δ_2 and Δ_1 are homothetic. The decomposition is illustrated in the first row of Figure 7.

In order to decompose all lattice polygons $T_{m,n}$ we distinguish the following cases:

- $m \geq 1, n = 0$
  In this case $T_{1,0}$ is a triangle with vertices $\{(0,0),(m,0),(0,1)\}$ and its edge sequence is $\{(m,0),(-m,1),(0,1)\}$. It very easy to check that this triangle is irreducible with respect to Minkowski sum. We can reach the same result if we use the approach of [5, Th. 8] by checking that $\gcd(0,1,m) = 1$.
- $m = 0, n \geq 1$

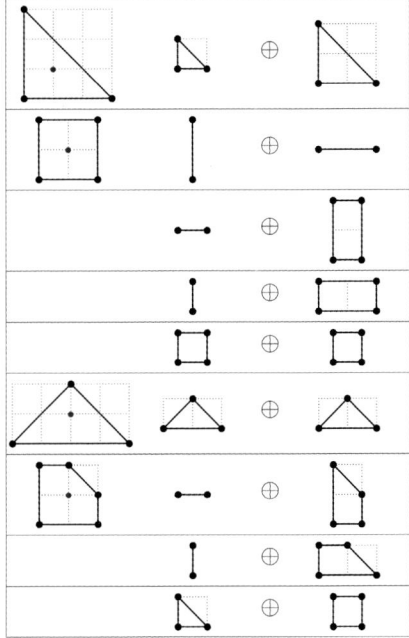

Fig. 5. Continued: Minkowski decomposition of lattice polygons, with one interior lattice point.

In this case, $T_{0,n}$ is a rectangular with vertices $\{(0,0), (n,0), (n,1), (0,1)\}$ and its edge sequence is $\{(n,0), (0,1), (-n,0), (0,-1)\}$. The Minkowski decomposition of this polygon is either two line segments (this is the first equality of the second row of Figure 7) or a line segment and a rectangle (this is the second equality of the second row of Figure 7, where $1 \leq k \leq n$). The last equality of the second row of Figure 7 presents the unique Minkowski decomposition of $T_{0,n}$ to $n+1$ irreducible summands.

- $m \geq 1, n \geq 1$
  In this case, $T_{m,n}$ is a trapezoid with vertices $\{(0,0), (m+n,0), (n,1), (0,1)\}$ and its edge sequence is $\{(m+n,0), (-m,1), (-n,0), (0,-1)\}$. We can decompose $T_{m,n}$, either to a Minkowski sum of a line segment and a trapezoid (this is the first equality of the third row of Figure 7, where $1 \leq k \leq n$) or to a Minkowski sum of triangle and a line segment (this is the second equality of the third row of Figure 7). The last equality of the third

row of Figure 7 presents the unique Minkowski decomposition of $T_{0,n}$ to irreducible summands.

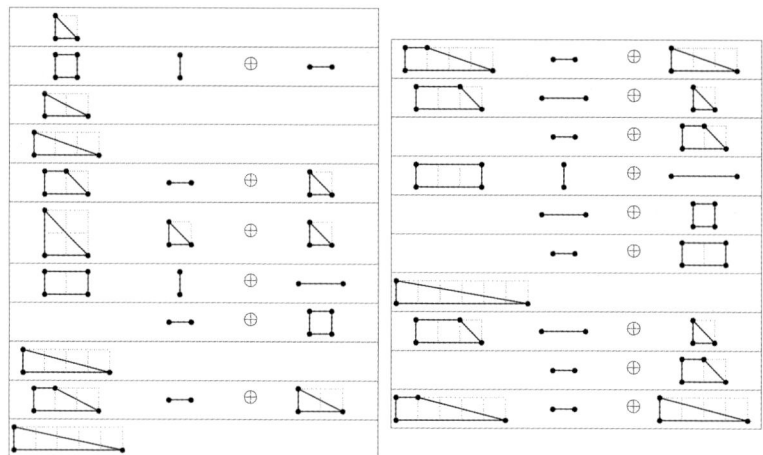

Fig. 6. Minkowski decomposition of lattice polygons, with zero interior lattice points and area less than or equal to 3.

5 Improving the general decomposition algorithm

In this section we return to the general problem, MINKOWSKI-DECOMPOSITION, and propose a different approach than the one by Gao and Lauder ([5]). This shall improve the asymptotic complexity. More importantly, we expect our approach to lead to efficient implementations in practice and to permit randomized and approximation algorithms.

The main idea is that it suffices to find combinations of vectors such that their sum is zero. Note that the sum of a subset of the vectors is zero iff both the sum of $x-$coordinates and the sum of $y-$coordinates is zero.

We need the definition of the SUBSET-SUM problem, which is an NP-complete problem:

Problem 18. SUBSET-SUM

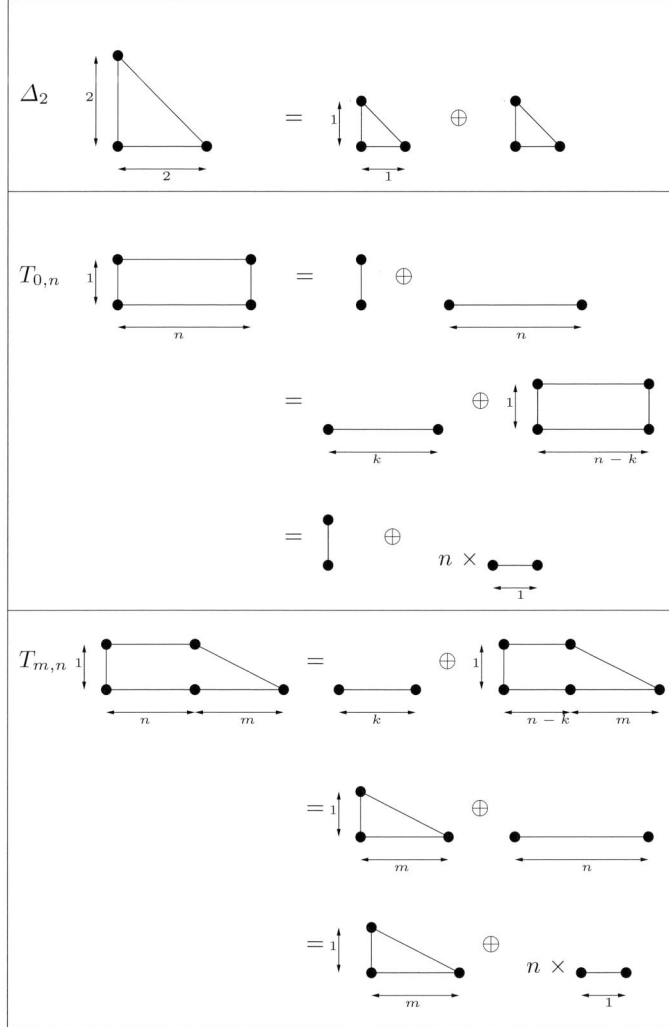

Fig. 7. Minkowski decompositions of all lattice polygons without interior lattice points.

Given a set of n positive integers and a goal sum S, decide whether there exists a subset, such that its elements add up to S.

We use the following transformation:

Lemma 19. *An instance of a* MINKOWSKI-DECOMPOSITION *problem can be transformed to an instance of a* SUBSET-SUM *problem, such that the instance of* MINKOWSKI-DECOMPOSITION *admits a solution if and only if the instance of* SUBSET-SUM *admits a solution.*

Proof. Let Q be a lattice polygon with n vertices, and also let DE be the maximum integer length of its coordinates. We compute the primitive edge sequence $\mathcal{E}$. We consider the coordinates of the primitive vectors e_i and we associate to every vector the positive number $a_i = e_{ix} + Le_{iy} + DE$, where $1 \leq i < n$ and L sufficiently large, for example $L = nDE$. We add the quantity DE to every a_i so that $a_i > 0, 1 \leq i < n$. We consider d_i copies of each a_i, thus the total number of them is $\sum_{i=1}^n d_i = O(nD)$.

Now our transformation is complete since the polygon Q is decomposable if and only if there is a subset of the a_i's that sums up to zero. The key idea is that if an a_i belongs to a subset that sums up to zero then its corresponding edge belongs to a summand of the polygon and vice versa.

Notice that the transformation takes $O(nD)$ time, which is also the size of the instance of the SUBSET-SUM problem. □

The time for solving SUBSET-SUM via dynamic programming is $O(N^2W)$ (see for example [9], [2]) where N is the cardinality of the set and W is an upper bound on the absolute value of every element. In our case $N = O(nD)$ and $W = O(nDE^2)$, thus the total complexity of the algorithm is $O(n^3D^3E^2)$. The complexity of this algorithm is the same as the complexity of the algorithm in [5].

However, our approach may use the dynamic programming paradigm and is completely different from the one in [5], since we completely avoid the computation of the interior lattice points of the polygon. Moreover, if we use a balancing algorithm ([16]) for solving the corresponding SUBSET-SUM problem, which is the best available and has time complexity $O(NW)$, then the complexity of our algorithm is $O(n^2D^2E^2)$. Thus, we improve the complexity by a factor nD.

So we have the following theorem

Theorem 20. *There is an algorithm for the decision* MINKOWSKI DECOMPOSITION *problem that has $O(n^2D^2E^2)$ arithmetic complexity.*

6 Future work

In order to enumerate all possible summands of a polygon, following the approach of Section 5, we can use the various algorithms for the PARTITION problem (refer to [9]). However, a detailed experimental analysis is needed in order to decide the best approach.

Additionally, our approach of section 5 can easily lead to a randomized algorithm. We pick $L = nDE$. The quantities are of the form $a_i = e_{ix} + Le_{iy}$, of max-value $E(L+1)$, and there are $d_i \leq D$ copies of each a_i. So the maximum sum value is $nDE(L+1) = O((nDE)^2)$.

In order to check if a sum vanishes mod p, where $p > 0$ is a random integer, we have to bound the probability $Pr[failure]$, that random sum $S \in [0, nDE(L+1)]$, vanishes mod p, when $S \neq 0$, where p is a prime uniformly distributed in $[2,\ldots,x]$. We can do this using the randomized algorithm for verifying equality of strings from [14].

Lemma 21. *[14] Let a, b be two numbers with τ bits each. If $a \neq b$, then*

$$Pr[failure] = Pr[a = b \mod p] < \frac{1}{2}$$

where p is a prime uniformly distributed in $[2,\ldots,4\tau^2]$.

In our case $a = 0$ and $b = S$, so we need $\tau \simeq 2\lg(nDE)$ bits to encode them. Thus we can use the previous lemma, if we choose a prime in $[2,\ldots,4\tau^2]$, to obtain $Pr[failure] < \frac{1}{2}$.

Last, but not least, the reduction to the SUBSET-SUM problem, can lead to approximate algorithms for the Minkowski decomposition. The first step towards this direction may be the adoption of the first fully polynomial-time approximation scheme for the SUBSET-SUM problem that Ibbara and Kim suggested in [10] or the adoption of the best known, so far, approximation algorithm of Kellerer et al ([11]).

Acknowledgments

Both authors acknowledge partial support by Kapodistrias, project 70/4/6452 of the Research Council of National University of Athens and by PYTHAGO-RAS, project 70/3/7392 under the EPEAEK program funded by the Greek Ministry of Educational Affairs and EU.

References

1. CGAL: Computational geometry algorithms library. http://www.cgal.org.
2. T. H. Cormen, C. E. Leiserson, R. L. Rivest, and C. Stein. *Introduction to Algorithms*. MIT Press, Cambridge, MA, 2nd edition, 2001.
3. J. Erickson. Lower bounds for satisfiability problems. *Chicago J. of Theoretical Computer Science*, 8, 1999. http://cjtcs.cs.uchicago.edu/articles/1999/8/contents.html
4. A. Gajentaan and M. H. Overmars. On a class of $O(n^2)$ problems in computational geometry. *Computational Geometry, Theory and Applications*, 5(3):165–185, October 1995.
5. S. Gao and A. G. B. Lauder. Decomposition of polytopes and polynomials. *Discrete and Computational Geometry*, 26:89–104, 2001.
6. S. Gao and A. G. B. Lauder. Fast absolute irreducibility testing via Newton polytopes. *preprint*, 2004.
7. R. Goldman. *Pyramid Algorihtms: A dynamic approach to curves and surfaces for geometric modeling*. Morgan Kaufmann, 2002.
8. J. E Goodman and J. O' Rourke. *Handbook of computational geometry*. Elsevier science, Amsterdam, 1995.

9. E. Horowitz and S. Shani. Computing partitions with applications to the knapsack problem. *Journal of ACM*, 21(2):277–292, April 1974.
10. O. Ibarra and C. Kim. Fast approximation algorithms for the knapsack and sum of subset problems. *J. ACM*, 22(4):463–468, 1975.
11. H. Kellerer, R. Mansini, U. Pferschy, and M. G. Speranza. An efficient fully polynomial approximation scheme for the subset-sum problem. *J. Comput. Syst. Sci.*, 66(2):349–370, 2003.
12. R. Krasauskas. Toric surface patches: Advances in geometrical algorithms and representations. *Adv. Comput. Math.*, 17(1-2):89–113, 2002.
13. R. Krasauskas and R. Goldman. Toric Bezier Patches with Depth. In R. Goldman and R. Krasauskas, editors, *Topics in Geometric Modeling and Algebraic Geometry*, volume 334, pages 65–91. AMS Mathematics of Computation, 2003.
14. R. Motwani and P. Raghavan. *Randomized Algorithms*. Cambridge University Press, 1995.
15. A. M. Ostrowski. Über die Bedeutung der Theorie der konvexen Polyeder für die formale Algebra. *Jahresberichte Deutsche Marth. Verein* 30 (1921), 98–99.
16. D. Pisinger. *Algorithms for the Knapsack problems*. PhD thesis, Department of Computer Science, University of Kopehagen, February 1995.
17. F. P. Preparata and M. I. Shamos. *Computational Geometry: An Introduction*. Springer-Verlag, 3rd edition, October 1990.
18. S. Rabinowitz. A census of convex lattice polygons with at most one interior lattice point. *Ars Combinatorica*, 28:83–96, 1989.
19. J. Schicho. Simplification of surface parametrizations - a lattice polygon approach. *J of Symbolic Computation*, 36:535–554, 2003.
20. G. Woeginger. Open problems around exact algorithms. Manuscript, TU Eindhoven, 2004.
21. G. Woeginger. Exact algorithms for NP-hard problems: A survey. In M. Juenger, G. Reinelt, and G. Rinaldi, editors, *Combinatorial Optimization - Eureka! You shrink!*, volume 2570, pages 185–207. LNCS, Springer, 2003.
22. C.K. Yap. *Fundamental Problems of Algorithmic Algebra*. Oxford University Press, New York, 2000.
23. S. Zube. The n-sided toric patches and the A-resultants. *Computer Aided Geometric Design*, 17:695–714, 2000.

Reducing the number of variables of a polynomial

Enrico Carlini

Politecnico di Torino, Corso Duca degli Abruzzi 24, 10129 Torino, Italia
enrico.carlini@polito.it

Summary. In this paper, we consider two basic questions about presenting a homogeneous polynomial f: how many variables are needed for presenting f? How can one find a presentation of f involving as few variables as possible? We give a complete answer to both questions, determining the minimal number of variables needed, $N\text{ess}(f)$, and describing these variables through their linear span, $\text{EssVar}(f)$. Our results give rise to effective algorithms which we implemented in the computer algebra system **CoCoA** [4].

1 Introduction

Polynomials, also seen as symmetric tensors, are ubiquitous in Applied Mathematics. They appear in Mechanics ([8]), Signal and Image Processing ([3]), Algebraic Complexity Theory ([1]), Coding and Information Theory ([9]), etc..

One of the main open issue is to manipulate polynomials in order to obtain presentations suiting the special needs of the application at hand.

In Mechanics, it is often useful to *separate variables*. Given a polynomial $f(x_1,\ldots,x_n)$, one splits the set of variables in two pieces, e.g. $\{x_1,\ldots,x_r\}$ and $\{x_{r+1},\ldots,x_n\}$, and a presentation of f is searched of the following type

$$f(x_1,\ldots,x_n) = g(x_1,\ldots,x_r) + h(x_{r+1},\ldots,x_n)$$

for some polynomials g and h.

Separating variables is a well established technique and the search for splitting methods in general is very active (see [8]).

In Signal Processing, homogeneous polynomials (also known as *quantics* from ancient Invariant Theory) are of crucial importance. The main interest is in the so called *sum of powers presentations*, where a homogeneous polynomial f of degree d is presented as

$$f = l_1^d + \ldots + l_s^d$$

where $l_1, \ldots, l_s$ are linear forms.

Sum of powers presentations are treated in connection with quantics in [3], while a more general approach relating them to Polynomial Interpolation and Waring Problem can be found in [2].

In this paper, we consider two basic questions about presenting a homogeneous polynomial (from now on referred to as a *form*) in a "easier" way. Given a form f, how many variables are needed for presenting it? How can one find a presentation of f involving as few variables as possible?

Even if these problems are so natural, we are not aware of a complete solution existing in the literature. In this paper, we give a complete answer to both questions. Our results give rise to effective algorithms which we implemented in the computer algebra system **CoCoA** (freely available at cocoa.dima.unige.it).

After the paper was completed, A. Iarrobino let us notice that the part of Proposition 10 about N_{ess} is completely equivalent to Lemma 1.22 in [7, p. 15]. We wish to thank him for the kind remark. Our Propostion is less formal than the one in [7] and also contains a characterization of the essential variables which allows us to produce effective algorithms.

More precisely, given a form $f \in S = k[x_1, \ldots, x_n]$, k any field, we call *essential number of variables of f* the smallest integer r for which there exists a set of linear forms $\{y_1, \ldots, y_r\} \subset S$ such that

$$f \in k[y_1, \ldots, y_r];$$

the linear forms $y_1, \ldots, y_r$ are called *essential variables of f*. Then our main result is (see Definition 4, Definition 8 and Section 2 for the notation involved):

Proposition 10 *Let f be a homogeneous element in $S = k[x_1, \ldots, x_n]$ and $T = k[\partial_1, \ldots, \partial_n]$ denote the ring of differential operators. Then*

$$N_{\text{ess}}(f) = rk(\mathcal{C}_f),$$

i.e. the number of essential variables of f is the rank of its first catalecticant matrix, and

$$EssVar(f) = \langle D \circ f : D \in T_{d-1} \rangle,$$

i.e. the essential variables of f span the space of its $(d-1)^{th}$ partial derivatives.

In Section 2, we briefly recall some facts from Apolarity Theory which are the main tools of our analysis.

In Section 3, we use Apolarity and Catalecticant Matrices to obtain our main result. In Subsection 3.1, we give some examples of the use of our algorithms.

Remark 1. In this paper we work with forms, i.e. homogeneous polynomials. To apply our results to *any* polynomial f, it is enough to work with its homogenization f^h with respect to a new variable. Clearly, e.g., a presentation of f^h in essential variables readily produces a presentation of f in essential variables: it is enough to dehomogenize.

Remark 2. Throughout the paper k will denote a field of characteristic 0. Our results also hold in positive characteristic, but more advanced techniques are required for proving them, e.g. differentiations have to be replaced with contractions and divided powers have to be introduced (see [7]).

The author wishes to thank B. Reznick and C. Ciliberto for their ideas on the problem. The **CoCoA** Team in Genoa, and especially Anna Bigatti, were of great help in the implementation of the algorithms. The comments and criticisms of the two anonymous referees were of help in improving the presentation of the results.

2 Apolarity

In this section we will briefly recall some basic facts from *Apolarity Theory* or, in modern terms, *Inverse Systems Theory*. Comprehensive references are [5], mainly Lecture 2, 6 and 8, and [7].

Consider the polynomial rings

$$S = k[x_1, \ldots, x_n] \text{ and } T = k[\partial_1, \ldots, \partial_n],$$

where k denotes a field of characteristic 0, and give S a T-module structure via differentiation, i.e. we will think of T as the ring of differential operators acting on S. We denote this action with "$\circ$", e.g. $\partial_j \circ f = \frac{\partial}{\partial x_j} f$ for $f \in S$.

There is a natural perfect paring between homogeneous pieces of the same degree of S and T, namely

$$S_i \times T_i \longrightarrow k$$
$$(f, D) \mapsto D \circ f$$

is a perfect pairing for all i; in particular, S_i and T_i are dual to each other. Given subspaces,

$$V \subseteq S_i \text{ and } W \subseteq T_i$$

we denote by

$$V^\perp \subseteq T_i \text{ and } W^\perp \subseteq S_i$$

their orthogonal with respect to this pairing; notice, e.g., that $\dim_k V + \dim_k V^\perp = \dim_k S_i = \dim_k T_i$.

Given a form $f \in S_d$, the ideal

$$f^\perp = \{D \in T : D \circ f = 0\}$$

is a homogeneous ideal of T and it is called the *orthogonal* ideal of f.

Orthogonal ideals play a central role in the theory: they contain all the differential operators annihilating a given form and even more information, as it is shown by the following Lemma (for a proof see [5], Proposition 8.10).

Lemma 3. *Let f be a degree d form in S, then $D \in T_i, i < d$, is such that*
$$D \circ f = 0$$
if and only if
$$D \circ (D' \circ f) = 0$$
for all $D' \in T_{d-i}$. In other terms, for $0 < i < d$, $(f^\perp)_i$ is orthogonal to the k-vector space spanned by the $(d-i)^{th}$ partial derivatives of f.

Orthogonal ideals can be easily described introducing *ad hoc* matrices. In this paper it will be enough to describe the degree one part of a given orthogonal ideal, but similar descriptions exist in each degree.

Definition 4. *Let $f \in S_d$ and fix the standard monomial basis, e.g. with respect to lex order, $\{M_1, \ldots, M_N\}$ of the k-vector space S_{d-1}. For $i = 1, \ldots, n$, consider the first partials*
$$\partial_i \circ f = c_{i1} M_1 + \ldots + c_{iN} M_N.$$
The first catalecticant matrix *of f is*
$$(\mathcal{C}_f)_{ij} = c_{ij},$$
$i = 1, \ldots, n, j = 1, \ldots, N$.

For a general treatment of catalecticant matrices and their applications see [6], [5] and [7].

Example 5. *Let $f = x_1 x_2 x_3 \in k[x_1, x_2, x_3]$ and consider the monomial basis*
$$\{\, x_1^2, x_1 x_2, x_1 x_3, x_2^2, x_2 x_3, x_3^2 \,\}$$
of the space of degree two forms. Then
$$\mathcal{C}_f = \begin{pmatrix} 0 & 0 & 0 & 0 & 1 & 0 \\ 0 & 0 & 1 & 0 & 0 & 0 \\ 0 & 1 & 0 & 0 & 0 & 0 \end{pmatrix}.$$

Catalecticant matrices determine the degree one part of orthogonal ideals readily:

Lemma 6. *Let $f \in k[x_1, \ldots, x_n]$ be a form, then*
$$(a_1 \partial_1 + \ldots + a_n \partial_n) \circ f = 0$$
if and only if the vector $(a_1, \ldots, a_n)$ is in the left kernel of $\mathcal{C}_f$. In particular, $\dim_k (f^\perp)_1 = n - \mathrm{rk}(\mathcal{C}_f)$.

Proof. The statement simply follows writing down the action of $a_1\partial_1 + \ldots + a_n\partial_n$ on f componentwise and considering the corresponding linear system of equations.

Remark 7. Let $l \in k[x_1,\ldots,x_n]$ be a linear form and consider its d-th power $f = l^d$. Let L be a linear differential operator and notice that $L \circ f = 0$ if and only if $L \circ l = 0$ which is a linear equation in the coefficients of L. Hence $\text{rk}(\mathcal{C}_{l^d}) = 1$ (actually, even the converse is true). In particular, this means that the form of Example 5 is not a pure power.

3 How many variables?

In this section we will use apolarity to answer our two basic questions: how many variables do we need to present a given form? How can we find a presentation involving as few variables as possible?

In what follows, we will work with the polynomial ring $S = k[x_1,\ldots,x_n]$, where k is *any* field such that $\text{char}(k) = 0$ (in positive characteristic similar results hold, but, in this paper, we decided to avoid the technical difficulties involved).

Lets introduce some definitions:

Definition 8. *Given a form f in S, the* number of essential variables *of f, $N\text{ess}(f)$, is the smallest integer r such that there exist linear forms $y_1,\ldots,y_r \in S$ for which $f \in k[y_1,\ldots,y_r]$. We call* essential variables *of f any set of generators of the k-vector space $\text{EssVar}(f) = \langle y_1,\ldots,y_r\rangle$.*

Roughly speaking, given a form $f \in S$, $N\text{ess}(f)$ tells us how many variables are necessary for presenting f, while $\text{EssVar}(f)$ tells us how we can find such variables. In particular, it is clear that, if

$$N\text{ess}(f) = r \text{ and } \text{EssVar}(f) = \langle y_1,\ldots,y_r\rangle,$$

then there exists $g \in k[y_1,\ldots,y_r] \subset S$ such that $f = g$.

Example 9. Consider the form $f = f(x_1,x_2,x_3) = (x_1+x_2)(x_1-x_3)^2$ in $k[x_1,x_2,x_3]$. Clearly f is an element of the subring $k[y_1,y_2]$, where $y_1 = x_1+x_2$ and $y_2 = x_1 - x_3$. Hence $N\text{ess}(f) \leq 2$ and equality holds by Remark 7, as $\text{rk}(\mathcal{C}_f) \neq 1$ and f is not a pure power. Also, notice that $\text{EssVar}(f) = \langle x_1+x_2, x_1-x_3\rangle$ and sets of possible essential variables are: $\{x_1+x_2, x_1-x_3\}$, $\{x_2+x_3, 2x_1+x_2-x_3\}$, etc..

Using apolarity we can effectively determine $N\text{ess}$ and EssVar for a given form:

Proposition 10. *Let f be a homogeneous element in $S = k[x_1,\ldots,x_n]$ and $T = k[\partial_1,\ldots,\partial_n]$ denote the ring of differential operators. Then*

$$\mathrm{Ness}(f) = \mathrm{rk}(\mathcal{C}_f),$$

i.e. the number of essential variables of f is the rank of its first catalecticant matrix, and

$$\mathrm{EssVar}(f) = \langle D \circ f : D \in T_{d-1} \rangle,$$

i.e. the essential variables of f span the space of its $(d-1)^{th}$ partial derivatives.

Proof. If $\mathrm{Ness}(f) = r$, then $f \in k[y_1, \ldots, y_r]$ for some linear forms $y_1, \ldots, y_r$ in S. Let

$$\langle y_1, \ldots, y_r \rangle^\perp = \langle L_1, \ldots, L_{n-r} \rangle \subset T_1$$

and notice that $(f^\perp)_1 \supseteq \langle L_1, \ldots, L_{n-r} \rangle$. Thus, by Lemma 6, we have $\mathrm{rk}(\mathcal{C}_f) \leq \mathrm{Ness}(f)$.

If $\mathrm{rk}(\mathcal{C}_f) = t$, then $(f^\perp)_1 = \langle D_1, \ldots, D_{n-t} \rangle$. Complete this to a basis of T_1

$$\langle D_1, \ldots, D_{n-t}, Y_1, \ldots, Y_t \rangle$$

and consider the dual basis of S_1 defined by the apolarity perfect pairing

$$\langle z_1, \ldots, z_{n-t}, y_1, \ldots, y_t \rangle.$$

Hence, after a linear change of variables, we have $f = f(z_1, \ldots, z_{n-t}, y_1, \ldots, y_t)$. But D_j annihilates all the elements of the chosen basis of S_1 but z_j. As $(f^\perp)_1 = \langle D_1, \ldots, D_{n-t} \rangle$ we conclude that

$$f \in k[y_1, \ldots, y_t]$$

and $\mathrm{rk}(\mathcal{C}_f) \geq \mathrm{Ness}(f)$.

To conclude the proof, notice that the prefect pairing $S_1 \times T_1 \to k$ induces a well defined perfect pairing of k-vector spaces

$$V \times \left(\frac{T}{f^\perp}\right)_1 \longrightarrow k$$

where

$$V = \left((f^\perp)_1\right)^\perp = \langle l : l \in S_1, L \circ l = 0 \text{ for all } L \in (f^\perp)_1 \rangle$$

and, with the notations above, $\left(\frac{T}{f^\perp}\right)_1 = \langle Y_1, \ldots, Y_t \rangle$ and hence $V = \mathrm{EssVar}(f)$. The result follows applying Lemma 3 ($i = 1$ case) which yields

$$V = \langle D' \circ f : D' \in T_{d-1} \rangle.$$

Example 11. Given the form

$$f = x_1^3 + x_1^2 x_2 - 2x_1^2 x_3 - 2x_1 x_2 x_3 + x_1 x_3^2 + x_2 x_3^2$$

we want to determine $\mathrm{Ness}(f)$ and $\mathrm{EssVar}(f)$. In order to apply Proposition 10, we compute the first catalecticant matrix of f

$$C_f = \begin{pmatrix} 3 & 2 & -4 & 0 & -2 & 1 \\ 1 & 0 & -2 & 0 & 0 & 1 \\ -2 & -2 & 2 & 0 & 2 & 0 \end{pmatrix}.$$

Hence $N_{\text{ess}}(f) = \text{rk}(C_f) = 2$ and f can be presented as a form in two variables. To determine the essential variables of f, it is enough to compute the span of the second partial derivatives of f:

$$\text{EssVar}(f) = \langle x_2 + x_3, x_1 - x_3 \rangle.$$

Summing these up, we see that there exists a degree 3 form $g(y_1, y_2) \in k[y_1, y_2]$ such that

$$g(x_2 + x_3, x_1 - x_3) = f(x_1, x_2, x_3),$$

but how can we find g?

To complete our analysis, we want to present a form f as a polynomial only involving essential variables: this can be done almost tautologically, but the notation are quite involved. We begin with an example.

Example 12. Consider the form $f \in S = k[x_1, x_2, x_3]$ in Example 11. We already showed that there exists $g \in k[y_1, y_2] \subset S$ such that $f = g$. To determine $g(y_1, y_2)$, consider $\text{EssVar}(f) = \langle x_2 + x_3, x_1 - x_3 \rangle$ and complete its basis to a basis of S_1: we choose $\{y_1 = x_2 + x_3, y_2 = x_1 - x_3, z_1 = x_1\}$. Hence we have a linear change of variables given by

$$\begin{cases} x_1 = z_1, \\ x_2 = y_1 + y_2 - z_1, \\ x_3 = z_1 - y_2. \end{cases}$$

The basic requirement of the form $g(y_1, y_2)$ is to satisfy the relation

$$g(x_2 + x_3, x_1 - x_3) = f(x_1, x_2, x_3).$$

From this, changing variables, we get

$$g(y_1, y_2) = f(z_1, y_1 + y_2 - z_1, z_1 - y_2) = y_1 y_2^2 + y_2^3,$$

which is the desired presentation in essential variables. As a byproduct, we readily see that

$$f = (x_2 + x_3)(x_1 - x_3)^2 + (x_1 - x_3)^3$$

which is quite surprising considering the original presentation

$$f = x_1^3 + x_1^2 x_2 - 2x_1^2 x_3 - 2x_1 x_2 x_3 + x_1 x_3^2 + x_2 x_3^2.$$

The procedure showed in the previous Example works in general. Given a form $f = f(x_1, \ldots, x_n) \in S$, we compute $\text{Ness}(f) = r$ and we choose a basis for $\text{Ness}(f) = \langle y_1, \ldots, y_r \rangle$; to avoid triviality, assume $r < n$. Now, our goal is to determine $g = g(y_1, \ldots, y_r) \in k[y_1, \ldots, y_r] \subset S$ such that $f = g$. To do this, complete the basis of $\text{Ness}(f) \subset S_1$ to a basis of S_1

$$S_1 = \langle y_1, \ldots, y_r, z_1, \ldots, z_{n-r} \rangle.$$

As $S_1 = \langle x_1, \ldots, x_n \rangle$, the completed basis yields a linear change of variables

$$(\dagger) \begin{cases} x_1 = x_1(y_1, \ldots, y_r, z_1, \ldots, z_{n-r}), \\ \vdots \\ x_n = x_n(y_1, \ldots, y_r, z_1, \ldots, z_{n-r}). \end{cases}$$

Notice that $y_1, \ldots, y_r$ are linear forms in S and hence there exist linear functions such that $y_i = y_i(x_1, \ldots, x_r), i = 1, \ldots, r$. Moreover, the following identities hold by construction of $(\dagger)$

$$y_i = y_i(x_1(y_1, \ldots, y_r, z_1, \ldots, z_{n-r}), \ldots, x_r(y_1, \ldots, y_r, z_1, \ldots, z_{n-r}))$$

for $i = 1, \ldots, n$.

To determine g, it is enough to consider the desired relation

$$f(x_1, \ldots, x_n) = g(y_1(x_1, \ldots, x_r), \ldots, y_r(x_1, \ldots, x_r))$$

and to apply the linear change of variables $(\dagger)$. Thus we obtain $g(y_1, \ldots, y_r)$:

$$g(y_1, \ldots, y_r) =$$
$$= g(y_1(x_1(y_1, \ldots, y_r, z_1, \ldots, z_{n-r}), \ldots, x_r(y_1, \ldots, y_r, z_1, \ldots, z_{n-r})), \ldots$$
$$\ldots, y_r(x_1(y_1, \ldots, y_r, z_1, \ldots, z_{n-r}), \ldots, x_r(y_1, \ldots, y_r, z_1, \ldots, z_{n-r}))) =$$
$$= f(x_1(y_1, \ldots, y_r, z_1, \ldots, z_{n-r}), \ldots, x_n(y_1, \ldots, y_r, z_1, \ldots, z_{n-r})).$$

Notice that, as f and the functions $x_i(y_1, \ldots, y_r, z_1, \ldots, z_{n-r}), i = 1, \ldots, n$, are *explicitly* known, we have completely determined g as an element in $k[y_1, \ldots, y_r]$.

Remark 13. As a straightforward application of the theory, we consider the detection of cylinders (i.e. algebraic surfaces ruled by a family of parallel lines moving along a fixed curve). Suppose you are given the polynomial equation of a surface $\mathcal{F} : f(x, y, z) = 0$ in three space and you want to decide whether $\mathcal{F}$ is a cylinder or not. It is well known that $\mathcal{F}$ is a cylinder if and only if its defining equation is a function of two planes, i.e. there exist linear forms $m(x, y, z)$ and $l(x, y, z)$ such that $f(x, y, z) = g(m, n)$ for some polynomial g. Hence, we readily have an effective procedure for cylinder detection:

$$\mathcal{F} \text{ is a cylinder if and only if } \text{Ness}(f^h) \leq 3,$$

where f^h denotes the homogenization of f (see Example 16). Clearly, the method applies in any dimension for deciding whether a given hypersurface is a cylinder or not.

3.1 Using a computer

The results of our analysis can be easily translated into algorithms and we wrote down procedures to be used with the Computer Algebra system **CoCoA**.

We begin with reporting a **CoCoA** session illustrating the use of our algorithms to work out the expository Examples 11 and 12.

Example 14. First we define the form we want to study

```
F := x^3 + x^2y -2x^2z -2 xyz + xz^2 +yz^2
```

To compute the number of essential of variables, use the function NEssVar(F):

```
NEssVar(F);
2
```

To determine a choice of essential variables, use the function EssVar(F):

```
EssVar(F);
[y + z, x - z]
```

Finally, NewPres(F) produces a presentation of the form involving the essential variables $y[1] = y + z, y[2] = x - z$:

```
NewPres(F);
y[1]y[2]^2 + y[2]^3
```

Usually, a given polynomial $f(x_1, \ldots, x_n)$ will essentially involve n variables, i.e. $\mathcal{N}\mathrm{ess}(f) = n$. Hence our algorithms *do not* help in solving the polynomial equation $f = 0$. Nevertheless, our procedure should be used as a pre-processing tool. In fact, *if* the number of variables can be decreased, then the numerical solution of the equation can be performed much more efficiently. We illustrate this with the following "extreme" example.

Example 15. We consider the degree three polynomial in four variables

$$f(x, y, z, t) = f_0(x, y, z, t) + f_1(x, y, z, t) + f_2(x, y, z, t) + f_3(x, y, z, t),$$

where

$$\begin{aligned}
f_0 &= 3 \\
f_1 &= -x - y + 2z + 3t \\
f_2 &= 5x^2 + 10xy + 5y^2 - 20xz - 20yz + 20z^2 - 30xt - 30yt + 60zt + 45t^2 \\
f_3 &= x^3 + 3x^2y + 3xy^2 + y^3 - 6x^2z - 12xyz - 6y^2z + 12xz^2 + 12yz^2 + \\
&\quad -8z^3 - 9x^2t - 18xyt - 9y^2t + 36xzt + 36yzt - 36z^2t + 27xt^2 + 27yt^2 + \\
&\quad -54zt^2 - 27t^3.
\end{aligned}$$

In order to solve the equation $f(x, y, z, t) = 0$, we apply our algorithms to the degree 2 and 3 pieces of f:

```
EssVar(F2);
[x + y - 2z - 3t]

NewPres(F2);
5y[1]^2
```

and hence $f_2(x, y, z, t) = 5y_1^2$, where $y_1 = x + y - 2z - 3t$. Similarly

```
EssVar(F3);
[x + y - 2z - 3t]

NewPres(F3);
y[1]^3
```

which yields $f_3(x, y, z, t) = y_1^3$. In conclusion, to solve the equation $f(x, y, z, t) = 0$, it is enough to solve the equation in one variable

$$y_1^3 + 5y_1^2 - y_1 + 3 = 0$$

and to apply some linear algebra to find all the solutions.

We conclude with a Geometric example about cylinder detection.

Example 16. Consider the degree five surface in three space $\mathcal{F} : f(x, y, z) = 0$, where

$$f = f_0 + f_2 + f_5$$

and

$$f_0 = -1, f_2 = x^2 - xy - 2y^2 - 3yz - z^2,$$

$$f_5 = x^5 + 2x^4y - 2x^3y^2 - 8x^2y^3 - 7xy^4 - 2y^5 + 3x^4z - 18x^2y^2z - 24xy^3z + \\ -9y^4z + 2x^3z^2 - 12x^2yz^2 - 30xy^2z^2 - 16y^3z^2 - 2x^2z^3 - 16xyz^3 + \\ -14y^2z^3 - 3xz^4 - 6yz^4 - z^5.$$

In order to decide whether $\mathcal{F}$ is a cylinder or not, we follow Remark 13. Introduce a new variable t and consider the homogenization of f, $f^h = t^5 f_0 + t^3 f_2 + f_5$. Using **CoCoA** and denoting by FH the form $f^h(x, y, z, t)$, we get:

```
NEssVar(FH);
3

EssVar(FH);
[t, y + 2/3z, x + 1/3z]

NewPres(FH);
-y[1]^5 - 2y[1]^3y[2]^2 - 2y[2]^5 - y[1]^3y[2]y[3] -
7y[2]^4y[3] + y[1]^3y[3]^2 - 8y[2]^3y[3]^2 - 2y[2]^2y[3]^3 +
2y[2]y[3]^4 + y[3]^5
```

In conclusion, $f^h(x,y,z,t) = g(y_1, y_2, y_3)$ where g is the output of the function `NewPres(FH)` and

$$\begin{cases} y_1 = t \\ y_2 = y + \frac{2}{3}z \\ y_3 = x + \frac{1}{3}z \end{cases}.$$

Hence, we have the polynomial equality $f(x,y,z) = g(1, y_2, y_3)$ and $\mathcal{F}$ is a cylinder ruled by lines parallel to the line $y_2 = y_3 = 0$.

References

1. P. Bürgisser, M. Clausen, and M.A. Shokrollahi. *Algebraic Complexity Theory*, volume 315 of *Grund. der Math. Wiss.* Springer, Berlin, 1997.
2. Ciro Ciliberto. Geometric aspects of polynomial interpolation in more variables and of Waring's problem. In *European Congress of Mathematics, Vol. I (Barcelona, 2000)*, volume 201 of *Progr. Math.*, pages 289–316. Birkhäuser, Basel, 2001.
3. P. Comon and B. Mourrain. Decomposition of quantics in sums of power of linear forms. *Signal Processing*, 53(2):93–107, 1996. Special issue on High-Order Statistics.
4. CoCoATeam. CoCoA: a system for doing Computations in Commutative Algebra. Available at http://cocoa.dima.unige.it, 2004.
5. Anthony V. Geramita. Inverse systems of fat points: Waring's problem, secant varieties of Veronese varieties and parameter spaces for Gorenstein ideals. In *The Curves Seminar at Queen's, Vol. X (Kingston, ON, 1995)*, volume 102 of *Queen's Papers in Pure and Appl. Math.*, pages 2–114. Queen's Univ., Kingston, ON, 1996.
6. Anthony V. Geramita. Catalecticant varieties. In *Commutative algebra and algebraic geometry (Ferrara)*, volume 206 of *Lecture Notes in Pure and Appl. Math.*, pages 143–156. Dekker, New York, 1999.
7. Anthony Iarrobino and Vassil Kanev. *Power sums, Gorenstein algebras, and determinantal loci*, volume 1721 of *Lecture Notes in Mathematics*. Springer-Verlag, Berlin, 1999.
8. Robert I. McLachlan and G. Reinout W. Quispel. Splitting methods. *Acta Numer.*, 11:341–434, 2002.
9. S. Roman. *Coding and Information Theory*. Springer, New York, 1992.

Index

A

acyclic, 31, 34
algebraic geometry, 1, 6, 16, 23, 55
apolarity, 238
approximate
    implicitization, 7, 38, 39, 42, 45, 49, 72
    techniques, 71
arithmetically Cohen-Macaulay, 34

B

B-spline, 40, 45
base point, 12, 17, 18, 56, 57, 60, 121
Bernstein
    basis, 73
    polynomial, 137, 148
    theorem, 4, 6, 17
Betti number, 88, 90, 96–98, 100, 106–108, 111, 112
Bézier, 148
    curve, 135–140, 142, 143, 146
    patch, 135, 136, 138–140, 142, 144, 146–149
      toric, 135, 136, 147–149
    quadrangle, 137
    rectangle, 136, 138
    representation, 142
    subdivision, 5, 16, 18
    tensor-product, 213
    triangle, 136–139, 141–143
Bezout resultant, 8, 10, 18
Bezoutian matrix, 10
blossoming, 218
blowup, 121

C

canal surface, 170, 173, 177, 181
Castelnuovo-Mumford regularity, 28
catalecticant matrix, 240
Čech complex, 28
class, 121
classification, 123
Clebsch cubic surface, 129
cluster, 206
COCOA, 32, 245
$\mathcal{Z}$-complex, 26, 29
condition number, 74
    algorithm, 75
configuration
    banana, 208
    hinge, 208
constraint
    graph, 207
    solver, 205
control points, 136, 139, 144, 145, 147–149
convex
    hull, 79, 139, 144
    polygon, 217
Coxeter diagram, 130
critical
    point, 64, 68
    value, 64, 68
cubic
    B-spline, 45

surface, 119
curve
  bicorn, 76
  cardioid, 76
  offset, 82
  tacnode, 76
  trifolium, 76
cyclic covering, 97, 99, 100
cylinder, 244

D

decomposable, 219
degree of freedom, 207
Descartes' dream, 1
desingularization, 122
determinant of a complex, 30
differential operator, 239
discriminant variety, 66, 68
distance, 77
Dixon resultant, 8, 9, 18
double point, 121
dual basis, 242

E

Eckardt points, 129
eigenvalue, 42
eigenvector, 42
elimination, 55, 64, 65
equations, 124
  nice, 124
error measurement, 45
essential variables, 238
evaluation map, 73

F

Fitting ideal, 31

G

generic resultant, 18
generically rigid system, 206
genus, 90, 93–95, 104, 114, 156, 165–167
geometric
  extraneous component, 174
  modeling, 1, 6, 16, 18, 23
graph of constraint, 207
grid, 186, 189–191, 193–195, 197, 198
Gröbner basis, 2, 5, 16, 18, 37, 53, 54, 66

H

Hausdorff distance, 72, 77
Hodge
  decomposition, 109
  number, 109
  structure, 109
  theory, 109, 110, 112
homography, 154
homotopy continuation, 2, 5, 16
hyperplane arrangement, 98

I

implicit, 161
  approximation, 42
  curve, 71
  degree, 145, 155, 158–160, 165, 166
  equation, 7, 11, 12, 17, 55, 64, 151–154, 157, 161–163, 165, 166, 172, 181
  surface, 45, 47, 71
  variety, 64
implicitization, 7, 16, 18, 23, 38, 42, 49, 53, 59, 154, 161–163, 171, 183, 186, 218
  error, 80
index of a parameterization, 57, 66, 68
instability, 63, 64
intersection, 64
inverse system, 238
isolated singularity, 121

J

Jacobian, 186, 189
  ideal, 92

K

Koszul complex, 29
Kronecker's method, 14

L

local complete intersection, 17, 25, 26, 32
localization, 55
looped patch, 46
lower bound, 221

M

M-variety, 106
MACAULAY 2, 32, 34
Macaulay resultant, 8
MAPLE, 214
Milnor, 92, 107
   fiber, 91, 92, 94, 98, 99
   lattice, 92, 93, 95, 96
   number, 91–93, 98, 99
minimal gradient algorithm, 81
Minkowski
   decomposition, 219, 232
   sum, 217
ML, 42, 47
moduli, 151, 152, 154–161
moving
   curve, 37
   hyperplane, 25
   hypersurface, 25
   plane, 11, 13
   quadric, 11, 13
   surface, 7, 11, 13, 37

N

nested nodal surface, 46
Newton polytope, 218
normal form, 151–153, 155–159, 161, 163, 164, 166
NP-complete problem, 211, 220, 228
number of variables, 237
numerical
   error, 71
   stability, 63
NURBS surface, 37, 38

O

offset curve, 82
orthogonal ideal, 239

P

parameterization, 64
   of a variety, 56
parametric
   variety, 65
   curve, 71
   surface, 71

parametrization, 146, 152, 154–161, 163–166, 185, 186, 188
   degree, 138, 139, 149
   extension algorithm, 146
   extension problem, 135–138
   universal rational, 135
partial elimination, 210
partition problem, 234
perfect pairing, 239
perturbation, 82
pipe surface, 173, 177
polynomial, 73
   equation, 212
   factorisation, 218
   interpolation, 238
   problem, 228
   system, 205
PPL, 39, 47–49
PPS, 40, 47–49
prime decomposition, 66
problem
   NP-complete, 211, 220, 228
   of partition, 234
   of Waring, 238
   polynomial, 228
   subset-sum, 232
   summand, 221
projection, 55, 64, 68
proper parameterization, 57, 58, 66, 67
PS, 41, 47

Q

quadric surface, 6
quartic surface, 46

R

real parameterization, 58
reality index, 122
realization, 206
residual resultant, 7
resolvant, 14, 17, 18
resultant, 2, 5, 8, 13, 14, 16, 18, 37, 173, 180
   sparse, 217
revolution surface, 170, 175, 181
rigid configuration, 205
ringed surface, 171, 179
ruled surface, 171, 183

S

saturation
  of a module, 27
  of an ideal, 26, 27, 32
self-intersection, 151, 185–187, 189, 195–198, 200, 202
simple singularity, 93, 94
singular
  locus, 54, 66, 67
  value, 42
    decomposition, 41, 74
SINGULAR, 32, 132
singularity, 130
  isolated, 121
Smith
  bound, 108
  exact sequence, 105, 106
  inequality, 113
  theory, 105, 106, 108, 110
solid modeling, 169
solution, 206
solver
  algebraic, 205
  by subdivision, 213
  numeric, 205
  of constraints, 205
sparse resultant, 7, 217
spectral sequence, 105, 106
spline, 38
stability, 68
statistics, 63
Stewart mechanism, 213
Sturm-Habicht sequence, 175
subdivision, 191, 192, 195
  solver, 213
subset-sum problem, 232
sum of powers, 237
summand, 217, 220
  problem, 221
surface
  cubic, 119
    of Clebsch, 129
SURFEX, 123, 131
Sylvester resultant, 8, 9, 18, 172
SYNAPS, 213
syzygy, 11, 18, 25, 26, 28, 33
  matrix, 29

T

tangent cone, 123
tensor-product
  Bernstein polynomial, 41
  B-spline, 39, 40
  surface, 153
topological type, 123
toric, 148, 149
  Bézier patch, 217
  map, 148
  surface, 135, 136, 144, 145, 148
    abstract, 145
    almost, 145–147

V

M-variety, 106, 107, 109, 111

W

Waring problem, 238
weight, 207
Whitney umbrella, 94, 164, 166

Z

zero-dimensional, 32, 61, 65

Printing: Krips bv, Meppel
Binding: Stürtz, Würzburg